D1582243

Please return on or before the date below.

Non-loan returns

Clifton Library
Rotherham College of Arts & Technology,
Eastwood Lane, Rotherham, S65 1EG
√19
**Need to renew, book a PC or book a help session?
Call: (01709) 722869 or 722738**

Rotherham College of Arts and Technology

R097853

Programmable Logic Controllers

CLIFTON LIBRARY
ACC NO
DATE
CLASS

About the Authors

Khaled Kamel, Ph.D., is currently a professor of computer science at Texas Southern University. Previously, he was for 22 years a professor and the chair of the Department of Computer Engineering and Computer Science at the University of Louisville Engineering School. He also worked as an instrumentation engineer at GE Jet Engine, and served as the founding dean of the College of Information Technology at United Arab Emirates University and of the College of Computer Science and Information Technology at Abu Dhabi University. Dr. Kamel received a B.S. in electrical engineering from Cairo University, a B.S. in mathematics from Ain Shams University (Egypt), an M.S. in computer science from the University of Waterloo, and a Ph.D. in electrical and computer engineering from the University of Cincinnati.

Eman Kamel, Ph.D., holds a B.S. in electrical engineering from Cairo University, an M.S. in electrical and computer engineering from the University of Cincinnati, and a Ph.D. in industrial engineering from the University of Louisville. She has extensive experience in process automation at companies including Dow Chemical, GE Jet Engine, Philip Morris, VITOK Engineers, Evana Tools, and PLC Automation. She designed and implemented PLC-based automation projects in application areas including tobacco manufacturing, chemical process control, wastewater treatment, plastic sheets processing, and irrigation water level control. Dr. Kamel has wide-ranging expertise in Siemens and Allen Bradley PLC programming, instrumentation, communication, and user interfaces. She has taught classes in the areas of PLCs, computer control, and automation at several universities.

CLIFTON LIBRARY
ACC NO. R097853
DATE 10/5/18
CLASS 629.89 KAM

Programmable Logic Controllers

Industrial Control

Khaled Kamel, Ph.D.

Eman Kamel, Ph.D.

New York   Chicago   San Francisco
Athens   London   Madrid
Mexico City   Milan   New Delhi
Singapore   Sydney   Toronto

Library of Congress Cataloging-in-Publication Data

Kamel, Khaled.
    Programmable logic controllers : industrial control / Khaled Kamel,
Eman Kamel.
        p.   cm.
    ISBN 978-0-07-181045-6
    1. Programmable controllers.   2. Automatic control.   I. Kamel, Eman.
    II. Title.
    TJ223.P76K36   2014
    629.8′955115—dc23

                                                                        2013025308

Copyright © 2014 by McGraw-Hill Education. All rights reserved. Printed in the United States of America. Except as permitted under the United States Copyright Act of 1976, no part of this publication may be reproduced or distributed in any form or by any means, or stored in a data base or retrieval system, without the prior written permission of the publisher.

1 2 3 4 5 6 7 8 9 0   DOC/DOC   1 9 8 7 6 5 4 3

ISBN 978-0-07-181045-6
MHID 0-07-181045-5

Sponsoring Editor
  Michael Penn

Editing Supervisor
  Stephen M. Smith

Production Supervisor
  Pamela A. Pelton

Acquisitions Coordinator
  Amy Stonebraker

Project Manager
  Nidhi Chopra,
  Cenveo® Publisher Services

Copy Editor
  James K. Madru

Proofreaders
  Manish Tiwari and Yamini Chadha,
  Cenveo Publisher Services

Art Director, Cover
  Jeff Weeks

Composition
  Cenveo Publisher Services

Printed and bound by RR Donnelley.

McGraw-Hill Education books are available at special quantity discounts to use as premiums and sales promotions or for use in corporate training programs. To contact a representative, please visit the Contact Us page at www.mhprofessional.com.

This book is printed on acid-free paper.

McGraw-Hill Education, the McGraw-Hill Education logo, and related trade dress are trademarks or registered trademarks of McGraw-Hill Education and/or its affiliates in the United States and other countries and may not be used without written permission. All other trademarks are the property of their respective owners. McGraw-Hill Education is not associated with any product or vendor mentioned in this book.

Information contained in this work has been obtained by McGraw-Hill Education from sources believed to be reliable. However, neither McGraw-Hill Education nor its authors guarantee the accuracy or completeness of any information published herein, and neither McGraw-Hill Education nor its authors shall be responsible for any errors, omissions, or damages arising out of use of this information. This work is published with the understanding that McGraw-Hill Education and its authors are supplying information but are not attempting to render engineering or other professional services. If such services are required, the assistance of an appropriate professional should be sought.

Contents

v

Preface

Programmable Logic Controllers: Industrial Control offers readers an introduction to PLC programming with a focus on real industrial process automation applications. The Siemens S7-1200 PLC hardware configuration and the Totally Integrated Automation (TIA) Portal are used throughout the book. A small and inexpensive training setup with a Siemens power supply, processor, processor-integrated discrete inputs-outputs, processor-integrated two-point analog inputs, processor one-output analog signal board, eight ON/OFF switch plug-in simulator, human-machine interface (HMI), four-port Ethernet switch module, and programming laptop is described and used to illustrate all programming concepts and the implementation of parts of automation projects completed by the authors in the past 15 years. The authors greatly appreciate the generous support of Siemens during the production of this book, including an expert technical review of the book conducted by the company.

At the end of each chapter is a set of homework questions and small laboratory design, programming, debugging, or maintenance projects. A comprehensive capstone design project is detailed in Chap. 9, the final chapter. All programs and system configurations used in this book are fully implemented and tested. The book's website, www.mhprofessional.com/ProgrammableLogicControllers, contains a Microsoft PowerPoint multimedia presentation with several interactive simulators. Readers are encouraged to go through the presentation and practice with the simulators to fully understand the PLC programming fundamentals covered.

An introduction to the concepts of process control and automation is provided in Chap. 1. Chapter 2 details the fundamentals of relay logic programming. It also covers the architecture and operation of PLCs. Configuration, operation, and the programming of timers and counters are the focus of Chap. 3. The book's website contains very useful simulators for the different types of PLC timers: ON-DELAY, OFF-DELAY, and retentive timers. The website also contains additional simulators illustrating concepts covered in the book's first three chapters, including motor start/stop and forward/reverse control.

Chapter 4 is dedicated to the coverage of mathematical, logic, and commonly used command operations, with emphasis on their use in real-time industrial applications. Ladder programming for both PLC ladder logic and HMIs is discussed in detail in Chap. 5. Modular structured programming design is presented with emphasis on industrial standards and safety. Coverage is specific for the Siemens S7-1200 processor, the SIMATIC basic-panel HMI, and the PROFINET Ethernet protocol, but the concepts are applicable to other systems.

System checkouts and troubleshooting are typically the most challenging and time-consuming tasks in industrial automation process-control applications. Chapter 6 contains common design and troubleshooting techniques. It also addresses critical issues of validation, hazards, safety standards, and protection against hardware/software failures or malfunctions. Analog programming and associated instrumentation are covered in Chap. 7. Configuration, interface, scaling, calibration, and associated user interfaces are briefly covered.

Chapter 8 presents a comprehensive introduction to open- and closed-loop digital process control. Topics covered include sensors, actuators, ON/OFF control, feedback control, PID tuning, and measures of good control. This chapter is intended to provide users with an understanding of the "big picture" of a control system in terms of system tasks, requirements, and overall expectations. It can best serve advanced engineering/technology, computer science, or information technology students as a prerequisite for the fundamentals and hands-on activities covered in the first seven chapters of the book. It also can serve other readers as a cap for the skills learned in previous chapters.

The book concludes with a comprehensive case study in Chap. 9. The case details the specifications of an irrigation-canal downstream water-level control. Coverage proceeds from the specification level to the final system design/implementation with associated documentation. The project is a small part of a much larger project implemented by the authors in an African country more than 10 years ago. All implementations are redone using the Siemens S7-1200 PLC system.

Recent advances in industrial process control have produced more intelligent and compact PLC hardware than the one we adopted in this book, the Siemens S7-1200 system. These advances also have made available extremely friendlier development software for structured ladder programming, communication, configuration, modular design, documentation, and overall system troubleshooting, and have created many opportunities for challenging and rewarding careers in the areas of PLC technology and process automation. This book is intended for a two-quarter sequence or one four-credit semester course in an academic setting with the expectation of weekly hands-on laboratory work activities. It also can be used for a two-week industrial training sequence in a small-group setting with adequate training setup for users. Successful career opportunities in the demanding field of PLC control and automation require acquisition of the skills in this book along with adequate hands-on experience.

Khaled Kamel, Ph.D.
Professor, CS Department
Texas Southern University

Eman Kamel, Ph.D.
Senior Control Engineer
PLC Automation

Introduction to PLC Control Systems and Automation

This chapter is an introduction to the world of PLCs and their evolution over the past 50 years as the top choice and most dominant among all systems available for process-control and automation applications.

Chapter Objectives

▶ Understand the concepts of process control.

▶ Learn the history of PLCs and relay logic.

▶ Understand the PLC hardware architecture.

▶ Learn hardwired and PLC system characteristics.

A programmable logic controller (PLC) is a microprocessor-based computer unit that can perform control functions of many types and varying levels of complexity. The first commercial PLC system was developed in the early 1970s to replace hardwired relay controls used in large manufacturing assembly plants. The initial use of PLCs covered automotive assembly lines, jet engines, and large chemical plants. PLCs are used today in many tasks including robotics, conveyors system, manufacturing control, process control, electrical power plants, wastewater treatment, and security applications. This chapter is an introduction to the world of PLCs and their evolution over the past 50 years as the top choice and most dominant among all systems available for process-control and automation applications.

1.1 Control System Overview

A *control system* is a device or set of structures designed to manage, command, direct, or regulate the behavior of other devices or systems. The entire control system can be viewed as a multivariable process that has a number of inputs and outputs that can affect the behavior of the process. Figure 1.1 shows this functional view of control systems. This section is intended as a brief introduction to control systems. Additional material will be covered in more detail in Chap. 7.

1.1.1 Process Overview

In the industrial world, the word *process* refers to an interacting set of operations that lead to the manufacture or development of some product. In the chemical industry, *process* means the operations necessary to take an assemblage of raw materials and cause them to react in some prescribed fashion to produce a desired end product, such as gasoline. In the food industry, *process* means to take raw materials and operate on them in such a manner that an edible high-quality product results. In each use, and in all other cases in the process industries, the end product must have certain specified properties that depend on the conditions of the reactions and operations that produce them. The word *control* is used to describe the steps necessary to ensure that the regulated conditions produce the correct properties in the product.

A process can be described by an equation. Suppose that we let a product be defined by a set of properties; $P_1, P_2, \ldots, P_n$. Each of these properties must have a certain value for the product to be correct. Examples of properties are such things as color, density, chemical composition, and size. The process can be assumed to have m variables characterizing its unique behavior. Some of these

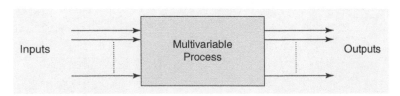

Figure 1.1 Control systems functional view.

variables also can be categorized as input, output, process property, and internal or external system parameters. The following equations express a process property and a variable as a function of process variables and time:

$$P_i = F(v_1, v_2, \ldots, v_m, t)$$

$$v_i = G(v_1, v_2, \ldots, v_m, t)$$

where P_i = the ith process property
v_i = the ith process variable
t = time

To produce a product with the specified properties, some or all the m process variables must be maintained at specific values in real time. Figure 1.2 shows free water flow through a tank, similar to rain flow in a home gutter system. The tank acts in a way to slow the flow rate through the piping structure. The output flow rate is proportional to the water head in the tank. Water level inside the tank will rise as the input flow rate increases. At the same time, output flow rate will increase with a noticeable increase in the tank water level. Assuming a large enough tank, level stability will be reached when the flow in is equal to the flow out. This simple process has three primary variables: FLOW IN, FLOW OUT, and the tank level. All three variables can be measured and, if desired, also can be controlled. The tank level is said to be a *self-regulated variable*.

Some of the variables in a process may exhibit the property of self-regulation, whereby they will naturally maintain a certain value under normal conditions. Small disturbances will not affect the tank level stability because of the self-regulation characteristic. A small increase in tank inflow will cause a slight increase in the water level. An increase in water level will cause an increase in tank outflow, which eventually will produce a new stable tank level. Large disturbances in tank input flow may force undesired changes in the tank level. Control of variables is

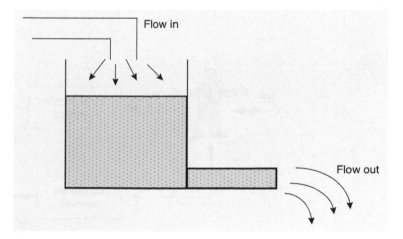

Figure 1.2 Water flow tank process.

necessary to maintain the properties of the product, the tank level in our example, within specification. In general, the value of a variable v actually depends on many other variables in the process as well as on time.

1.1.2 *Manual Control Operation*

In a manual control system, humans are involved in monitoring the process and carrying out necessary decisions to bring about desired changes in the process. Still, computers and advanced digital technologies may be used to automate a wide variety of process operation, status, command, and decision-support functions. Sensors and measurement instruments are used to monitor the status of different process variables sconditions, whereas final control elements or actuators are used to force changes in the process. As shown in Fig. 1.3, humans close the control loop and establish the connection between measured values, desired conditions, and the needed activation of the final control elements.

Manual control is widely available and can be effective for simple and small applications. The initial cost of such systems might be relatively smaller than that of automated ones, but the long-term cost is typically much higher. It is difficult for operators to achieve the same control/quality because of varying levels of domain expertise as well as unexpected changes in the process. The cost of operation and training also can become a burden unless certain functions are automated. Most systems start by using manual control or existed previously with manual operation. System owners acquire and accumulate process-control experience over time and use this knowledge eventually to make process improvements and eventually automate the control system.

The introduction of digital computers into the control loop has allowed the development of more flexible control systems, including higher-level functions and advanced algorithms. Furthermore, most current complex control systems could not be implemented without the application of digital hardware. However, the simple sequence of sensing, control, and actuation for the classic feedback control becomes more complex as well. A real-time system is one in which the correctness of a result depends not only on the logical correctness of the

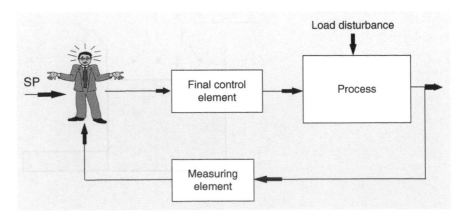

Figure 1.3 Manual control systems.

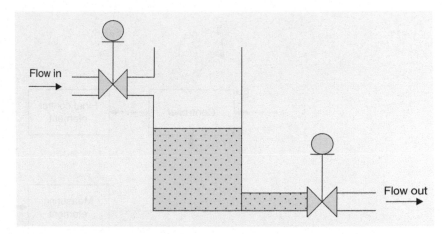

Figure 1.4   Tank level manual controls.

calculation but also on the time at which the different tasks are executed. Time is one of the most important entities of the system, and there are timing constraints associated with system tasks. Such tasks normally have to control or react to events that take place in the outside world that are happening in *real time*. Thus a real-time task must be able to keep up with external events with which it is concerned.

Figure 1.4 shows a simple manual control system. The level in the tank shown varies as a function of the flow rate through the input valve and the flow rate through the output valve. The level is the control or controlled variable, which can be measured and regulated through valve control and adjustment at the input or output flow or both. The two valves can be motorized and activated from an easy-to-use operator interface. Valve position variations are achieved through an operator input based on observed real-time process conditions. We will see next that the operator can easily be eliminated.

1.1.3   *Automated-System Building Blocks*

The closed control loop shown in Fig. 1.5 consists of the following five blocks:

- Process
- Measurement
- Error detector
- Controller
- Control element

In manual control, the operator is expected to perform the task of error detection and control. Observations and actions taken by operators can lack both consistency and reliability. The limitations of manual control can be eliminated through the implementation of closed-loop systems and the associated process-control strategies. Details of such strategies will be provided in Chap. 7. Figure 1.5 shows a block diagram of a single-variable closed-loop control.

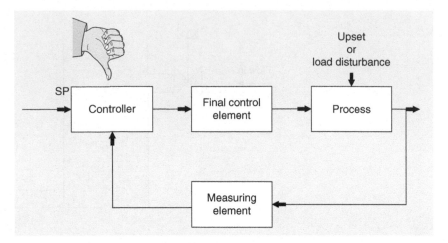

Figure 1.5 Closed-loop control.

The controller can be implemented using variety of technologies, including hardwired relay circuits, digital computers, and more often PLC systems.

It is impossible to achieve perfect control, but in the real world, it is not needed. We can always live with small errors within our acceptable quality range. An oven with a desired temperature of 500°F can achieve the same results at 499.99°F. In most cases we are limited by the precision and cost of the actual sensors. There is no good justification for spending more money to achieve unwanted/unnecessary gains in precessions.

Errors in real time are used to judge the quality of the system design and its associated controller. The errors can be measured in three ways, as explained by the following definitions:

Absolute error = set point − measured value

Error as percent of set point = absolute error/set point × 100

Error as percent of range = absolute error/range × 100

Range = maximum value − minimum value

Errors are commonly expressed as a percentage of range and occasionally as a percentage of set point but rarely as an absolute value. Also, most process variables are also commonly quantified as percentage of the defined range. This quantification allows for universal input-output PLC computer interfaces regardless of the physical nature of the sensory and actuating devices. A PLC analog input module having several input slots can accommodate and process temperature, pressure, motor speed, viscosity, and many other measurements in exactly the same way. Later chapters will detail the PLC hardware and software as applied to real-world industrial control applications. Even though the implementation focus will be on the Siemens S7-1200 system, the concepts covered will apply to other PLCs with no or very little modification. International standards and the success of open-system architectures are the main reasons for the universal nature of today's PLC technology and its compatibility.

1.2 Hardwired Systems Overview

Prior to the widespread use of PLCs in process control and automation, hardwired relay control systems or analog single-loop controllers were used. This section will briefly introduce relay systems and the logic used in process control. It is important that you understand relay fundamentals so that you have a full appreciation of the role of PLCs in replacing relays, simplifying process-control design/implementation and enhancing process quality at much lower system overall cost. Coverage in this section is limited to functionality and applications without much detail of either electrical or mechanical characteristics.

1.2.1 *Conventional Relays*

This section shows how a relay actually works. A *relay* is an electromagnetic switch that has a coil and a set of associated contacts, as shown in Fig. 1.6. Contacts can be either normally open or normally closed. An electromagnetic field is generated once voltage is applied to the coil. This electromagnetic field generates a force that pulls the contacts of the relay, causing them to make or break the controlled external circuit connection. These electrically actuated devices are used in automobiles and industrial applications to control whether a high-power device is switched on or off. While it is possible to have a device such as a large industrial motor or ignition system powered directly by an electric circuit without the use of a relay, such a choice is neither safe nor practical. For example, in a factory, a motor control may be placed far away from the high-voltage electric motor and its power source for safety reasons. In this case, it is more practical to have a low-power electrical relay circuit control the high-power relay contacts than to wire a high-power electric switch directly from the control area to the motor and its independent power supply.

Figure 1.6 Typical industrial relays.

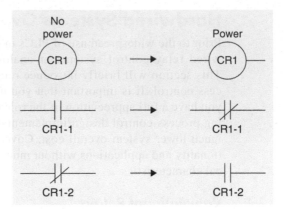

Figure 1.7 Relay with two contacts normally open and normally closed.

Figure 1.7 shows a control relay CR1 with two contacts normally open (CR1-1) and normally closed (CR1-2). On the left side of the figure, power is not applied to the coil (CR1), and the two contacts are in the normal state. On the right side of the figure, the power is applied to the coil, and the two contacts switch state.

Figure 1.8 shows a simple relay circuit for controlling a bell using a single-pole, single-through (SPST) switch. Pressing the switch causes the bell to sound. A relay is typically used to control a device that requires high voltage or draws large current. The relay allows full power to the device without needing a mechanical switch that can carry the high current. A switch is normally used to control the low-power side, the relay-coil side. Notice that there are two separate circuits: the bottom uses the direct-current (dc) low power, whereas the top uses the alternating-current (ac) high power. The two circuits are only connected through electromagnetic field coupling. The low-power dc side is connected to the coil, whereas the high-power ac side in this example is located in the field away from the control room. The two sides are normally powered from two independent sources in a typical industrial facility automated application. Of course, it is not cost-effective to replace the relay in this example with a PLC, but it does for a real application with hundreds or thousands of input-output (I/O) devices.

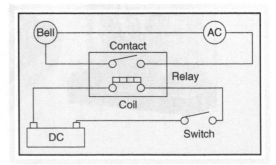

Figure 1.8 Simple relay circuit.

Relay Logic System

Relay logic systems are control structures appropriate for both industrial and municipal applications. The operations/processes that will be controlled by relay logic systems are hardwired, unlike programmable logic control systems. These systems are inflexible and can be difficult to modify after deployment. Because the operation of relay logic controllers is built directly into the device, it is easy to troubleshoot the system when problems arise. Such control systems are developed with fixed features for specific applications. Typically, large pumps and motors will be equipped with hardwired relay control to protect them against damage under overloads and other undesired working condition. PLC systems provide needed flexibility and allow for future continuous quality improvements in the process.

Figure 1.9 shows two relay circuits for implementing two inputs: AND and OR logic functions, respectively. Each relay has two magnetic coils and an associated normally closed (NC) set of contacts. The two inputs are connected to one side of each of the two coils, and the other end of the coil is connected to ground. The contacts are connected in a predefined manner to produce the desired output as a function of the two inputs. Input A and input B can be at either the ground level (0/low logic/false logic) or the +V level (1/high logic/true logic). The AND arrangement produces the +V logic (high logic) only when the two inputs are high, whereas the OR configuration produces the ground logic (low logic) only when the two inputs are low. Notice that the relay operation involves electrical (coils and power supply) and mechanical (moving contacts) components.

Schematic diagrams for relay logic circuits are often called *logic diagrams*. A relay logic circuit is an electrical diagram consisting of lines/networks/rungs in which each must have continuity to enable the intended output device. A typical circuit consists of a number of networks, with each controlling an output.

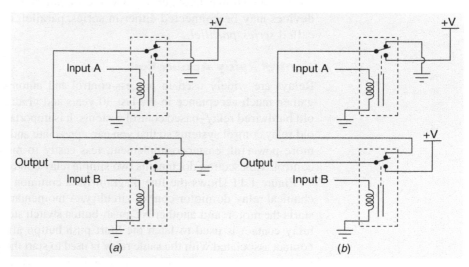

Figure 1.9 (*a*) AND logic function. (*b*) OR logic function.

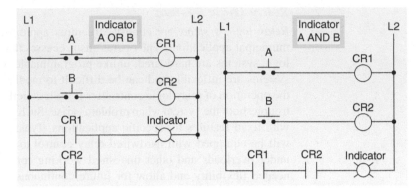

Figure 1.10 Relay logic line diagrams.

Each output is controlled through a combination of input or output conditions (e.g., switches and control relays) connected in series, parallel, or series-parallel to obtain the desired logic to drive the output. Relay logic diagrams represent the physical interconnection of devices. It is possible to design a relay logic diagram directly from the narrative description of a process-control event sequence. In ladder-logic diagrams, an electromechanical relay coil is shown as a circle and the contacts actuated by the coil as two parallel lines. Given this notation, the relay logic line diagrams for the AND and OR logic functions are shown in Fig. 1.10.

The L1 and L2 designations in this logic diagram refer to the two poles of a 120-Vac supply. L1 is the hot side of the supply, and L2 is the ground/neutral side. Output devices are always connected to L2. Any device overloads that are to be included must be shown between the output device and L2; otherwise, the output device must be the last component before L2. Input devices are always shown between L1 and the output device. Relay contact control devices may be connected either in series, parallel, or a combination of both called *series-parallel*.

1.2.3   *Control Relay Application*

Relays are widely used in process-control and automation applications. PLCs gained much acceptance in the last 30 years and gradually replaced most of the old hardwired relay–based control systems. It is important that you understand the old relay control systems so that you can appreciate and make the transition to the more powerful, easier-to-implement, less costly to maintain, and reliable PLC control. This section documents two simple relay control applications.

Figure 1.11 shows the line diagram for a common application of electromechanical relay dc motor-control circuitry. A momentary NO push-button switch starts the motor, and another NC push-button switch stops the motor. The control relay contact is used to latch the Start push button after it is released. Another contact associated with the same relay is used to start the motor. Pressing the Stop push button at any time will interrupt the flow of electricity to the motor and cause it to stop.

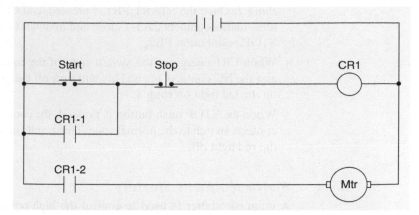

Figure 1.11   Dc motor controls.

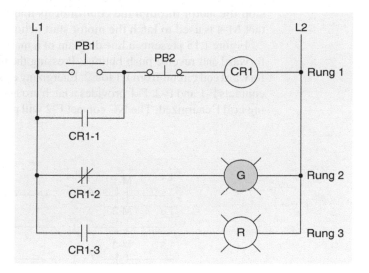

Figure 1.12   Relay controlling two pilot lights.

Another application is shown in Fig. 1.12. The line diagram illustrates how a hardwired relay is used to control two pilot lights. The desired control is accomplished using two push-button switches; PB1 starts the operation, and PB2 terminates it at any time.

Below are the critical steps for this example:

• With no power applied to the control relay, the contacts are in a normal state. The NO contacts are open, and the NC contacts are closed. The green pilot light receives power and turns on as indicated by the green fill light. The red pilot light is off as shown.

- *Rung 1:* Once the START PB1 is pressed, coil CR1 becomes energize; this, in turn, makes contacts CR1-1 close and maintains power to CR1 through the NC STOP push button PB2.
- When CR1 energizes the switch state of the contacts, the NO contacts close, and the NC contacts open. This will turn off the green light on rung 2 and turn on the red light on rung 3.
- When the STOP push button is pressed, the control relay loses power, and the contacts switch to the normal state. This results in turning the green light on and the red light off.

1.2.4 *Motor Magnetic Starters*

A magnetic starter is used to control the high power to the motor, as shown in Fig. 1.13. Three of the motor magnetic starter contacts are used to connect the three phases of the high-voltage supply. In addition, overload relays are physically attached for the motor overload protection. Figure 1.14 shows the low-power motor starter circuit. START and STOP push-button switches start and stop the motor through the control of its magnetic starter. Magnetic starter contact M-4 is used to latch the motor start action.

Figure 1.15 presents a line diagram of a magnetic reversing starter controlled by forward and reverse push buttons. Pressing the forward push button completes the forward coil circuit from L1 to L2. Energizing coil F, in turn, energizes two auxiliary contacts F-1 and F-2. F-1 provides a latch around the forward push button maintaining coil F energized. The NC contact F-2 will prevent the motor from running in the

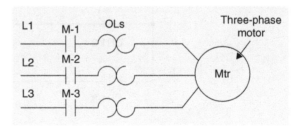

Figure 1.13 High-power motor circuit.

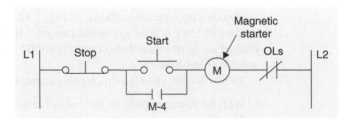

Figure 1.14 Low-power motor starter circuit.

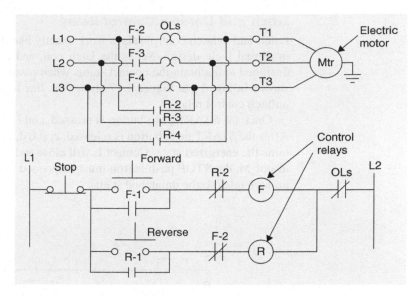

Figure 1.15   Control of reversing motor starter.

reverse direction if the reverse push button is pressed before the Stop push button while the motor is running in the forward direction. The lower part of Fig. 1.15 presents a line diagram of the magnetic reversing starter controlled by forward and reverse push buttons.

Pressing the reverse push button completes the reverse-coil circuit from L1 to L2. Energizing coil R, in turn, energizes two auxiliary contacts R-1 and R-2. R-1 provides a latch around the reverse push button maintaining coil R energized. The NC R-2 contact will prevent the motor from running in the forward direction if the reverse push button is pressed before the Stop push button while the motor is running in the reverse direction. Please refer to the book's website, www .mhprofessional.com/ProgrammableLogicControllers, for an interactive simulator illustrating the forward/reverse motor operation.

Reversing the motor running direction is accomplished by switching two of the motor input voltage phases. When coil R energizes R-2, R-3 and R-4 are closed; L1 connects to T3, L3 to T1, and L2 to T2, causing the motor to run in the reverse direction.

Vertical gate control for downstream water level regulation is one such application that makes use of this reversal of motor running direction. A desired increase in downstream water level requires running the motor in a certain direction, which causes the gate to move upward. Running the motor in the opposite direction will cause the downstream water level to decrease. Movements in both directions are accomplished by a single motor. These motors are heavy-load, high-power devices/actuators with wide use in industrial process-control and automation applications. Typical cost for each such motor is high, and they come ready equipped with a magnetic starter with all the needed instrumentation and protective gear, such as overload relay contacts.

1.2.5   *Latch and Unlatch Control Relay*

Latch and unlatch control relays work exactly like the set-reset flip-flop covered in digital logic design. Set is the latch coil, and reset is the unlatch coil. It is designed to maintain the contact status when power is removed from the coil, as shown in Fig. 1.16. Figure 1.17 shows the line logic diagram for the latch and unlatch control relay.

Once the START push button is pressed, coil L receives power and energizes. After the START push button is released, coil L does not receive power but maintains the energized state. Contact L will close and cause motor M to run. To stop motor M, the STOP push button must be pressed to switch the state of the latch-unlatch relay to the unlatched status.

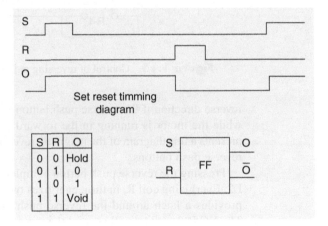

Figure 1.16   Latch and unlatch operation.

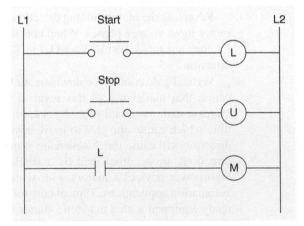

Figure 1.17   Latch-unlatch control line diagram.

The START and STOP push-button switches are interlocked through hardwiring. Either action can be activated at any time, but never both at the same time. The START (latch) and the STOP (unlatch) can be generated through program logic events instead of the two push-button switches shown, such as the temperature in a chemical reactor exceeding a certain range or the level in a boiler drum being below a certain threshold.

1.3 PLC Overview

This section is intended as a brief introduction to PLCs, their history/evolution, hardware/software architectures, and the advantages expected from their use relative to other available choices for process control and automation.

1.3.1 What Is a PLC?

A *programmable logic controller* (PLC) is an industrial computer that receives inputs from input devices and then evaluates those inputs in relation to stored program logic and generates outputs to control peripheral output devices. The I/O modules and a PLC functional block diagram are shown in Fig. 1.18. Input devices are sampled and the corresponding PLC input image table is updated in real time. The user's program, loaded in the PLC memory through the programming device, resolves the predefined application logic and updates the output internal logic table. Output devices are driven in real time according to the updated output table values.

Standard interfaces for both input and output devices are available for the automation of any existing or new application. These interfaces are workable with all types of PLCs regardless of the selected vendor. Sensors and actuators allow the PLC to interface with all kinds of analog and ON/OFF devices through the use of digital I/O modules, analog-to-digital (A/D) converters, digital-to-analog (D/A) converters, and adequate isolation circuits. Apart from the power supply input and the I/O interfaces, all signals inside the PLC are digital and low voltage. Details of PLC hardware and interfaces will be discussed later in Chaps. 2, 5, 7, and 8.

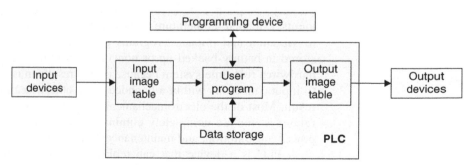

Figure 1.18 Input-output (I/O) PLC architecture.

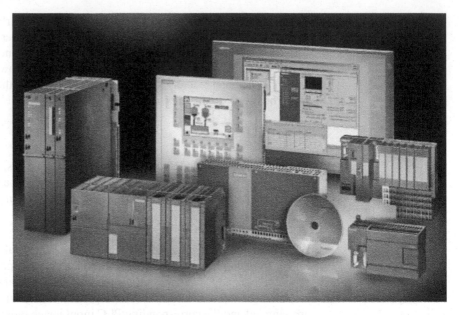

Figure 1.19 Typical industrial PLCs.

Since the first deployment of PLCs four decades ago, old and new vendors have competed to produce more advanced and easier-to-use systems with associated user-friendly development and communications tools. Figure 1.19 shows a sample of actual industrial and popular PLCs. You should notice the diversity of sizes and obviously associated capabilities, thus not only allowing cost accommodation but also enabling the design and implementation of complex distributed control systems. Most vendors allow the integration of other PLCs as part of a networked distributed control system. It is also possible to implement extremely large system control on one PLC system with large number of interconnected chassis and modules.

Wikipedia states that "a *programmable logic controller* (PLC) or *programmable controller* is a digital computer used for automation of electromechanical processes, such as control of machinery on factory assembly lines, amusement rides, or light fixtures." PLCs are used in many industries and machines. Unlike general-purpose computers, the PLC is designed for multiple input and output arrangements, extended temperature ranges, immunity to electrical noise, and resistance to vibration and impact. Programs to control machine operation are typically stored in battery-backed-up or nonvolatile memory. A PLC is an example of a hardwired real-time system because output results must be produced in response to input conditions within a bounded time; otherwise, unintended operation will result. Most of the electromechanical components needed for hardwired control relay systems are completely eliminated, resulting in great reduction in space, power consumption, and maintenance requirements.

A PLC is a device that can replace the necessary sequential relay circuits needed for process control. The PLC works by sampling its inputs and, depending on their state, actuating its outputs to bring about desired changes in the controlled system. The user enters a program, usually via software, that allows control systems to achieve the desired result. Programs are typically written in ladder logic, but higher-level development environments are also available. The International Electrotechnical Commission (IEC) 1131-3 Standard (global standard for industrial control programming) has tried to merge PLC programming languages under one international standard. We now have PLCs that are programmable in function block diagrams, instruction lists, C computer language, and structured text all at the same time! Personal computers (PCs) are also being used to replace PLCs in some applications.

PLCs are used in a great many real-world applications. The evolution of the competitive global economy mandated industries and organizations to commit to investments in digital process control and automation using PLCs. Wastewater treatment, machining, packaging, robotics, materials handling, automated assembly, and countless other industries are using PLCs extensively. Those who are not using this technology are wasting money, time, quality, and competitiveness. Almost all application that use electrical, mechanical, or hydraulic devices have a need for PLCs.

For example, let's assume that when a switch turns on, we want to turn on a solenoid for 15 seconds and then turn it off regardless of the duration of the switch ON position. We can accomplish this task with a simple external timer. What if our process includes 100 switches and solenoids? We would need 100 external timers to handle the new requirements. What if the process also needed to count how many times the switches individually turned ON? We'd have to employ a large number of external counters along with the external timers. All this would require extensive wiring, energy, and space and expensive maintenance requirements. As you can see, the bigger the process, the more of a need there is for PLCs. You can simply program the PLC to count its inputs and turn the solenoids on for the specified time.

1.3.2 *History of PLCs*

Prior to the introduction of PLCs, all production and process-control tasks were implemented using relay-based systems. Industrialists had no choice but to deal with this inflexible and expensive control system. Upgrading a relay-based machine-control production system means a change to the entire production system, which is very expensive and time-consuming. In 1960s, General Motors (GM) issued a proposal for the replacement of relay-based machines. PLC history started with an industrialist named Richard E. Morley, who was also one of the founders of Modicon Corporation in response to the GM proposal. Morley finally created the first PLC in 1977 and sold it to Gould Electronics, which presented it to General Motors. This first PLC is now safely kept at company headquarters.

The website plcdev.com shows the history (reproduced in the following figures) of the development of the PLC by different manufacturers. It spans the period from 1968 to 2005. The new S7-1200 microcontroller was introduced by Siemens in 2009. It was designed to provide an easy-to-use and scalable infrastructure for small and large distributed control applications. Details of the S7-1200 and associated interfaces, including hardware, software, human-machine interfaces (HMIs), communication, and networking, along with industrial-control application implementation using this Siemens infrastructure, will be the focus of this book. Reduction in size, lower cost, larger capabilities, standard interfaces, open communication protocols, user-friendly development environment, and HMI tools are the trend in the evolution of PLCs, as shown in timeline.

The history of PLCs is displayed in time categories starting from the early systems introduced from 1968 to 1971 (Fig. 1.20). This is followed by a span of 6 years labeled as the first PLC generation. The second generation started in 1979 and covered a period of 7 years, ending in 1986. This period showed a greater number of vendors mostly from existing U.S. companies in addition to German and Japanese firms. The early third generation started 1987 and lasted for 10 years, followed by a lasting period of continued growth and advancement in both hardware and software tools, which led to a wide deployment of PLCs in most manufacturing automation and process-control activities.

1.3.3 *PLC Architecture*

A typical PLC mainly consists of a central processing unit (CPU), power supply, memory, communication module, and appropriate circuits to handle I/O data. The PLC can be viewed as an intelligent box having hundreds or thousands of

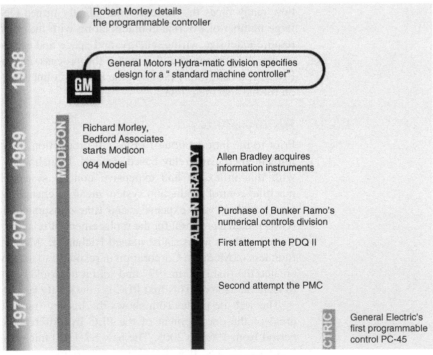

1968 to 1971 Early PLC Systems

Figure 1.20 PLC history chart (R. Morley, father of PLCs).

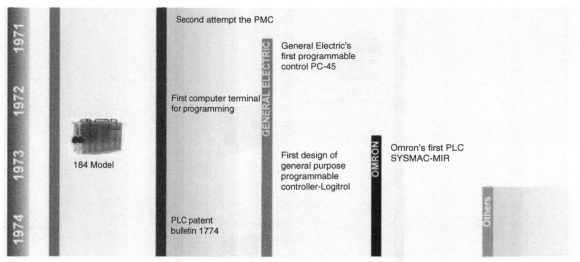

1971 to 1974 First-Generation PLC Systems

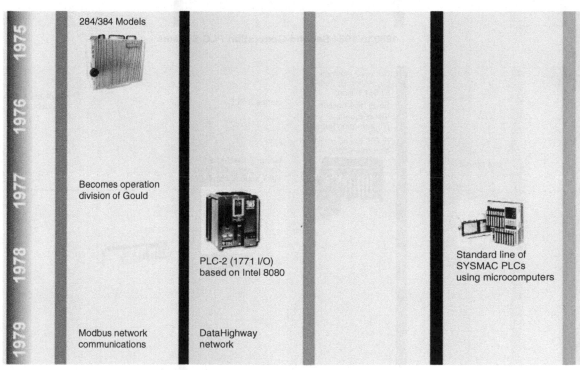

1975 to 1979 Early Second-Generation PLC Systems

Figure 1.20 (*Continued*)

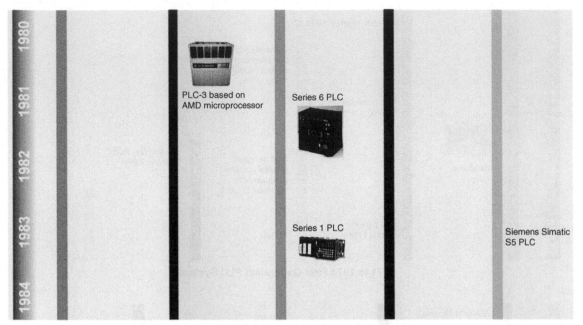

1980 to 1984 Second-Generation PLC Systems

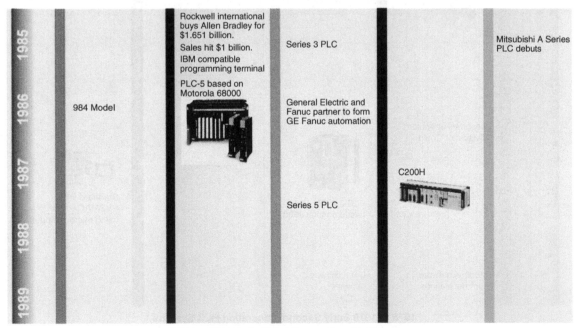

1985 to 1989 Early Third-Generation PLC Systems

Figure 1.20 (*Continued*)

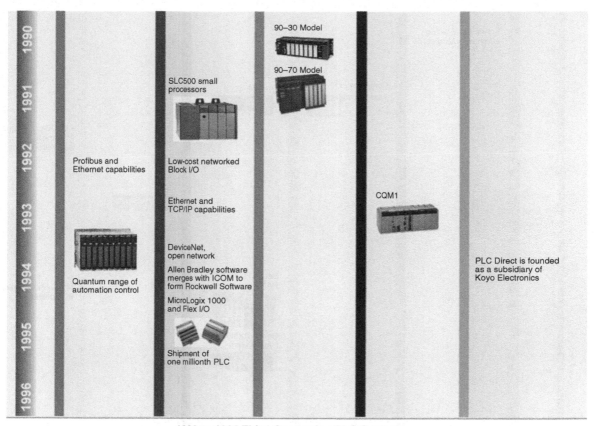

1990 to 1996 Third-Generation PLC Systems

Figure 1.20 (*Continued*)

separate relays, counters, timers, and data-storage locations. These counters, timers, and relays do not exist physically, but they are software-simulated internal entities. The internal relays are simulated through bit locations in memory registers. Figure 1.21 shows a simplified block diagram of a typical generic PLC hardware architecture.

PLC input modules are typically implemented using transistors and do exist physically. They receive signals from external switches and sensors through contacts. These modules allow the PLC to interface with and get a real-time sense of the process status. Output modules are typically implemented using transistors and use triodes for alternating current (TRIACs) to switch the connected power to the output coil when the output reference bit is TRUE. They send ON/OFF signals to external solenoids, lights, motors, and other devices. These modules allow the PLC to interface with and regulate in real time the controlled process. Counters are software-simulated and do not exist physically. They can be programmed to count up, down, or both up and down events/pulses. These simulated counters are limited in their counting speed but suitable for most real-time applications.

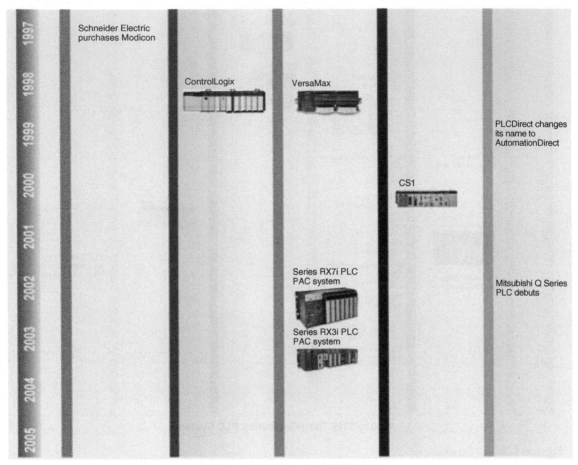

1997

Schneider Electric
purchases Modicon

1998

ControlLogix

VersaMax

1999

PLCDirect changes
its name to
AutomationDirect

2000

CS1

2001

2002

Series RX7i PLC
PAC system

Mitsubishi Q Series
PLC debuts

2003

Series RX3i PLC
PAC system

2004

2005

1997 to 2005 Compact PLC Systems

Figure 1.20 (*Continued*)

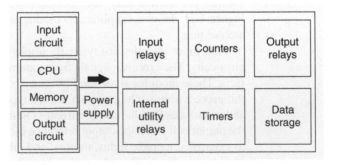

Input circuit	Power supply	Input relays	Counters	Output relays
CPU				
Memory		Internal utility relays	Timers	Data storage
Output circuit				

Figure 1.21 PLC architecture.

Most PLC vendors provide high-speed counter modules that are hardware-based and can accommodate extremely fast events. Typical counters include up-counters, down-counters, and up/down-counters. Timers are also software-simulated and do not exist physically. The most common types are the on-delay, off-delay, and retentive timers. Timing increments vary but typically are larger than 1/1000 of a second. Most process-control applications make extensive use of timers and counters in a variety of ways and applications that will be detailed in Chap. 3.

Data storage is a high-speed memory/register assigned to simply store data. These registers are usually used in math or data manipulation as temporary storage. They also are used to store values associated with timers, counters, I/O signals, and user-interface parameters. Communication buffers and related networking and user-interface tasks also make use of high-speed storage. They also typically can be used to store data and programs when power is removed from the PLC. On power-up, the same contents that existed before power was removed will still be available.

1.3.4 *Hardwired System Replacement*

As stated in the preceding section, PLCs were introduced to replace hardwired relays. In this section we will introduce the process of replacing the relay logic control with a PLC. The example we will use to demonstrate this replacement process may not be very cost-effective for the use of a PLC, but it will demonstrate the fundamental concepts. As shown earlier, the first step is to create the process ladder-logic diagram /flowchart. PLCs do not understand these schematic diagrams, but most vendors provide software to convert ladder logic diagrams into machine code, which shields users from actually having to learn the PLC processor-specific code. Still we have to translate all process logic into the standard symbols the PLC recognizes. Terms such as *switch, solenoid, relay, bell, motor,* and other physical devices are not recognized by PLCs. Instead, *input, output, coil, contact, timer, counter,* and other terms are used.

Ladder-logic diagrams use standard symbols and associated addresses to uniquely represent different elements and events. Two vertical bars, representing L1 and L2, span the entire diagram and are called the *power/voltage bus bars*. All networks/rungs start at the far left, L1, and proceed to the right, ending at L2. Power flows from left to right through available closed circuits. Inputs such as switches are assigned the contact symbol of a relay as shown in Fig. 1.22. Output such as the bell is assigned the coil symbol of a relay. The ac/dc supply is an

A contact symbol **A coil symbol**

Figure 1.22   Contact and coil symbols.

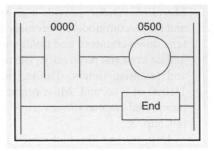

Figure 1.23 Bell logic diagram.

external power source and thus does not show in the ladder-logic diagram. The PLC executes the logic and turns an output ON or OFF using a TRIAC switching interface without any regard to the physical device connected to that output.

The PLC must know the location of each input, output, or other element used in the application. For example, where are the switch and the bell going to be physically connected to the PLC? The PLC has prespecified I/O addresses in a wide variety of signal forms and sizes to interface with all types of devices. For now, assume that the input (the push-button switch) will be labeled "0000" and the output (the bell) will be labeled "0500." The final step converts the schematic into a logical sequence of events telling the PLC what to do when certain real-time events or conditions are satisfied. In the example, we obviously want the bell to sound when the push-button switch is pressed. An electrical power connection to the bell is made while the push-button switch is being pressed. Once the push button is released, the electrical power connection to the bell is removed. The only requirement for this small system to work is to have the push button connected to the PLC input module and to have the bell wired to the PLC output module, as will be shown later. Figure 1.23 shows the logic diagram for this simple example. More industrial control examples and extensive discussion will illustrate this concept in subsequent chapters.

1.3.5 *PLC Ladder Logic*

PLCs use a ladder-logic program, which is similar to the line diagram used in hardwired relay control systems. Figure 1.24 describes the control circuit for a ladder-logic program rung, which is composed of three basic sections: the signal,

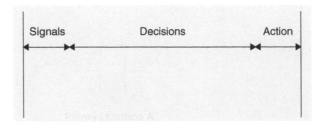

Figure 1.24 Ladder rung/network.

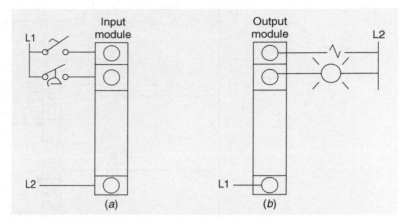

Figure 1.25 I/O terminal connection.

the decision, and the action. The PLC input module scans the input signals, and the CPU executes the ladder-logic program in relation to the input status and makes a decision. The output module updates and drives all output devices. The following sections show the I/O terminal connection and describe the digital I/O addressing format.

As shown in Fig. 1.25(*a*), the input devices are connected to the input module through the hot (L1), whereas the neutral is connected directly to the input module. Figure 1.25(*b*) shows the outputs wired to the output terminal module, the outputs are wired to the output terminal module, and the neutral (L2) is connected to the output devices. The figure shows two digital inputs, a foot switch and a pressure switch, and two outputs, a solenoid and a pilot light.

1.3.6 *Manual/Auto Motor Control Operation*

Figure 1.26 shows a manual/auto control of a three-phase induction motor. While the M/A switch is being held in the manual position, pressing the Start push button energizes the motor magnetic starter M1. Because the Start push button is an NO momentary switch, the power to the magnetic starter is maintained through the latch with the auxiliary contact M1-1 around the Start push button. When the M/A switch is placed in the AUTO position, the digital output module receives the hot line (L1) through the AUTO switch. When rung logic in the software for the output M1 is TRUE, switching of L1 occurs, and the magnetic starter is energized, causing the motor to run. Motor status can be monitored with NO contacts M1-2 wired between the hot line (L1) and the digital input module. The neutral phase is connected directly to the input module. The motor overload conditions, which typically is deployed in the motor for protection and safety operation, are combined and shown in the PLC wiring connection in Fig. 1.26. These safety and protection features are part of the standard safety requirement for most industrial motors.

Figures 1.27 and 1.28 show the results of converting a hardwired control relay to a PLC ladder-logic control. The first example implements a simple motor control

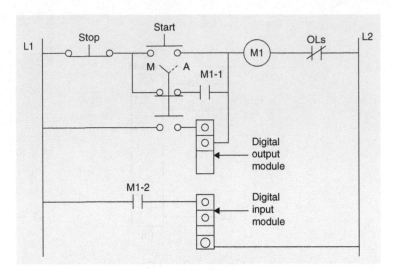

Figure 1.26 Manual/auto motor control PLC connection.

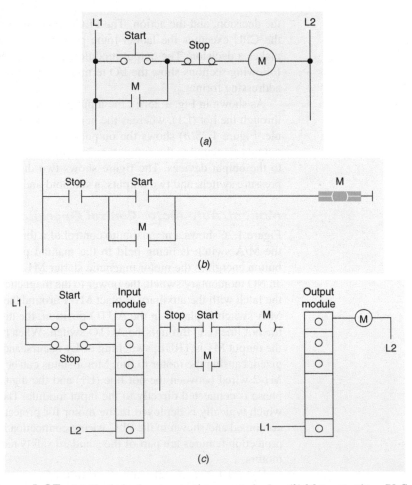

Figure 1.27 (a) Hardwired motor start/stop control relay. (b) Motor start/stop PLC ladder-logic control. (c) Motor start/stop PLC ladder logic in relation to I/O modules.

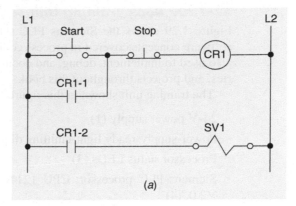

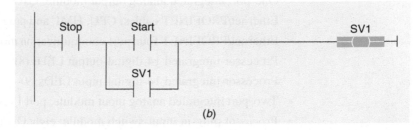

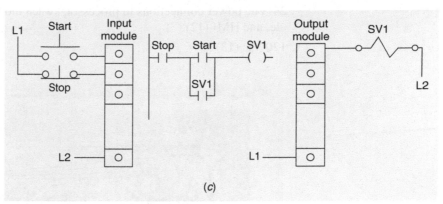

Figure 1.28 (*a*) Hardwired solenoid valve relay control. (*b*) Solenoid valve PLC ladder-logic control. (*c*) Solenoid valve PLC ladder-logic control in relation to I/O modules.

using momentary START and STOP push buttons. The START push button is a NO contact that closes when the switch is pressed and opens when it is released. The STOP push button is a NC contact that opens when the switch is pressed and closes when it is released. The second example shows a simple solenoid-valve control using START and STOP momentary push buttons. The solenoid valve is activated when the START push button is pressed and deactivated through the STOP switch action.

1.3.7 *S7-1200 Book Training Unit Setup*

Figure 1.29 shows the Siemens PLC setup used to demonstrate hardware and software concepts covered in process control and automation. The training unit is also used to implement, debug, and document all examples, homework, laboratories, and projects throughout this book.

The training unit shown in this figure consists of the following items:

- 24-V power supply (1)
- Power-supply-ready light-emitting diode (LED) (2)
- Processor status LEDs (3)
- Siemens PLC processor; CPU 1214C DC/DC/DC, 6ES7 214-1AE30-0XB0 V2.0. (4)
- One-port integrated analog output module: QW80 (5)
- Ethernet/PROFINET cables; CPU, HMI, and programming computer (6)
- Ethernet/PROFINET four-port communication module (7)
- Processor-integrated 14-digital-output LEDs (8)
- Processor-integrated 14-digital-input LEDs (9)
- Two-port integrated analog input module: port 1 (IW64) and port 2 (IW66) (10)
- Processor plug-in input-switch module: eight ON/OFF switches (11)
- 24-Vdc power connections to processor, switch module, communication module, and HMI (12)
- 120 Vac (13)

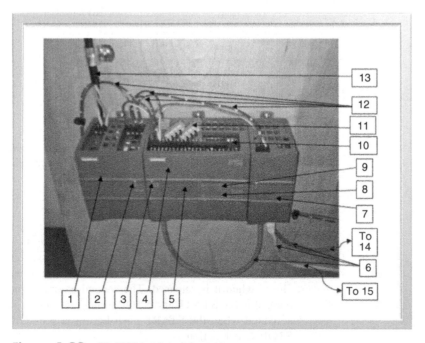

Figure 1.29 S7-1200 book training unit.

- SIMATIC HMI basic color with six function keys: KTP600 basic PN (14)
- Windows programming computer/laptop (15)

The eight ON/OFF switches (I0.0 to I0.7) are used to simulate and test discrete input devices. The corresponding LEDs indicate their individual status. The remaining six digital inputs (I1.0 to I1.5) can be wired to additional discrete input devices. The 14 ON/OFF outputs (Q0.0 to Q0.7 and Q1.0 to Q1.5) are used to simulate and test discrete output devices. The corresponding LEDs indicate their individual status. The two analog inputs (IW64 and IW66) are connected to two potentiometers (0 to 10 V) that simulate analog variable signals. The analog output is connected to a small voltmeter (0 to 10 V). These are the tools and configurations used to establish the trainer in this book. Other configurations can be implemented and used.

1.3.8 *Process-Control Choices*

PLCs are not the only devices available for controlling a process or automating a system. Control relays and PCs can be used to implement the same control. Each choice may be of benefit depending on the control application. This debate has been going on for a long time while a mix of technologies has advanced at an incredible rate. With the continuing trend of declining PLC prices, size shrinking, and performance improving, the choice has become less of a debate in favor of PLCs . Yet system owners and designers have to ask themselves if a PLC is really overkill for an intended process-control or automation application. Table 1.1

Table 1.1 PLC and Control Relay Comparison

Issue of PLC and Control Relay Comparison	PLCs	Control Relays
Control logic changes	Changes in logic can be easily implemented in software.	Changes require more complex hardware modifications.
Deployment on different systems	Easier to customize and download software.	Requires construction of new control panels.
Future expansion	New I/O modules, expansion chassis, HMIs, and software patches can be added. Networked control systems can be used.	Expansion is possible but at higher cost.
Reliability	PLCs are more robust, and redundancy is available.	Less reliable because of the use of individual components.
Down time	Troubleshooting/changes can be made online with no downtime.	Changes or troubleshooting often requires the system to go offline.
Space requirement	Space requirement rapidly decreases as the number of relays increase.	Huge space requirement for a system with large number of relays.
Data acquisition and communication	PLCs support data collection, analysis, and communication.	Not directly or easily possible.
Maintenance and speed of control	Less maintenance and faster speed of control.	Mechanical parts require more maintenance and reduce speed of control.
Cost	Effective cost and performance for a wide range of process-control applications.	Can be cost-effective for very small systems.

summarizes a brief comparison between PLCs and control relays and addresses important issues to be considered:

A *dedicated controller* is a single instrument that is dedicated to controlling one process variable such as temperature for a heating control. Dedicated controllers typically use *proportional integral derivative* (PID) control and have the advantage of an all-in-one package, typically with display and buttons. These controllers can be an excellent tool to use in simple applications. PLCs can compete pricewise and functionally with these controllers, especially if several controllers are needed. PLCs offer greater degree of flexibility and can be programmed to handle existing and future scenarios.

PCs also can be fitted with special hardware and software for use in process-control applications. PCs can provide an advantage in certain control tasks relative to PLCs, but their use is not as widespread as PLCs. Hybrid networked system of PLCs and PCs is in wide use in large distributed control applications. Table 1.2 shows a brief comparison between PLCs and PCs and addresses important issues to be considered.

Table 1.2   PLC and PC Control System Comparison

Comparison Issues	PLCs	PCs
Environment	PLCs are specifically designed for harsh conditions with electrical noise, magnetic fields, vibration, and extreme temperatures or humidity.	Common PCs are not designed for harsh environments. Industrial PCs are available but at a much higher cost.
Ease of use	By design, PLCs are friendlier to technicians because they are programmed in ladder logic and have easy connections.	Operating systems such as Windows, UNIX, and Linux are common. Connecting I/O to the PC is not always easy.
Flexibility	PLCs in rack format are easy to exchange and expand. They are designed for modularity and expansion.	Typical PCs are limited by the number of special cards they can accommodate and are not easily expandable.
Speed	PLCs execute a single program in sequential order and have a better ability to handle real-time events.	PCs are designed to handle multitasks. Real-time operating systems can handle real-time events.
Reliability	A PLC rarely crashes over long periods of time.	Locking up and crashing are more frequent with PCs.
Programming languages	PLC languages used are typically ladder logic, function block, or structured text.	PCs are very flexible and powerful in providing a wide variety of programming tools.
Data management	Memory is limited in its ability to store and analyze large amounts of data.	PCs excel with any long-term data storage, modeling, simulation, and trending.
Cost	Hard to compare pricing with many variables like I/O counts, hardware needed, programming software, and so on.	

Homework Problems and Laboratory Projects

Problems

1.1 What is the meaning of the word *process* in the chemical industry?

1.2 Define the following:
   a. Self-regulated process
   b. Process variable
   c. Process set point
   d. Controlled variable
   e. Controlling variable
   f. The difference between manual and automated control
   g. Dead band

1.3 What is the difference between open- and closed-loop control?

1.4 Describe the difference between direct-acting and reverse-acting control.

1.5 List at least four advantages of PLC control over hardwired relay control.

1.6 Explain the advantages of using a logic diagram or flowchart in programming.

1.7 Explain the steps used in implementing single-variable closed-loop control.

1.8 Define the following:
   a. Absolute error
   b. Error as a percent of set point
   c. Error as a percent of range

1.9 If an oven set point = 210°C, measured value = 200°C, range = 200 to 250°C, answer the following.
   a. What is the absolute error?
   b. What is the error as a percent of set point?
   c. What is the error as a percent of what range?
   d. Repeat the preceding parts for a measured temperature value of 230°C, assuming that the same set point and range do not change.

1.10 Explain why the *National Electrical Code* demand users to control a motor's START/STOP using NO/NC momentary push-button switches instead of maintain switches.

1.11 Explain the following:
   a. The function of a process controller
   b. The function of the final control element
   c. The main objectives of process control

1.12 Study the circuit in Figure 1.30, and answer the following questions:
   a. What logic gate type does the indicator represent?
   b. What is the status (ON/OFF) of the indicator if push buttons A and B are pressed and released one time?
   c. What is the status of the indicator if push buttons A and B are pressed and maintained closed all the time?
   d. What is the status of the indicator if push button A or B is pressed one time?
   e. Show how you can modify the circuit to maintain the indicator status ON if push button A or B is pressed and maintained closed.
   f. Modify the circuit in the figure to maintain the indicator ON once the two push buttons are activated.
   g. Add a STOP push button to turn the indicator OFF and restart the process at any time.

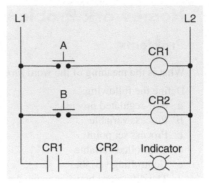

Figure 1.30

1.13   Figure 1.31 shows a line diagram for an auto/manual motor-control circuit. The correct circuit was discussed in this chapter (Fig. 1.26). The STOP push-button switch will stop the motor only in the manual mode. The START push-button switch will start the motor and maintain its running status in the manual mode through the magnetic starter contact M1-1. The START push button can only jog the motor in the auto mode. As shown, the circuit has an error. Correct the error, and explain why the circuit should be corrected.

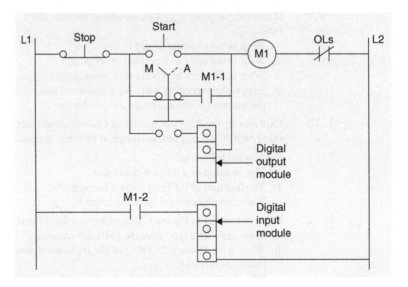

Figure 1.31   Auto/manual control.

1.14   What is the status of CR1, M1, and SV1 in Fig. 1.32 under the following conditions?
    a. PB1 is not pushed, and LS1 is open.
    b. PB1 is pushed, and LS1 is open.
    c. PB1 is pushed, and LS1 is close.

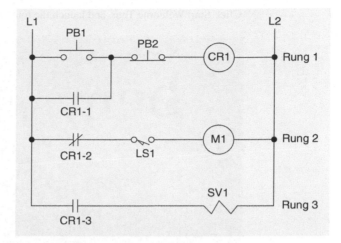

Figure 1.32 Hardwired control relay for motor and solenoid valve activation.

Projects

Laboratory 1.1: Getting familiar with Siemens S7-1200 PLC software

Start the Siemens TIA Portal to launch the project view shown in Fig. 1.33. Perform the following steps in offline mode:

a. Welcome tour and opening an existing application (My_First_Lab)
   Click Welcome Tour.

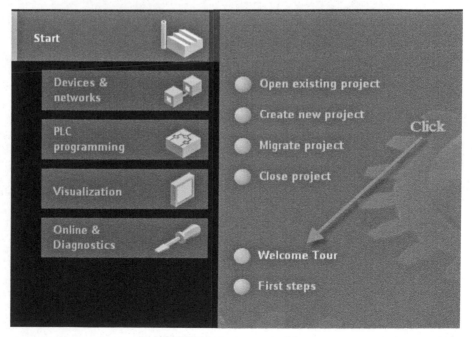

Figure 1.33 S7-1200 Portal project view.

Click Start Welcome Tour, and launch the tour as displayed in Fig. 1.34.

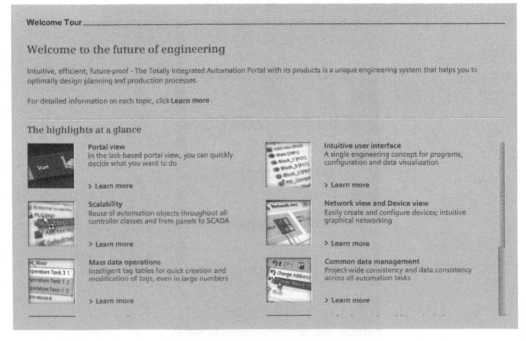

Figure 1.34 Navigating the welcome tour.

From your computer desktop find and click My_First_Lab as shown in Fig. 1.35. This file and shown associated folders are available from your instructor or can be downloaded from the book's website.

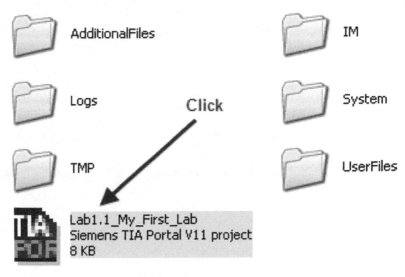

Figure 1.35 Loading My_First_Lab.

The project-tree functions look and behave like Windows Explorer:

- As with other Windows programs, a folder with a plus (+) sign can be expanded to show its contents.
- A folder with minus (−) sign can be collapsed to hide its contents.

Using the Windows tool bar, you can perform the following:

- Open files
- Delete files
- Copy files
- Rename files
- Create new file

Open file:

- After creating the application file (My_First_Lab), click Write PLC Program, and explore screen options.
- Explore the desired basic instructions, and use the Help menu for further clarification.

b. Laboratory requirements:

- Use an application file.
- Copy and save an application file.
- Navigate through the software.
- Navigate through the online help system.
- Get ready to create and edit a new PLC ladder program using the screen shown in Fig. 1.36.

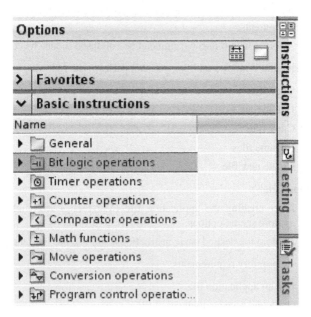

Figure 1.36 PLC ladder program creation.

Laboratory 1.2: Getting familiar with S7-1200 software, offline programming, and the help menu

Part A

Enter the network shown in Fig. 1.37, and do the following from the project tree (as illustrated in Fig. 1.38):

1. Drag and drop the NO contact.
2. Enter the address I0.0.
3. Click on the contact to rename Tag_1 to STOP.
4. Click Change.

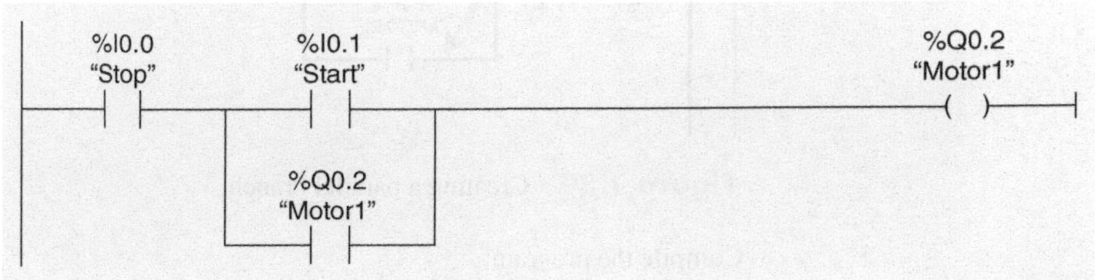

%I0.0 "Stop" — %I0.1 "Start" — %Q0.2 "Motor1"

%Q0.2 "Motor1"

Figure 1.37 Motor1 start network.

To complete Network 1, follow the preceding steps to enter START and MOTOR1 as illustrated in Fig. 1.39.

To enter the branch arround the start contact, do the following:

1. Drag and drop open branch.
2. Enter the NO contact MOTOR1.
3. Drag and drop close branch.

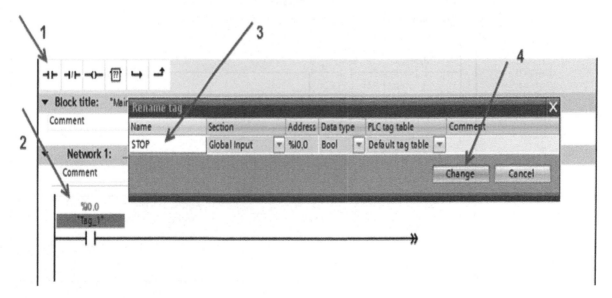

Figure 1.38 Motor1 start network creation steps.

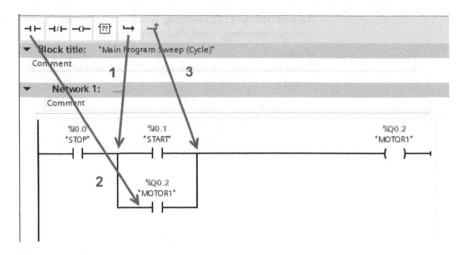

Figure 1.39 Creating a parallel branch.

Compile the program.

Notice that you have zero error and one warning after you compile the program as shown in Fig. 1.40. The warning is issued because the hardware is not configured. Hardware configuration will be covered in Chap. 2, Laboratory 2.2. Save the program.

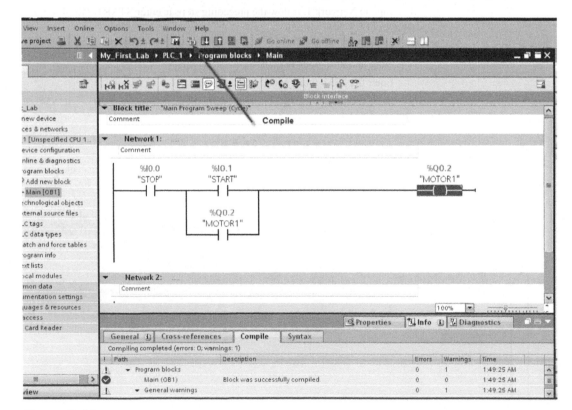

Figure 1.40 Compiling the program.

Part B

Enter the networks in Figs. 1.41 to 1.44 to implement the four combinational logics AND, OR, XOR, and XNOR. This diagram assumes two input switches (SW1 and SW2) and four output coils AND_LOGIC, OR_LOGIC, EXOR_LOGIC, and EXNOR_LOGIC.

Network 1

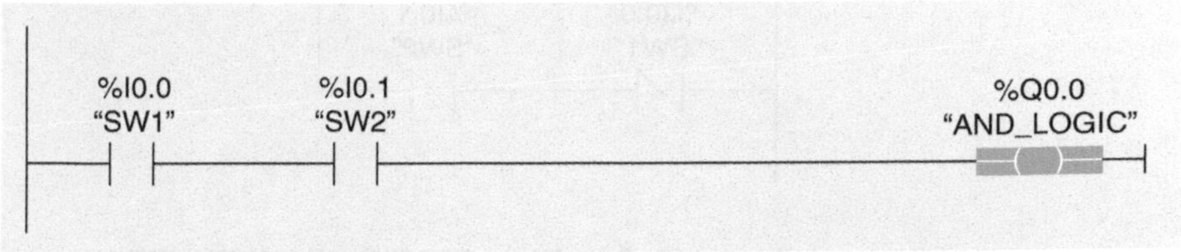

Figure 1.41   Logic AND network.

Network 2

Figure 1.42   Logic OR network.

Network 3

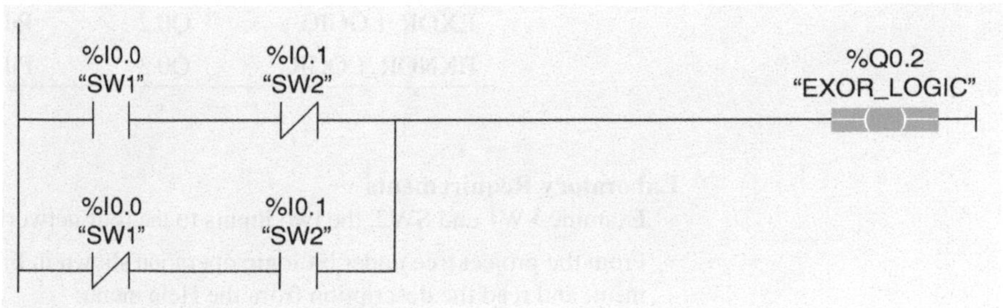

Figure 1.43   Exclusive OR (XOR/EXOR) logic.

Network 4

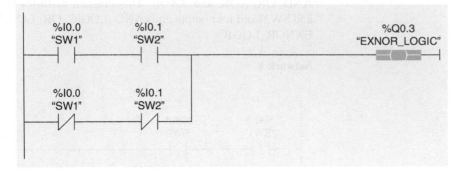

Figure 1.44 Exclusive NOR (XNOR/EXNOR) logic.

I/O addresses should be documented as follows:

System Input

Tag Name	Address	Comments
SW1	I0.0	Toggle switch
SW2	I0.1	Toggle switch

System Output

Tag Name	Address	Comments
AND_LOGIC	Q0.0	Pilot light1
OR_LOGIC	Q0.1	Pilot light2
EXOR_LOGIC	Q0.2	Pilot light3
EXNOR_LOGIC	Q0.3	Pilot light4

Laboratory Requirements

- Examine SW1 and SW2, the two inputs to the four network logics, and verify the logic.
- From the project tree under bit logic operation shown in Fig. 1.45, click Negate Assignment, and read the description from the Help menu.
- Edit and save Network 4, the XNOR logic network, to use the negate assignment and the XOR output instead of generating it from SW1 and SW2.
- Reload the program, and verify that new Network 4 logic works.
- Write the laboratory report to document all that you learned.

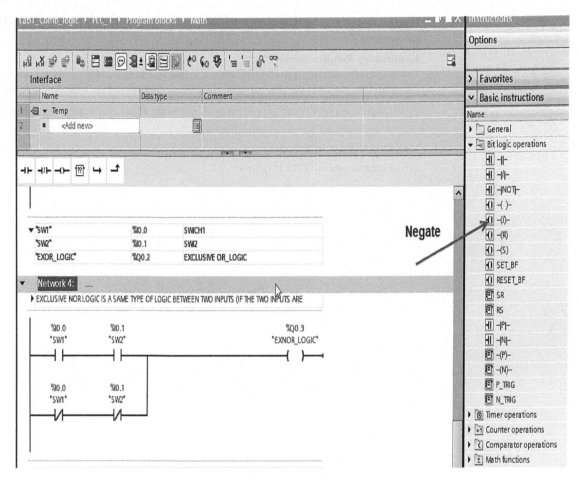

Figure 1.45 Bit logic operations.

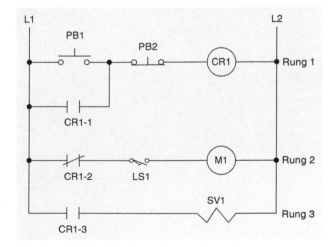

Figure 1.46

Laboratory 1.3: Converting hardwired control relay to a ladder-logic program

Laboratory Requirements

Using the information given in Laboratory 1.1, and assuming that PB1 and PB2 are NO and NC push buttons, respectively, wired hot in the line diagram in Fig. 1.46, do the following:

1. Convert the hardwired control relay shown in the figure to a ladder-logic program.

2. Assign and document all I/O addresses.

3. Document your program.

4. Compile and save the program.

2

Fundamentals of PLC Logic Programming

This chapter focuses on the PLC hardware fundamentals and logic programming for the Siemens S7-1200. Concepts covered are applicable to other types of PLCs.

Chapter Objectives

▶ Understand the PLC hardware and memory organization.

▶ Understand ladder-logic diagrams and programming.

▶ Understand combinational and sequential logic instructions.

▶ Use ladder programming in industrial process control.

A programmable logic controller (PLC) is a microprocessor-based computer unit that can perform control functions of many types and varying levels of complexity. PLC programming is a major task because in addition to requiring knowledge of the specific ladder-logic development environment and associated utilities, it assumes familiarity with the control application domain. Understanding of the PLC hardware details, human-machine interfaces (HMIs), and communication fundamentals is a must. This chapter focuses on the PLC hardware fundamentals and logic programming for the Siemens S7-1200. HMIs and communication will be addressed in Chap. 5

2.1 PLC Hardware

The S7-1200 PLC provides the flexibility and power to control a wide variety of devices in support of automation needs. The compact design, flexible configuration, and powerful instruction set combine to make the S7-1200 a perfect solution for controlling a wide variety of applications. The central processing unit (CPU) combines a microprocessor, an integrated power supply, input circuits, and output circuits in a compact housing to create a powerful PLC. After you download your program, the CPU contains the logic required to monitor and control the devices in your application. The CPU monitors the inputs and changes the outputs according to the logic of your user program, which can include Boolean logic, counting, timing, complex math operations, and communications with other intelligent devices.

2.1.1 S7-1200 Processor

The CPU provides a PROFINET port for communication over a PROFINET network. PROFINET uses the Ethernet network protocol as in offices and information technology (IT) departments. However, its capabilities have been enhanced to meet the far-tougher conditions encountered in factory automation, process automation, and other industrial applications. More coverage of communication, networking, and PROFINET is provided in Chap. 5. Communication modules are available for communicating over RS485 and RS232 networks. Figure 2.1 shows

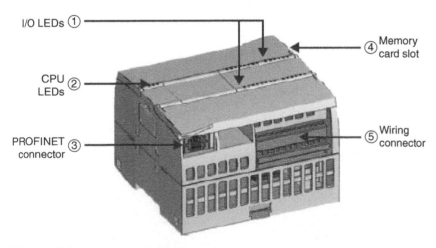

I/O LEDs ①

CPU LEDs ②

PROFINET ③ connector

④ Memory card slot

⑤ Wiring connector

Figure 2.1 Typical S7-1200 processor.

a typical Siemens S7-1200 processor. The SIMATIC S7-1200 system comes in three different models: CPU 1211C, CPU 1212C, and CPU 1214C. The following five areas are pointed out:

1. Status light-emitting diodes (LEDs) for the onboard input-output (I/O)
2. Status LEDs for the operational state of the CPU
3. PROFINET connector
4. Memory card slot (under door)
5. Removable user wiring connector

2.1.2 Operating Modes of the CPU

The CPU has three modes of operation: STARTUP, STOP, and RUN. The following are the characteristic of each of the three CPU modes:

- In STOP mode, the CPU is not executing the program. Projects cannot be executed in this mode.
- In STARTUP mode, the startup organizational blocks (OBs) are called by the operating system. OBs (if present) are executed once and usually contain setup instructions. Interrupt events are not processed during the startup phase.
- In RUN mode, the scan cycle is executed repeatedly in the processor memory, and outputs are activated according to the implemented program logics. The program cannot be downloaded in this mode.

2.1.3 Communication Modules

The S7-1200 family provides communication modules (CMs) that provide critical additional functionality to the system. There are two communication modules: RS232 and RS485. The CPU supports up to three CMs. Each CM connects to the left side of the CPU (or to the left side of another CM that is connected to the CPU). Figure 2.2 shows a typical Siemens S7-1200 communication module. The following areas are pointed out:

1. Status LEDs for the communication module
2. Communication connector

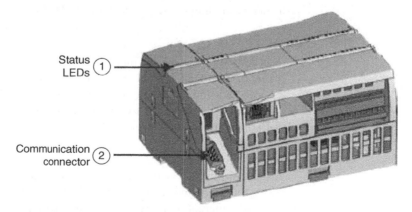

Figure 2.2   Communication module.

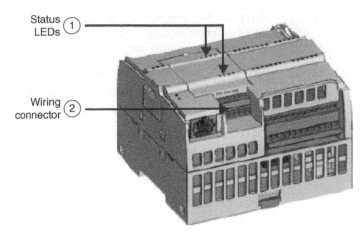

Status LEDs ①

Wiring connector ②

Figure 2.3 Signal board.

2.1.4   *Signal Boards*

A *signal board* (SB) allows user to add I/O module to the CPU. One SB can be added inside the front of any CPU to easily expand the digital or analog I/O without affecting the physical size of the controller. SBs can be connected to the right side of the CPU to further expand the digital or analog I/O capacity. CPU 1212C accepts two and CPU 1214C accepts eight SBs. Figure 2.3 shows an S7-1200 signal board.

The following are the two available SB types:

- SB with four digital I/O (2 × dc inputs and 2 × dc outputs)
- SB with one analog output

2.1.5   *Input-Output Modules*

I/O modules are of three types: digital, analog, and special. Digital I/O modules provide discrete ON/OFF voltage-type signals, and analog I/O modules provide variable (minimum to maximum value) voltage or current signals. An example of a special module is a high-speed pulse (HSP) counter or an ASCII module.

Digital Input Modules

As shown in Fig. 2.4, an input module performs four main tasks: it senses the presence of an input signal, maps the input signal (which is typically a 120 Vac or 24 Vdc to a low dc voltage), isolates the input signal from the mapped output signal, and outputs a direct-current (dc) signal to be sensed by the PLC processor (CPU) during the input scan cycle.

Digital Output Modules

An output module operates in the opposite manner as an input module. It acts as a triode for alternating current (TRIAC) switch connecting any selected input to the module's alternating-current (ac) or dc voltage, as shown in Fig. 2.5. The output

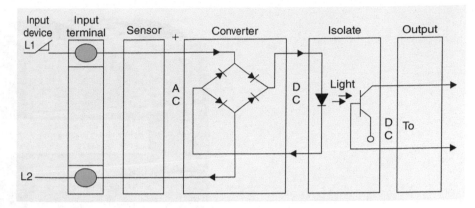

Figure 2.4   Digital input module.

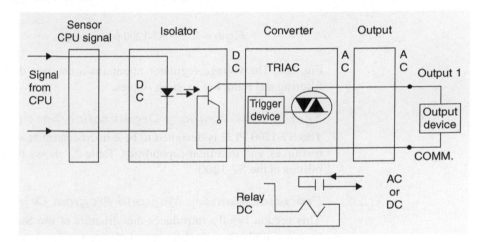

Figure 2.5   Digital output module.

module having the matched terminal address receives an output command from the CPU. The output module performs the voltage switching to the selected terminal output point.

TRIAC is a trade name for an electronic component that can conduct current in either direction when it is triggered (turned ON) and is formally called a *bidirectional triode thyristor* or *bilateral triode*. The bidirectional property makes TRIACs very convenient switches for ac circuits, also allowing them to control very large power flows.

2.1.6   *Power Supply*

The main function of the power supply is to convert the 120/220-Vac input to the 24 Vdc required for the PLC operation. The power supply has three main components: line conditioner, rectifier, and voltage regulator. The *line conditioner* purifies the input ac voltage waveform to a smoothed sine wave. The rectifier converts the stepped-down input ac voltage to the required dc voltage level, as shown in

Figure 2.6 S7-1200 power supply.

Fig. 2.6. The voltage regulator maintains a constant dc output voltage level by filtering and reducing existing ripples.

2.1.7 *S7-1200 PLC Memory Organization/Specifications*

The S7-1200 PLC is designed to be a microcontroller with compact size, limited resources, and excellent capabilities. Table 2.1 shows the specification and capabilities of the S7-1200.

2.1.8 *Processor Memory Map and Program Organization*

This section briefly introduces the structure of the Siemens S7-1200 processor memory. It also covers the concept of structured programming through the use of organization blocks, function blocks, and functions.

Memory Areas

The processor *memory area* is divided into three sections. As shown in Fig. 2.7, each memory area stores the user program, user data, and configuration. The following is a brief description of each section:

- *Load memory* is nonvolatile storage for the user program, data, and configuration.
- *Work memory* is a volatile storage work area for some elements of the user project while the user program is executing.
- *Retentive memory* is nonvolatile storage for a limited quantity of work memory values.

Memory Map

The processor *memory map* is divided into several data files, as shown in Fig. 2.8, where each data file consists of an operand and tags such as inputs, outputs, and bit memory. The CPU identifies these operands based on a numerical absolute address.

Table 2.1   S7-1200 PLC Specifications

Feature	CPU 1211C	CPU 1212C	CPU 1214C
Physical size (mm)	90 × 100 × 75		110 × 100 × 75
User memory			
• Work memory	• 25 Kbytes		• 50 Kbytes
• Load memory	• 1 Mbyte		• 2 Mbytes
• Retentive memory	• 2 Kbytes		• 2 Kbytes
Local on-board I/O			
• Digital	• 6 inputs/4 outputs	• 8 inputs/6 outputs	• 14 inputs/10 outputs
• Analog	• 2 inputs	• 2 inputs	• 2 inputs
Process image size	1024 bytes (inputs) and 1024 bytes (outputs)		
Signal modules expansion	None	2	8
Signal board	1		
Communication modules	3 (left-side expansion)		
High-speed counters	3	4	6
• Single phase	• 3 at 100 kHz	• 3 at 100 kHz 1 at 30 kHz	• 3 at 100 kHz 3 at 30 kHz
• Quadrature phase	• 3 at 80 kHz	• 3 at 80 kHz 1 at 20 kHz	• 3 at 80 kHz 3 at 20 kHz
Pulse outputs	2		
Memory card	SIMATIC memory card (optional)		
Real-time clock retention time	10 days, typical/6-day minimum at 40 degrees		
PROFINET	1 Ethernet communication port		
Real math execution speed	18 μs/instruction		
Boolean execution speed	0.1 μs/instruction		

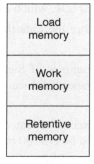

Figure 2.7   Memory areas.

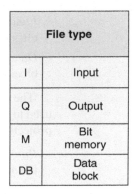

Figure 2.8   Memory map.

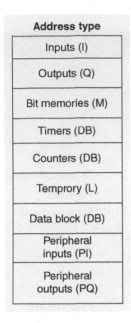

Address type
Inputs (I)
Outputs (Q)
Bit memories (M)
Timers (DB)
Counters (DB)
Temprory (L)
Data block (DB)
Peripheral inputs (PI)
Peripheral outputs (PQ)

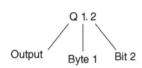

Figure 2.9   CPU memory addresses.   **Figure 2.10**   Discrete addressing format.

CPU Memory Addresses

The CPU can access elements addressed in the formats shown in Fig. 2.9. Also shown is the detailed format for a discrete output. Other address types follow the same pattern but use different number of bits per element.

Discrete Output Addressing Format

Figure 2.10 shows the discrete addressing format.

Code Blocks

The CPU supports the following types of code blocks that allow users to create an efficient modular program:

- *Organization blocks* (OBs) define the structure of a program.
- *Functions* (FCs) and *function blocks* (FBs) contain the program code that corresponds to a particular task, which can be executed frequently or as needed.
- *Data blocks* (DBs) store the data that can be used by the different program blocks.

The following are examples of Siemens programming blocks that are often used to structure and enhance the documentation of uses of ladder process-control projects:

- *Program-cycle OBs* execute repeatedly as long as the CPU is running. OB1 is the default; others must be OB200 or greater.
- *Startup OBs* execute one time when the operating mode of the CPU changes from STOP to RUN. All codes that need to be executed only one time, such as

initialization for certain parameters or the configuration of hardware modules, should be placed in startup OBs.

- *Time-delay OBs* execute at a specified interval after an event is configured by the start interrupt (`SRT_DINT`) instruction.
- *Cyclic-interrupt OBs* execute at a specified interval. A cyclic-interrupt OB will interrupt cyclic program execution at user-defined intervals.

A *subroutine* (also known in different programming languages as a *procedure, function, routine, method,* or *subprogram*) is a portion of code within a larger program that performs a specific task and is relatively independent of the remaining code. It is similar to the previously defined OBs and FCs for the Siemens system. A subroutine is often coded so that it can be started/called several times from different places during a single execution of the program and then branch back/return to the next instruction after the call point once its task is concluded. Examples of subroutine usage include the following:

- Loading and executing a specific recipe when needed
- Initializing a system startup
- Performing a common calculation at different points in a program
- Updating alarms and displays
- Updating communication data and protocol parameters

Functions are logic blocks without memory. After the function has been executed, the data in the temporary variables therefore are lost. Figure 2.11 shows a simple function, and the following provides details about its initiation and coding.

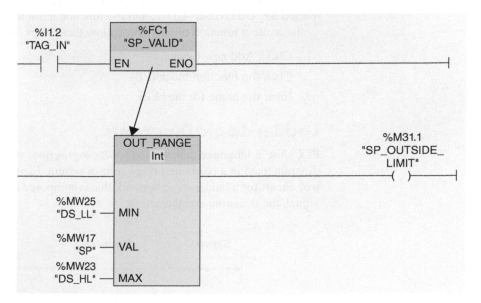

Figure 2.11   A simple function.

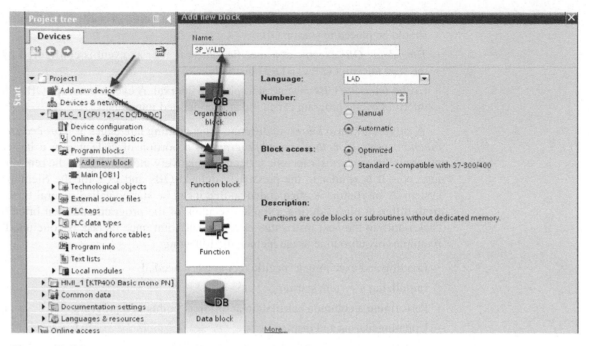

Figure 2.12 Creating a function block.

If `TAG_IN` is TRUE, the function `SP_VALID` is executed. The function is shown in the figure with only one network. At initiation of the function, program cyclic execution transfers to the `OUT_RANGE` instruction, which compares the set point (SP) to the `MIN` value, with tag name `DS_LL` and `MAX` value, and with tag name `DS_HL`. If the SP is outside the low to high limit, output energizes the tag named `SP_OUTSIDE_LIMT`, and the function terminates.

To create a function block (FB), follow these steps, as illustrated in Fig. 2.12:

1. Click "Add new block."
2. Click the function block (FB).
3. Enter the name for the block.

2.2 Ladder-Logic Diagrams

PLCs use a language called a *ladder-logic program*, which is similar to the line diagram used in a hardwired relay control system. Figure 2.13 describes the control circuit for a ladder-logic network that is composed of three basic sections: the signal, the decision, and the action.

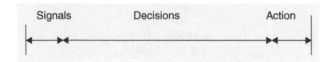

Figure 2.13 Ladder-logic diagram network/rung.

Table 2.2 PLC Program Scan Cycle

Event in Operating Cycle	Description
Input scan	The status of each input module is read, and the input image table in the processor is updated with the information
Program scan	The ladder program is executed
Output scan	The output image table information is transferred to the output module
Communication	Communication with computer and other devices takes place
Processor overhead	Internal housekeeping in the processor takes place, including updates of the status file and internal time base

The PLC input modules scan the input signals, and the CPU executes the ladder-logic program in relation to the input status and makes a decision. The output modules update and drive all output devices. The program scan process or what is referred to as *PLC events* in the operating cycle are summarized in Table 2.2. A description of each scan operation is also given. The following sections show the I/O terminal connection and describe the digital I/O addressing format.

2.2.1 *PLC Input/Output Terminal Connection*

As shown in Fig. 2.14(*a*), the input devices are connected to the input module via the hot line (L1), whereas the neutral line is connected directly to the input module. Figure 2.14(*b*) shows the output devices wired to the output terminal module through the neutral line (L2), whereas the hot line is connected directly to the output module.

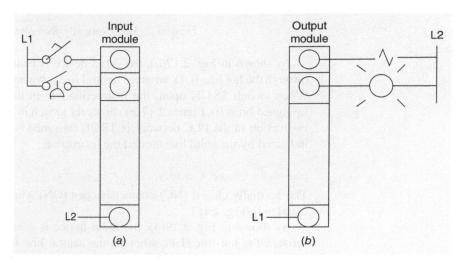

Figure 2.14 I/O hardwired connections.

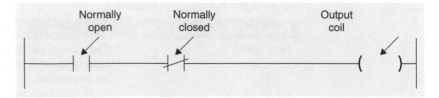

Figure 2.15 Ladder-logic network example.

Figure 2.15 shows a ladder-logic network. This is very similar to the line diagram used in a hardwired relay. Every instruction is examined if TRUE (TRUE means the bit in the PLC memory assigned the value of 1). If TRUE, continuity of the rung is maintained or power flow is maintained. When all input instructions are TRUE, the output will be set to 1 (ON). The figure shows an example of such network. The network has two input elements (representing decision) and a coil (representing action), as shown previously in Fig. 2.13. A brief description of the three commonly used instructions will be provided in the next section. The Siemens S7-1200 system notation, hardware, and software development tools will be assumed throughout this book and in all implemented examples and projects.

2.2.2   *PLC Boolean Instructions*

Normally Open Contact

The normally open (NO) contact is closed (ON) when the assigned bit value in memory is equal to 1 (Fig. 2.16).

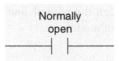

Figure 2.16 Normally open contact ON.

As shown in Fig. 2.17(*a*), the input device is connected to the input module through the hot line (L1), whereas neutral line is connected directly to the module. When switch SS1 is open, the instruction NO in the PLC network is FALSE (assigned bit is 0). Figure 2.17(*b*) shows the switch in the closed position, the NO instruction in the PLC network is TRUE (assigned bit is 1), and power flows as indicated by the solid line around the instruction.

Normally Closed Contact

The normally closed (NC) contact is open (ON) when the assigned bit value is equal to 0 (Fig. 2.18).

As shown in Fig. 2.19(*a*), the input device is connected to the input module through the hot line (L1), whereas the neutral line is connected directly to the module. When switch SS1 is open, the instruction NC in the PLC network is TRUE, and power flows through the network, as indicated by the solid line around

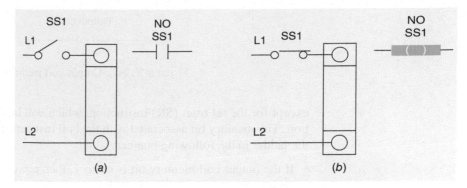

Figure 2.17 Hardwired connections and the associated instruction status.

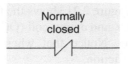

Figure 2.18 Normally closed contact ON.

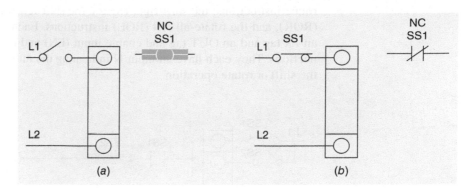

Figure 2.19 Hardwired connections and associated instruction status.

the instruction. Figure 2.19(*b*) shows the switch in the closed position; the NC instruction in the PLC network is FALSE, and power does not flow through the network.

Output Coil

The output-coil instruction writes a value for an output bit in the PLC memory based on the power-flow status preceding the instruction; if all preceding conditions are TRUE, the output instruction becomes TRUE. The output signals for the associated control actuators are wired to the output-coil terminals. Coils are assigned unique memory bit addresses. No two coils can have the same address

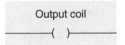

Figure 2.20 Output-coil memory bit.

except for the set reset (SR) instruction, which will be discussed in the next section. The memory bit associated with the coil instruction is updated every scan of the ladder in the following manner:

- If the output-coil memory bit is set to 1, then power flows through that output coil.
- If the output coil memory bit is set to 0, then power will not flow through that output coil (Fig. 2.20).

Figure 2.21 shows the three instructions: normally open (NO), normally close (NC), and output coil (OC) hardwired connections in relation to the PLC instructions. Notice that when an instruction is TRUE, power flows through the instruction.

2.2.3   *Shift and Rotate Instructions*

This section examines four of the commonly used memory-register shift and rotate instructions: the shift-right (SHR), the shift-left (SHL), the rotate-all-right (ROR), and the rotate-all-left (ROL) instructions. Each of these instructions has an IN tag and an OUT tag and enable input (EN) and enable output (ENO) connections. They each have an input N indicating the number of bits to be used for the shift or rotate operation.

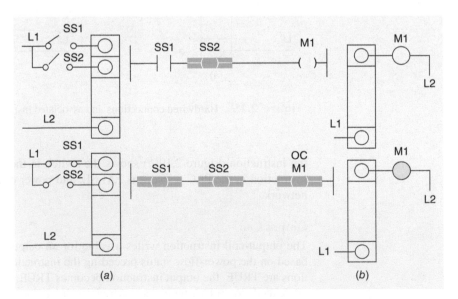

Figure 2.21 Hardwired connections and the associated instructions status.

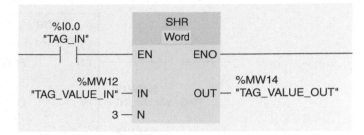

Figure 2.22 Shift-right instruction.

Shift-Right Instruction (SHR)

Figure 2.22 shows this instruction.

If `TAG_IN` is set, the shift right instruction is executed. The content of `TAG_VALUE_IN` is shifted 3 bit positions to the right. The result is sent to the `TAG_VALUE_OUT` output. If `TAG_VALUE_IN` = 0011 1111 1010 1111, then after the `SHR` instruction executes, the `TAG_VALUE_OUT` will equal 000 0 0111 1111 0101.

Shift-Left Instruction (SHL)

Figure 2.23 shows this instruction.

If `TAG_IN` is set, the shift-left instruction is executed. The content of `TAG_VALUE_IN` is shifted 4 bit positions to the left, as indicated by the unsigned integer `TAG_SHIFT_NUMBER`, which contains the value 4. The result is sent to the `TAG_VALUE_OUT` output. If `TAG_VALUE` = 0011 1111 1010 1111, then after the instruction SHL executes, the `TAG_VALUE_OUT` will equal 1111 1010 1111 0000.

Rotate-All-Right Instruction (ROR)

Figure 2.24 shows this instruction.

If `TAG_IN` is set, the rotate all right instruction is executed. The content of `TAG_VALUE_IN` is rotated 5 bit positions to the right, as indicated by the

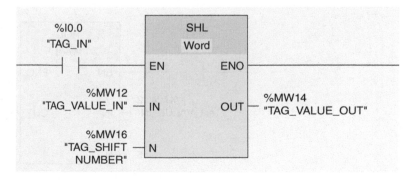

Figure 2.23 Shift-left instruction.

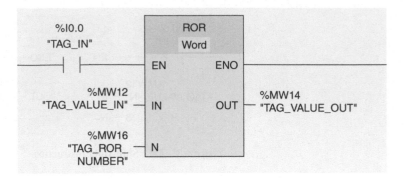

Figure 2.24   Rotate all right instruction.

unsigned integer `TAG_ROR_NUMBER`, which is assigned the value 5. The result is sent to the `TAG_VALUE_OUT` output. If `TAG_VALUE_IN` = 0000 1111 100**1 0011**, then after the ROR instruction executes, the `TAG_VALUE_OUT` will equal **1001 1**000 0111 1100.

Rotate-All-Left Instruction (ROL)

Figure 2.25 shows this instruction.

If `TAG_IN` is set, the rotate all left instruction is executed. The content of `TAG_VALUE_IN` is rotated 5 bit positions to the right, as indicated by the constant number 5 in the `TAG_ROR_NUMBER`. The result is sent to the `TAG_VALUE_OUT` output. If `TAG_VALUE_IN` = **1010 1**000 1111 0110, then after the ROR instruction executes, the `TAG_VALUE_OUT` will equal 0001 1110 110**1010 1**.

2.2.4 *Program-Control Instructions*

This section details two of the commonly used program-control instructions. These instructions divert the processor from the sequential network scan and execution under certain conditions. These conditions are synchronous events initiated

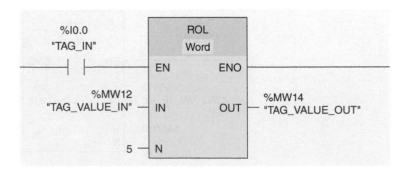

Figure 2.25   Rotate all left instruction.

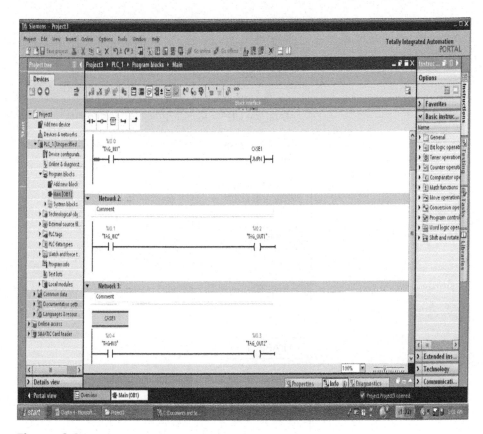

Figure 2.26 Jump and label instruction.

by the executing program, unlike asynchronous events such as interrupts. Jump and label, switch-jump distributor, and function calls are covered in this section.

Jump and Label Instructions

Program-control instructions are used to interrupt the linear execution flow of a program and resume it in another network. The destination network must be identified by a jump label (`LABEL`). These instructions are essential because they allow program scanning and decision making to be altered according to user's predefined conditions or real-time scenarios. The screen shown in Fig. 2.26 illustrates the jump and label instruction and its application, as documented in the Siemens S7-1200 PLC system.

If `TAG_IN1` is TRUE, the JMP instruction is executed. Linear execution of the program is interrupted and resumes execution in the network labeled `CASE1`. If `TAG_IN3` is set, output tag name `TAG_OUT2` is set, whereas the network 2 `TAG_OUT1` status will not update.

Switch-Jump Distributor

The switch-jump distributor allows program control to transfer to `DEST0`, `DEST1`, or the `ELSE` assigned labels. In the configured instruction in Fig. 2.27, we used `LABEL0`, `LABEL1`, and `LABEL2` for these assignments. The user can select the

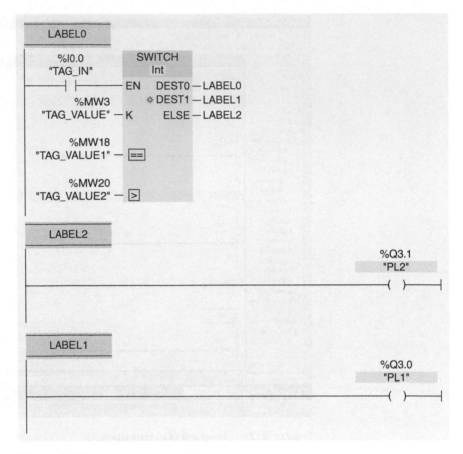

Figure 2.27 Switch-jump distributor.

=, >, or < quantifiers for each of the input tags. `TAG_VALUE1` and `TAG_VALUE2` are compared to the input value of `TAG_VALUE`. The comparison result determines whether program control goes to `LABEL0`, `LABEL1`, or `LABEL2`.

If the `TAG_INPUT` is set TRUE, the switch-jump distributor instruction is executed. If `TAG_VALUE` is equal to `TAG_VALUE1`, control goes to `LABEL0`, which is assigned to the same network in this example (no jump occurs). If `TAG_VALUE` is greater than `TAG_VALUE2`, then program execution jumps to `LABEL1` and PL1 turns ON. Otherwise, PL2 goes ON.

2.3 Sequential and Combinational Logic Instructions

Sequential and combinational logic instructions are discussed in this section. The set-reset (SR), the set (S), the reset (R), the positive-edge (P), and the negative-edge (N) instructions are covered. This will be followed by coverage of common combinational logic instruction with a brief review of digital-logic fundamentals.

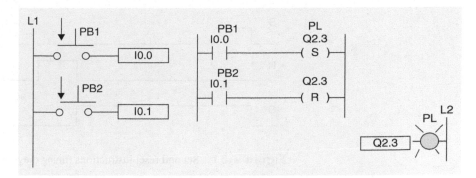

Figure 2.28   Set-reset hardwired connections and instruction status.

2.3.1   *The Set-Reset Flip-Flop Instructions*

The set-reset (SR) flip-flop logic is used to describe and document the LATCH and UNLATCH relay functions in the PLC. When both inputs PB1 and PB2 are at zero state, the output S is in a hold state, latching the previous output status. When PB1 is set to 1 and PB2 is reset to 0, the output S will be set to 1 (output R is reset). When PB1 is reset to 0 and PB2 is set to 1, the output S will be reset to 0 (output R is set). The state where PB1 is set to 1 and PB2 set to 1 is invalid (prohibited state for the SR flip-flop) and should not be used (hardwired switches should be interlocked). Figure 2.28 shows the SR logic and PLC connections. When PB1 is pushed, it sets S to 1, and R is reset. This will cause PL to turn ON, as indicated by the shaded circle. To turn the PL OFF, PB2 should be activated/pressed.

2.3.2   *Set and Reset Output Instructions*

Set-Output Instruction (S): Set 1 Bit

This instruction is executed only if the preceding logic for the same network is TRUE (power flows to the S coil); then S is activated. When the preceding network input (power flow to the coil) is FALSE, then S maintains the active status. S remains active until a reset action is executed (Fig. 2.29).

Reset-Output Instruction (R): Reset 1 Bit

This instruction is executed only if the preceding logic for the same network is TRUE (power flows to the R coil); then R is activated, which resets the S coil. When the preceding rung input (power flow to the coil) is FALSE, then S maintains the inactive status (Fig. 2.30).

Figure 2.29   Set output instruction.   **Figure 2.30**   Reset output instruction.

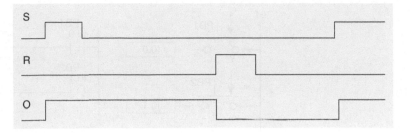

Figure 2.31 Set and reset instructions timing diagram.

Figure 2.31 shows the timing diagram for set-reset instructions. Notice that the two signals cannot be active at the same time. The figure assumes that O represents the logic status of input to the S or R coil network.

2.3.3 *Positive and Negative Edge Instructions*

Positive Edge Instruction

Figure 2.32 shows this instruction. The state of this contact is TRUE for one scan when a positive transition (OFF to ON) is detected on the assigned positive edge bit INPUT. Power flows in the network for one program scan from the time where a positive edge is triggered. Figure 2.33 shows the timing diagram for this instruction.

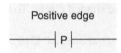

Figure 2.32 Positive edge instruction.

Negative Edge Instruction

The state of this contact is TRUE for one scan when a negative transition (ON to OFF) is detected on the assigned negative edge bit INPUT. Power flows in the network for one program scan from the time where edge is triggered. Figure 2.34 shows the operation of this instruction.

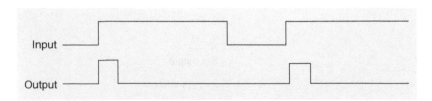

Figure 2.33 Positive edge timing diagram.

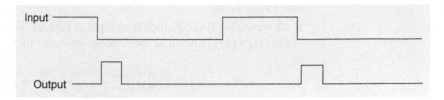

Figure 2.34 Negative edge timing diagram.

Notice that the address of the edge memory bit for the positive or the negative edge instruction must not be used more than once in the entire program; otherwise, the memory bit will be overwritten.

2.3.4 *Logic Gates and Truth Tables*

Digital systems logic elements are classified as *combinational* or *sequential*. The combinational logic includes AND, OR, NOT, NAND, NOR, XOR, and XNOR gates. Sequential logic instructions were covered in Sec. 2.3.3. These instructions will be used extensively throughout this book. The basic combinational logic operations are described as follows.

AND Logic Gate

If all inputs are TRUE, then the output becomes TRUE; otherwise, the output is FALSE (Fig. 2.35).

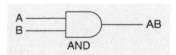

Figure 2.35 AND logic gate.

Truth tables show the output logic for all possible input logic conditions for a given gate or logic function. Table 2.3 is the logic table for a two-input AND gate. The product sign (.) is used as the Boolean operator for AND gate operation.

Table 2.3 Truth Table for Two-Input AND Gate

AND GATE LOGIC		
A	B	A.B
0	0	0
0	1	0
1	0	0
1	1	1

OR Logic Gate

If all inputs are FALSE, then the output is FALSE; otherwise, the output is TRUE. A plus sign (+) is known as the *Boolean operator* for logic OR (Fig. 2.36).

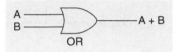

Figure 2.36 OR logic gate.

Table 2.4 is the truth table for a two-input OR gate.

Table 2.4 Truth Table for a Two-Input OR Gate

OR GATE LOGIC		
A	B	A+B
0	0	0
0	1	1
1	0	1
1	1	1

NOT Logic Gate

The output of the NOT (inverter) gate is the inverse of the input logic (Fig. 2.37).

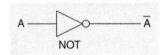

Figure 2.37 NOT logic gate.

Table 2.5 is the truth table for a NOT gate. If the input variable to the inverter is labeled **A**, then the inverted output is known as **A NOT**. This is also shown as **A'** or **A** with a bar over the top, as shown at the gate output.

Table 2.5 Truth Table for a NOT Gate

NOT GATE LOGIC	
A	A'
0	1
1	0

NAND Logic Gate

If all inputs are TRUE, then the NAND gate output turns FALSE; otherwise, the output is TRUE. This is a NOT-AND gate, which is equal to an AND gate followed by a NOT gate (Fig. 2.38).

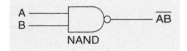

Figure 2.38   NAND logic gate.

Table 2.6 is the truth table for the NAND gate.

Table 2.6   Truth Table for a NAND Gate

NAND GATE LOGIC		
A	B	(A.B)′
0	0	1
0	1	1
1	0	1
1	1	0

NOR Logic Gate

If all inputs are FALSE, then the NOR gate output becomes TRUE; otherwise, the gate output is FALSE (Fig. 2.39).

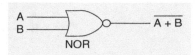

Figure 2.39   NOR logic gate.

Table 2.7 is the truth table for a NOR gate. This is a NOT-OR gate, which is equal to an OR gate followed by a NOT gate.

Table 2.7   Truth Table for an NOR Gate

NOR GATE LOGIC		
A	B	(A+B)′
0	0	1
0	1	0
1	0	0
1	1	0

XOR Logic Gate

If the two inputs are of different logic levels, then the gate output becomes TRUE; otherwise, the gate output is FALSE. An inside-circle plus sign ⊕ (Boolean operator) is used to show the XOR operation (Fig. 2.40).

Figure 2.40 XOR logic gate.

Table 2.8 is the truth table for an XOR gate.

Table 2.8 Truth Table for an XOR Gate

XOR GATE LOGIC		
A	B	$A \oplus B$
0	0	0
0	1	1
1	0	1
1	1	0

XNOR Logic Gate

If the two inputs are the same logic level, then the gate logic output is TRUE; otherwise, the gate output is FALSE. A negated $\oplus$ sign or bar over the top of the XOR output is used to show the XNOR operation (Fig. 2.41).

Figure 2.41 XNOR logic gate.

Table 2.9 is the truth table for an XNOR gate.

Table 2.9 Truth Table for an XNOR Gate

XNOR GATE LOGIC		
A	B	$(A \oplus B)'$
0	0	1
0	1	0
1	0	0
1	1	1

Table 2.10 is a summary truth table of the I/O combinations for the NOT gate together with all possible I/O combinations for the other gate functions. Note that n-bit inputs have 2^n rows. The names for XOR/EXOR and XNOR/EXNOR are interchangeably used in the literature.

Table 2.11 shows the symbolic representations of seven logic gates that are used to document PLC program-logic elements.

Table 2.10 Logic Gate Representation Using the Truth Table

NOT gate		INPUTS		OUTPUTS					
		A	B	AND	NAND	OR	NOR	EXOR	EXNOR
A	$\overline{A}$	0	0	0	1	0	1	0	1
0	1	0	1	0	1	1	0	1	0
1	0	1	0	0	1	1	0	1	0
		1	1	1	0	1	0	0	1

Table 2.11 Logic Gate Symbols

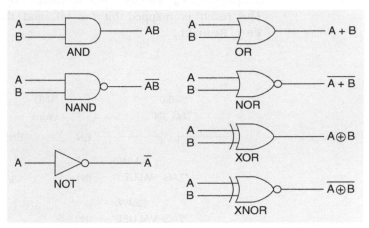

A *ladder diagram* (LD) is a graphical representation of Boolean equations using contacts (inputs) and coils (outputs). The ladder-diagram language allows these features to be viewed in a graphical form by placing graphic symbols into the program workspace similar to a relay-logic electrical diagram. Both ladder diagrams and relay- logic diagrams are connected on the left sides to the power rails.

2.3.5 *Combinational Logic Instructions*

Commonly used S7-1200 logic instructions are covered in this section. Please refer to the Siemens manual online for uncovered instruction or for more details.

NEGATE Assignment

The negate-assignment instruction inverts the result of a rung logic operation (RLO) and assigns it to the specified operand. When the RLO at the input of the coil is 1, the operand is reset. When the RLO at the input of the coil is 0, the operand is set to signal state 1. The instruction does not influence the RLO. The RLO at the input of the coil is sent directly to the output of the coil (Fig. 2.42).

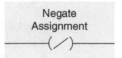

Figure 2.42 Negate-assignment instruction.

AND Logic Operation

The AND logic operation ANDs 2 bytes, words, or double words at the value **IN1** and **IN2** inputs bit by bit and outputs the result in the OUT output. Figure 2.43 shows the AND instruction. If operand **Tag_IN** has the signal state 1, the AND logic operation instruction is executed as shown in Table 2.12. The value of operand **Tag_Value1** and the value of operand **Tag_Value2** are ANDed. The result is mapped bit for bit, and the output is placed in operand **Tag_Result**.

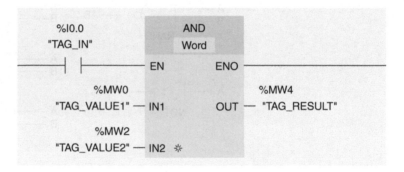

Figure 2.43 AND instruction.

Table 2.12 AND Instruction Example

Parameter	Tag Name	Value
IN1	Tag_Value1	01010101 01010101
IN2	Tag_Value2	00000001 00001111
OUT	Tag_Result	00000001 00000101

OR Logic Operation

The OR logic operation ORs 2 bytes, words, or double words at the value **IN1** and **IN2** inputs bit by bit and outputs the result at the OUT output. Figure 2.44 shows the OR instruction.

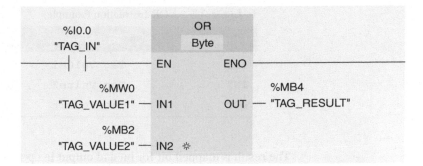

Figure 2.44 OR instruction.

Table 2.13 shows how the instruction works using specific operand values. If operand `Tag_IN` has the signal state 1, the OR logic operation instruction is executed. The value of operand `Tag_Value1` and the value of operand `Tag_Value2` are ORed. The result is in `Tag_Result`.

Table 2.13 OR Instruction Example

Parameter	Tag Name	Value
IN1	Tag_Value1	01010101
IN2	Tag_Value2	00001111
OUT	Tag_Result	01011111

XOR Logic Operation

The XOR logic operation XORs 2 bytes, words, or double words at the value `IN1` and `IN2` inputs bit by bit and outputs the result at the OUT output. Figure 2.45 shows the XOR instruction.

If operand `Tag_IN` has the signal state 1, the XOR logic operation instruction is executed. The value of operands `Tag_Value1` and `Tag_Value2` are XORed.

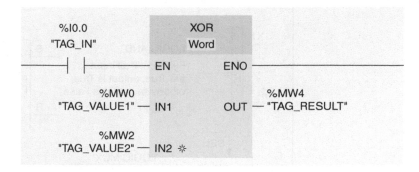

Figure 2.45 XOR instruction.

Table 2.14 XOR Instruction Example

Parameter	Tag Name	Value
IN1	Tag_Value1	00010001 01000101
IN2	Tag_Value2	00000001 01001101
OUT	Tag_Result	00010000 00001000

The result is mapped bit for bit and output in operand **Tag_Result**. The operation of the OR instruction is shown in Table 2.14 with values assigned in 2 bytes, which is equivalent to one memory word.

Table 2.15 summarizes the commonly used logic operations with the associated legend normally used in the specification and documentation of control logic. This logic legend notation will be used later in this book in documented industrial applications.

2.3.6 *Illustrative Ladder Examples*

This section presents a few implemented Siemens S7-1200 ladder-logic examples that illustrate the practical side of the discussions covered so far in this chapter. The reader is encouraged to practice these examples using a suitable training unit (similar to the one described in Fig. 1.29 and used in this book) and development software.

Table 2.15 Ladder-Logic Legend

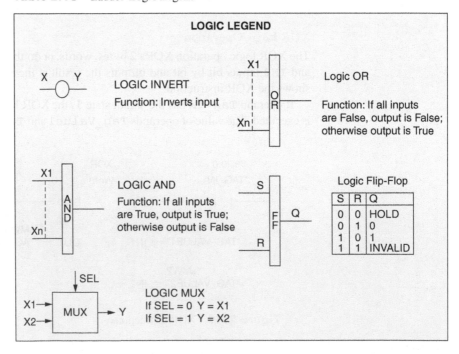

Example 2.1   Figure 2.46 shows a ladder-logic diagram for the three instructions normally open (NO), normally closed (NC), and output coil (OC). It implements the four combinational logics AND, OR, XOR, and XNOR. This diagram assumes two input switches (SW1 and SW2) and four output coils (AND_LOGIC, OR_LOGIC, XOR_LOGIC, and XNOR_LOGIC).

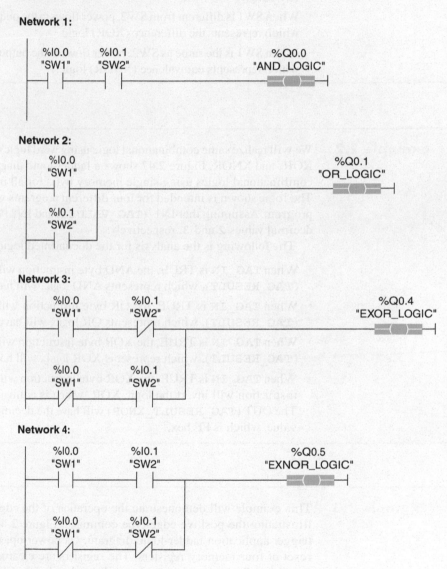

Network 1:

%I0.0   %I0.1   %Q0.0
"SW1"   "SW2"   "AND_LOGIC"

Network 2:

%I0.0   %Q0.1
"SW1"   "OR_LOGIC"

%I0.1
"SW2"

Network 3:

%I0.0   %I0.1   %Q0.4
"SW1"   "SW2"   "EXOR_LOGIC"

%I0.0   %I0.1
"SW1"   "SW2"

Network 4:

%I0.0   %I0.1   %Q0.5
"SW1"   "SW2"   "EXNOR_LOGIC"

%I0.0   %I0.1
"SW1"   "SW2"

Figure 2.46 Ladder logic for combinational logic using bit instructions.

The following is the analysis for the documented logic:

- When SW1 and SW2 are closed, the NO instructions are TRUE, and power flows to the output coil (**AND_LOGIC**), which represents series (AND) logic.
- When SW1 or SW2 or both switches are closed, the NO instructions are TRUE, and power flows to the output coil (**OR_LOGIC**), which represents parallel (OR) logic.
- When SW1 is different from SW2, power flows to the output coil (**EXOR_LOGIC**), which represents the difference (XOR) logic.
- When SW1 is the same as SW2, power flows to the output coil (**EXNOR_LOGIC**), which represents equivalence (XNOR) logic.

Example 2.2

We will realize same combinational logic using word logic operations for AND, OR, XOR, and XNOR. Figure 2.47 shows a ladder-logic diagram for each of the four combinational logics using single memory bytes for all inputs to the instructions. The logic shown is intended for four different programs with one network in each program. Assuming that IN1 (**TAG_VALUE1**) and IN2 (**TAG_VALUE2**) have the decimal values 2 and 3, respectively.

The following is the analysis for the documented logic:

- When **TAG_IN** is TRUE, the AND byte instruction will be executed. The OUT (**TAG_RESULT**), which represents AND logic, will have the decimal value 2.
- When **TAG_IN** is TRUE, the OR byte instruction will be executed. The OUT (**TAG_RESULT**), which represents OR logic, will have the decimal value 3.
- When **TAG_IN** is TRUE, the XOR byte instruction will be executed. The OUT (**TAG_RESULT**), which represents XOR logic, will have the decimal value 1.
- When **TAG_IN** is TRUE, the XOR byte instruction will be executed. The invert instruction will invert the logic XOR, which is equivalent to the logic XNOR. The OUT (**TAG_RESULT_XNOR**) will have the decimal value −2 signed 8 bits value, which is FE hex.

Example 2.3

This example will demonestrate the operation of the edge trigger instructions by illustrating the positive edge-type command. Figure 2.48 shows a positive edge trigger application ladder-logic diagram. On power-up, the program initializes a reset of four memory registers. The registers are cleared one time on power-up condition. The move instruction will be detailed in Chap. 4. Initialization is a common task in all computer systems, which takes place during power-up, reset, or restart conditions. The edge instruction triggers one scan.

Network 1:

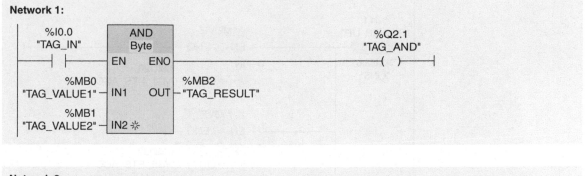

Network 2:

Network 3:

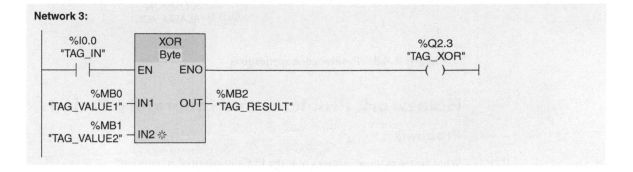

Network 4:

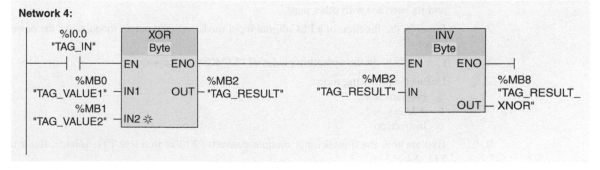

Figure 2.47 Combinational logic using word logic instructions.

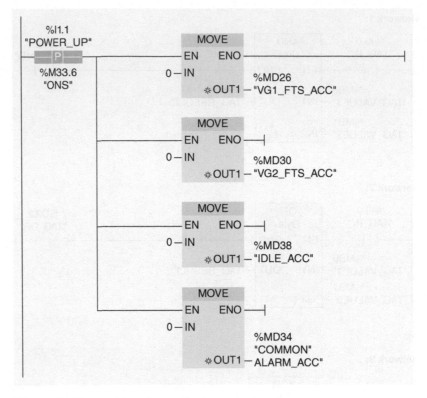

Figure 2.48   Positive edge applications.

Homework Problems and Laboratory Projects

Problems

2.1   What are the main advantages of using PLCs in industrial automation?

2.2   Draw a functional block diagram of a PLC, and describe briefly the role of each component and its interface with other parts.

2.3   Describe the function of a PLC digital input module, digital output module, and the power supply.

2.4   List and explain the operating modes of PLC (CPU/processor).

2.5   Define the following terms:
a. Program scan.
b. Address
c. Instruction

2.6   Explain how the digital input module converts 120 Vac to a low TTL voltage. Refer to Fig. 2.4.

2.7   What is the difference between the following?
a. Program-cycle organizational blocks (OBs) and the startup OB
b. Functions (FCs) and function blocks (FBs)

2.8   Explain how normally open, normally closed, and output energize instructions work.

2.9   What is the function of the power supply used in a PLC?

2.10   List the tasks performed by the CPU when the power supply is turned on?

2.11   Explain the difference between an output coil and the set output instruction. What reverses the status of the output in each case?

2.12   Write the Boolean equations for the following logic diagrams:
   a. Figure 2.49

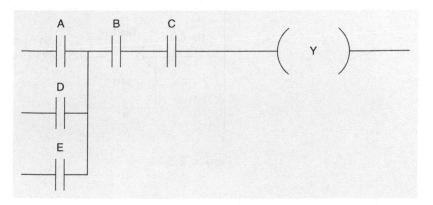

Figure 2.49

   b. Figure 2.50

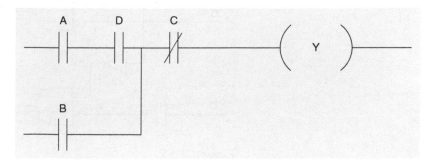

Figure 2.50

2.13   List the conditions required to turn ON motor M for the line diagram in Fig. 2.51.

2.14   Convert the following logic diagrams into a ladder-logic program.
   a. Figure 2.52
   b. Figure 2.53

2.15   Create a ladder-logic program for the following Boolean equation.

$$SV = (SW1 + SW2)(SW3)$$

where SV is a solenoid valve, and SW1, SW2, and SW3 are three ON/OFF switches.

2.16   Create a ladder-logic program for the following Boolean equation.

$$(SW1.SW2)' + (SW3) = PL1 \text{ (pilot light 1)}$$

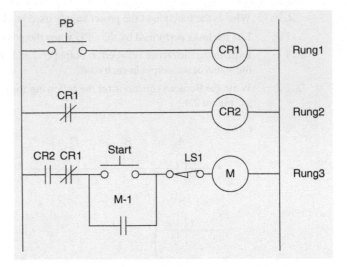

Figure 2.51

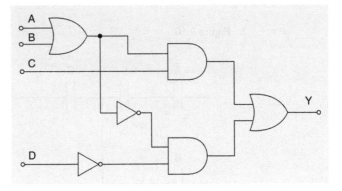

Figure 2.52

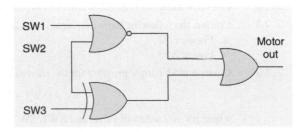

Figure 2.53

2.17   For the AND word instruction in Fig. 2.54, complete the table for the `Tag_Result` word below.

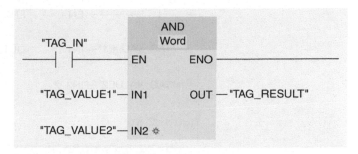

Figure 2.54

Parameter	Tag Name	Value
IN1	Tag_Value1	01010101 11010101
IN2	Tag_Value2	01010001 10101011
OUT	Tag_Result	

2.18   For the OR byte instruction in Fig. 2.55, complete the table for the `Tag_Result` byte below.

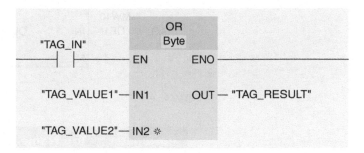

Figure 2.55

Parameter	Tag Name	Value
IN1	Tag_Value1	01010110
IN2	Tag_Value2	10011110
OUT	Tag_Result	

2.19   For the XOR word instruction in Fig. 2.56, complete the table for the `Tag_Result` word below.

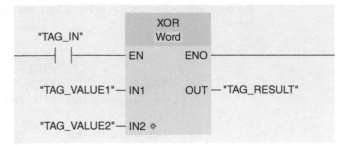

Figure 2.56

Parameter	Tag Name	Value
IN1	Tag_Value1	11010101 01010110
IN2	Tag_Value2	01000001 00101111
OUT	Tag_Result	

2.20   For the SHR word instruction in Fig. 2.57, complete the table for `Tag_Result` word below.

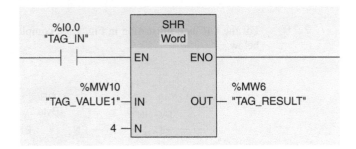

Figure 2.57

Parameter	Tag Name	Value
IN	Tag_Value1	11010101 01010110
N	4	
OUT	Tag_Result	

2.21   Repeat Problem 20 for the ROR and ROL instructions.

2.22   Show how you can program a network using a logic operation instruction to clear the most significant byte in memory word location MW5.

2.23   Show how you can program a network using a logic operation instruction to set the most significant bit in memory word location MW1.

2.24   Explain the set and reset output instructions, and complete the timing diagram in Fig. 2.58, assuming that the output (O) is initially high.

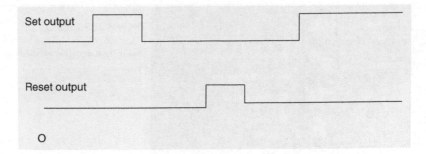

Figure 2.58

2.25   Explain the positive edge instruction, and complete the timing diagram in Fig. 2.59, assuming that the output is initially at low state.

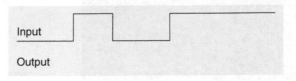

Figure 2.59

2.26   Show a network using bit logic instructions to implement logic NAND and logic EXCLUSIVE NOR.

Projects

Laboratory 2.1: Devise Configuration and Online Programming

The objective of this laboratory project is to familiarize the reader with the procedures used for PLC device configuration and online programming using the S7-1200 software-development portal.

a. Online mode. Follow the steps for device configuration online.

* From the portal view (Fig. 2.60) click on "Create New Project."
* Enter the name of the project, and then double-click "Create" (Fig. 2.61).
* Click "Write PLC Program" and then "Main" (Fig. 2.62).
* From the Project View, click "Add New Device," and choose the correct PLC series from the menu (Fig. 2.63)
* Under Catalog, choose the right digital input (DI) module (Fig. 2.64). The digital I/Os on the PLC processor module do not require configuration.
* Under Catalog, choose the right digital output (DO) module (Fig. 2.65). Do not configure the I/Os on the PLC processor module.
* Enter the following networks to implement the four combinational logics AND, OR, XOR, and XNOR. This diagram (Fig. 2.66) assumes two input switches (SW1 and SW2) and four output coils (AND_LOGIC, OR_LOGIC, EXOR_LOGIC, and EXNOR_LOGIC).

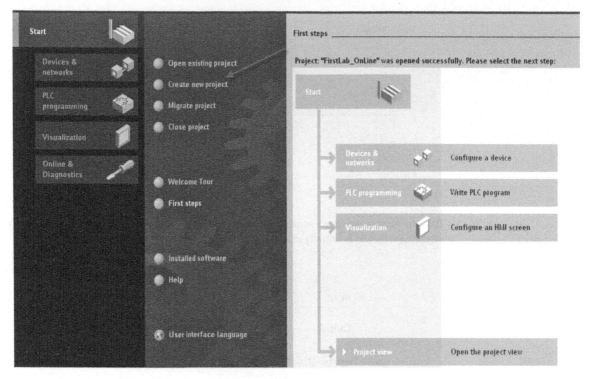

Figure 2.60

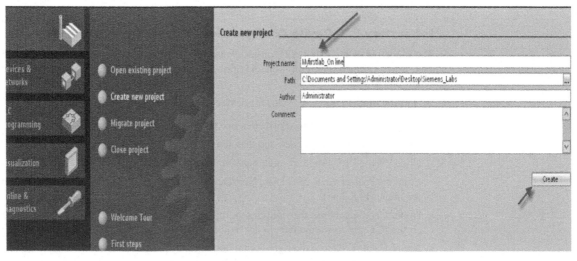

Figure 2.61

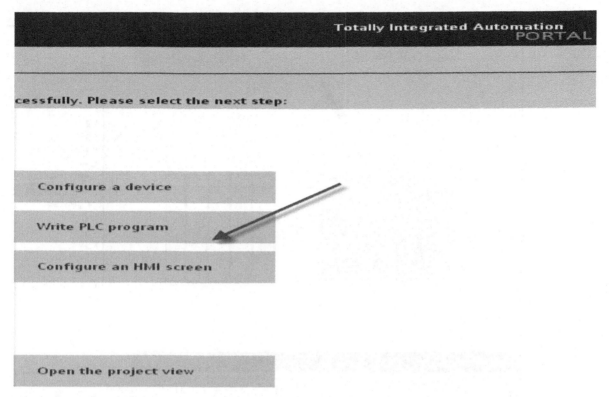

Figure 2.62

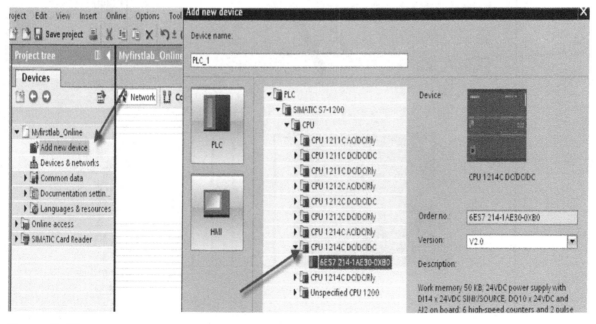

Figure 2.63

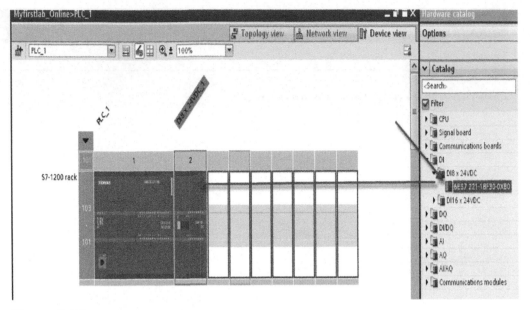

Figure 2.64

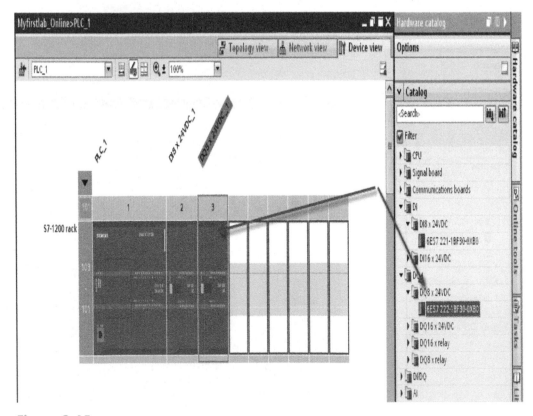

Figure 2.65

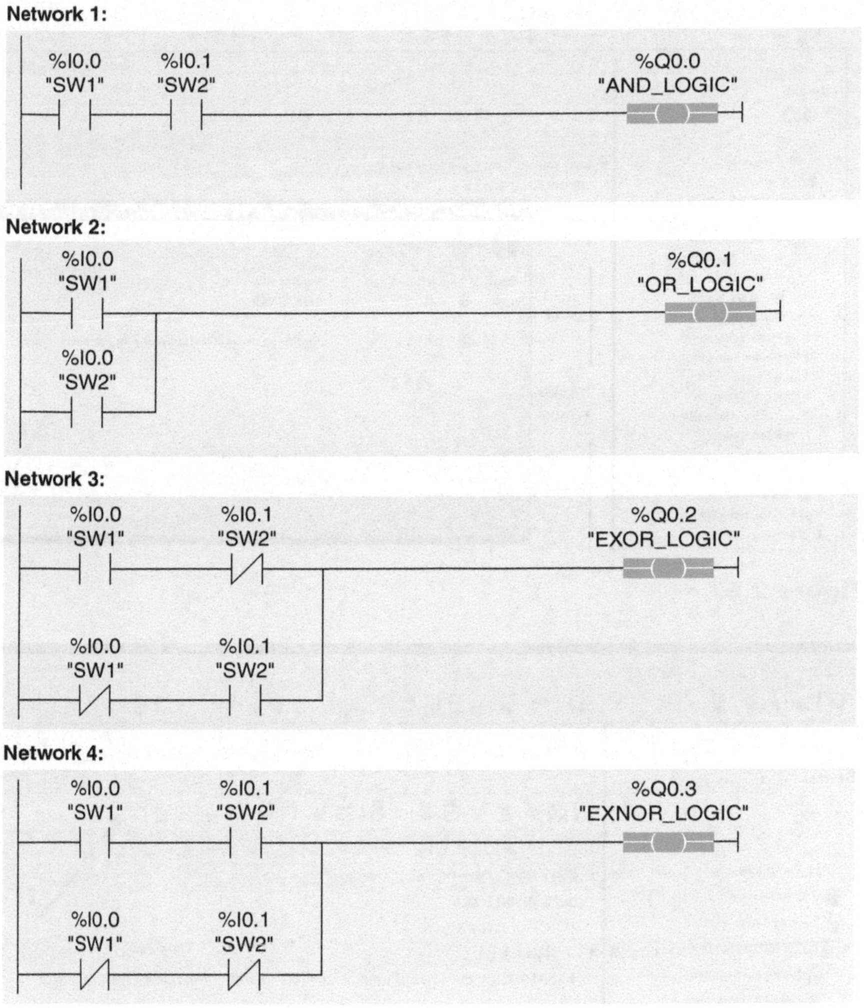

Network 1:

```
%I0.0      %I0.1                                    %Q0.0
"SW1"      "SW2"                                    "AND_LOGIC"
─┤ ├───────┤ ├──────────────────────────────────────( )──
```

Network 2:

```
%I0.0                                               %Q0.1
"SW1"                                               "OR_LOGIC"
─┤ ├──┬─────────────────────────────────────────────( )──
      │
%I0.0 │
"SW2" │
─┤ ├──┘
```

Network 3:

```
%I0.0      %I0.1                                    %Q0.2
"SW1"      "SW2"                                    "EXOR_LOGIC"
─┤ ├───────┤/├──┬────────────────────────────────────( )──
                │
%I0.0      %I0.1│
"SW1"      "SW2"│
─┤/├───────┤ ├──┘
```

Network 4:

```
%I0.0      %I0.1                                    %Q0.3
"SW1"      "SW2"                                    "EXNOR_LOGIC"
─┤ ├───────┤ ├──┬────────────────────────────────────( )──
                │
%I0.0      %I0.1│
"SW1"      "SW2"│
─┤/├───────┤/├──┘
```

Figure 2.66

b. **Online viewing.** In order to view and troubleshoot a program residing in PLC S7-1200 processor's memory, the programming terminal node must be communicating with the PLC processor. The following are the typical steps used in this task:

- Choose Main [OB1].
- Click "Download to Device."
- Push "Load" and "Finish" (Fig. 2.67).
- Go online.
- Push "Monitoring" ON/OFF (Fig. 2.68)
- Turn on SW1 on the training unit or Siemens simulator, and monitor the logic as shown in Fig. 2.69.
- Check out all logic using or simulating SW1 and SW2.

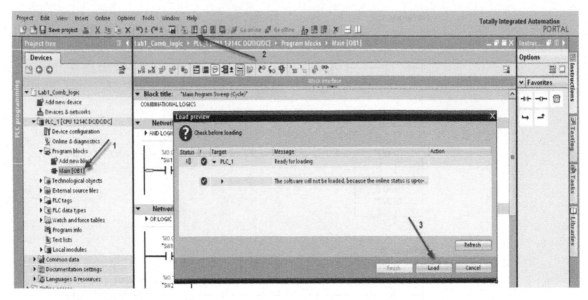

Figure 2.67

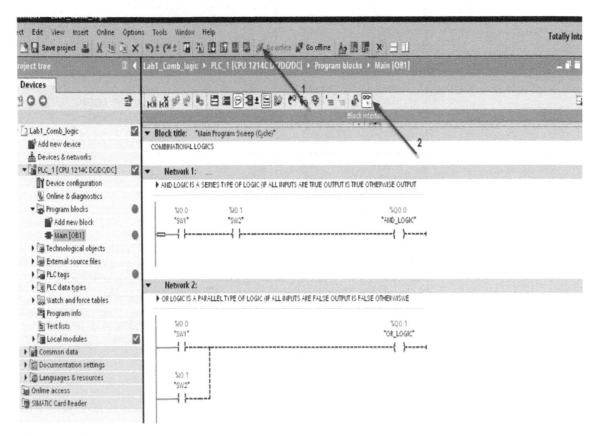

Figure 2.68

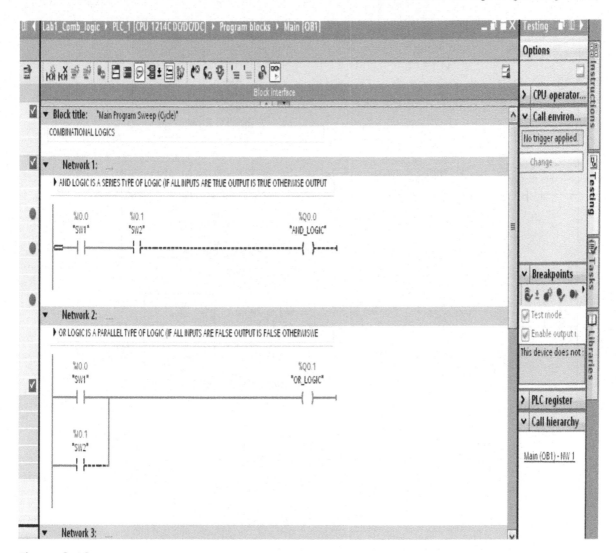

Figure 2.69

Lab Requirements

- Use SW1 and SW2 on the training unit or the Siemens simulator to simulate the inputs for the four network logics, and confirm that the logics work.

- Repeat the four logic operations using word logic instructions with 1 byte assigned to each of the two operands and the result.

NOTE After adding the rest of the functions in the second requirement, your networks should look like the program in Fig. 2.70 but implemented using four separate functions.

Notice that the same tag (TAG_RESULT) is used for all logic functions because the implemented interlock allows the user to select only one function at a time. The steps

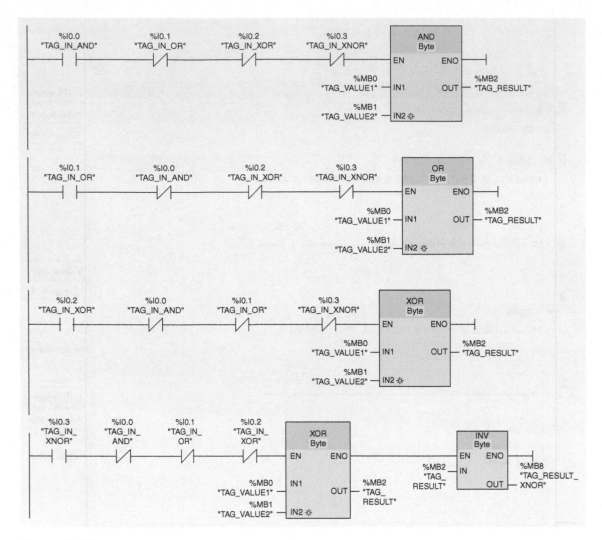

Figure 2.70 Ladder combinational logic using logic instructions.

and associated screen for this laboratory are shown in Fig. 2.71 for only the AND instruction.

Laboratory 2.2: Structured Programming

The objective of this laboratory is to realize the combinational logic functions used in Laboratory 2.1 using word logic operations in structured programming. To create a function (FC), follow the steps illustrated in Fig. 2.72:

- Click "Add New Block."
- Click the function block (FC).
- Enter the name for the block.

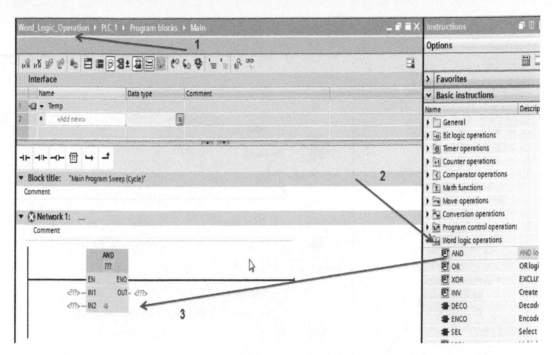

Figure 2.71

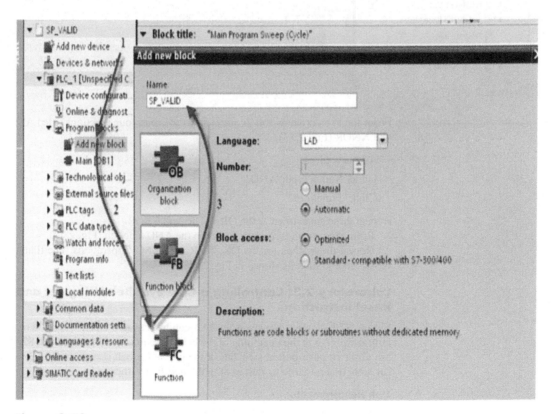

Figure 2.72

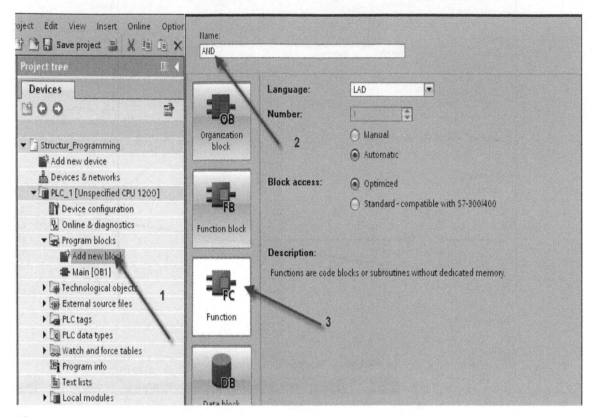

Figure 2.73

From the Project view, follow these steps to create four functions titled AND, OR, XOR, and XNOR (Fig. 2.73):

- Add new block.
- Under Name, type "AND."
- Click "Function."

Repeat these steps to enter the OR, XOR, and XNOR.

From the project tree, drag and drop the AND function (Fig. 2.74).

Repeat these steps for the OR, XOR, and XNOR operations. The final organizational blocks should look as shown in Fig. 2.75.

Laboratory 2.3: Controlling a Conveyor Belt Using Set and Reset Instructions

Figure 2.76 shows a conveyor belt that can be activated electrically. There are two push-button switches at the beginning of the belt (Location A): S1 for start and S2 for stop. There are also two push-button switches at the end of the belt (Location B): S3 for start and S4 for stop. It is possible to start or stop the belt from either end.

Lab Requirements

- Assign and document all I/O addresses.
- Enter the program.

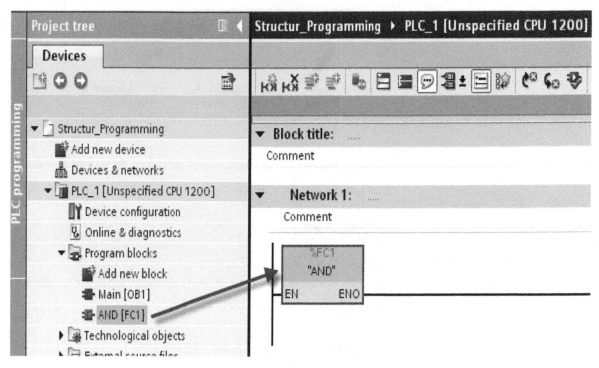

Figure 2.74

Figure 2.75

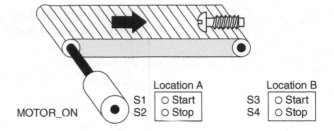

Network 1:
The conveyor belt motor is switched on when Start switch "S1" or "S3" is pressed.

```
   "StartSwitch_Left"   "MOTOR_ON"
   ─┤ ├───────────────────( S )───┤

   "StartSwitch_Right"
   ─┤ ├──────────────
```

Network 2:
The conveyor belt motor is switched off when Stop switch "S2" or "S4" is pressed.

```
   "StopSwitch_Left"   "MOTOR_ON"
   ─┤/├──────────────────( R )───┤

   "StopSwitch_Right"
   ─┤/├──────────────
```

Figure 2.76

- Download and go online.
- Provide system check as detailed in the preceding steps.
- Document your program.

Laboratory 2.4: Conveyor Belt Movement Direction

Figure 2.77 shows a conveyor belt that is equipped with two photoelectric barriers (PEB1 and PEB2). These are designed to detect the direction in which a package is moving on the

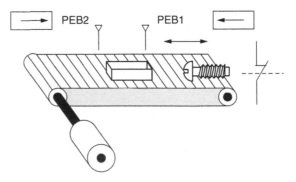

Figure 2.77

belt. Write a ladder-logic program to detect the belt movement direction. Two pilot lights (PL_R) and (PL_L) are used to indicate the direction status.

Laboratory Requirements

- Write a ladder program to detect the conveyor movement direction and activate one of the two pilot lights.

- Use positive-edge set and reset instructions to achieve the required task. Document your program.

- Download the program to your PLC hardware or trainer, or use the Siemens simulator. Check out and debug the program.

Timers and Counters Programming

This chapter will introduce common types of timers and counters used in PLC programming with an emphasis on the format used in the Siemens S7-1200 and other families of PLCs. Concepts will be enforced through the use of implemented real industrial application examples.

Chapter Objectives

▶ Understand timer types, operation, and implementation.

▶ Understand counter types, operation, and implementation.

▶ Understand core and special timing instructions.

▶ Use timers and counters in industrial process control.

As stated in Chap. 1, real-time system behavior is the core issue in most control applications. Timers and counters are critical elements used in programmable logic controllers (PLCs), hardwired controllers, and impeded systems to accommodate an application event and real-time constrains. This chapter will introduce common types of timers and counters used in PLC programming with emphasis on the format used in the Siemens S7-1200 and other families of PLCs. Concepts will be enforced through the use of implemented real industrial applications.

3.1 Timer Fundamentals

Siemens S7-1200 timers are available in four different forms: generate ON-DELAY (TON), generate OFF-DELAY (TOF), time accumulator (TONR), and generate pulse (TP). Table 3.1 shows the parameters for the TON, TOF, and TONR timers. Timer instruction preset/accumulated value uses the M (2 words, 16 bits), D (double words, 32 bits), or L (long, 64 bits) memory area, as shown in the data table. The same information for the TP timer is shown in Table 3.2.

3.1.1 ON-DELAY Timers (TONs)

This timer's main function is to delay the rising edge of output Q by the pre-defined period of the preset time (PT). The timer block is shown in Fig. 3.1 with appropriate tag names assigned. In the figure, these tags are displayed between double quotes. All timer-required variables are displayed using the standard system labels, which start with the percent (%) character. Timer preset-time (PT) input can be defined as displayed or as a constant value.

Table 3.1 TON, TOF, and TONR Parameters

Parameters	Declaration	Data Type	Memory Area	Description
IN	Input	BOOL	I, Q, M, D, L	Start input
PT	Input	TIME	I, Q, M, D, L, or constant	Duration of the on delay The value of the PT parameter must be positive
Q	Output	BOOL	I, Q, M, D, L	Output that is set when the time PT expires
ET	Output	TIME	I, Q, M, D, L	Current time value

Table 3.2 TP Parameters

Parameters	Declaration	Data Type	Memory Area	Description
IN	Input	BOOL	I, Q, M, D, L	Start input
PT	Input	TIME	I, Q, M, D, L, or constant	Duration of the pulse The value of the PT parameter must be positive

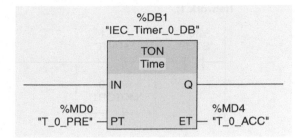

Figure 3.1 ON-DELAY timer (TON).

If rung input (IN) is TRUE, the timer accumulated value with tag name `T_0_ACC` will increment. When timer accumulated value is equal to the defined preset value with tag name `T_0_PRE`, the output Q will change status to ON, and the timer will stop timing. Figure 3.2 illustrates the timer timing diagram. Notice that the delay action applies to timer output Q. It turns ON after the prespecified preset time from the point where the timer input is enabled. If the timer enable input is lost before the prespecified preset time, the Q output will not turn ON. The timer input must stay ON during the entire preset time in order for the output to turn ON. The timer output will turn OFF once the input goes OFF. Please refer to the book's website, www .mhprofessional.com/ProgrammableLogicControllers, for an *interactive simulator* illustrating the operation of this timer.

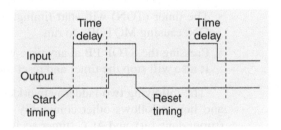

Figure 3.2 ON-DELAY timing diagram.

Figure 3.3 shows a ladder-logic diagram for the TON timer instruction. This diagram assumes a normally open START push button (PB), a normally open STOP push button, a timer preset (PT) value of 10 seconds, output MOTOR1 (Q0.0), and output MOTOR2 (Q0.1). The ladder diagram consists of two networks. The following are the critical events in the shown ladder-logic diagram:

- The first network initially during the first scan, I0.0, is TRUE because the STOP PB is wired high. Also, the START PB is FALSE because this switch is normally open.
- Once the START PB is pressed, I0.1 becomes TRUE; this, in turn, makes Q0.0 TRUE. Q0.0 is the output to MOTOR1 starter, which causes it to run. The next scan, Q0.0, will latch the START PB and maintain the Q0.0 TRUE status.

Network 1:

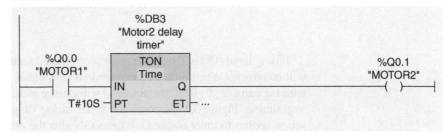

Network 2:

Figure 3.3   ON-DELAY ladder-logic diagram.

- The second network, Q0.0, is the input condition for the TON instruction, which is TRUE. This timer has a preset time value of 10 seconds.
- The timer (TON) will start timing. After 10 seconds of delay, Q0.1 will turn ON, causing MOTOR2 to run.
- Pressing the STOP PB at any time will cause MOTOR1 and MOTOR2 to stop. It also will stop the timer and reset its accumulated time register (ET).

The following two ladder networks explain how the ON-DELAY timer works and how it follows other commonly used notations: TT (timer timing bit), DN (timer done bit), and ACC (timer accumulated value).

- Network 1: When INPUT (I0.0) is TRUE, Timer0 starts trimming. Outputs Timer0_DN and OUTPUT are FALSE as long as Timer0_ACC did not reach 10.
- Network 2: When Timer0_DN is FALSE, the two compare instructions are executed. Timer0_TT is set, indicating that the timer is timing.
- When Timer0_ACC is greater than or equal to 10, the two outputs Timer0_DN and OUTPUT are TRUE. Network 2 input will be FALSE, and Timer0_TT goes OFF, indicating that the timer stopped timing.
- The timer timing (TT) bit is not directly available with Siemens ON-DELAY timers (TONs), but it can be generated using Network 2, as shown in Fig. 3.4. Other PLCs include the TT bit as part of the instruction for all supported timers. Instead, Siemens provides pulse timers (TBs), which initiate timing on an input pulse.

Network 1:

```
                    %DB1
                   "Timer0"
    %I0.0                                              %M0.0
   "INPUT"           TON                              "Timer0_DN"
                     Time
                 IN        Q                            ( )
    T#10S ─ PT                  %MD4
                          ET ─ "TIMER0_ACC"           %Q0.2
                                                      "OUTPUT"
                                                        ( )
```

Network 2:

```
    %M0.0          %MD4                                 %M0.1
  "TIMER0_DN"    "TIMER0_ACC"                         "Timer0_TT"
    ─▭▭╱▭─          > ┤                                 ( )
                    Time
                    T#0MS

                   %MD4
                 "TIMER0_ACC"
                    < ┤
                    Time
                    T#10S
```

Figure 3.4   Timer timing (TT) bit.

3.1.2   *OFF-DELAY Timers (TOFs)*

This timer's main function is to delay the falling edge of output Q by the pre-defined period of the preset time (PT). If the input to the instruction (IN) is TRUE, the output Q is set true. When input turns OFF, the timer starts timing. It resets the output Q when the timer accumulated value ET with tag name `T_0_ACC` is equal to the timer preset value PT with assigned tag name `T_0_PRE`. Figure 3.5 shows the timer, and Fig. 3.6 illustrates the TOF timing diagram. Please refer to

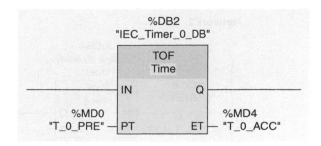

Figure 3.5   OFF-DELAY timer (TOF).

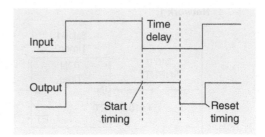

Figure 3.6 OFF-DELAY timing diagram.

the book's website for an *interactive simulator* illustrating the operation of this timer.

Figure 3.7 shows a ladder-logic diagram for the TOF timer instruction. This diagram assumes a normally open START push button (PB), a normally closed STOP push button, a timer preset (PT) value of 10 seconds, output MOTOR1 (Q0.0), and output MOTOR2 (Q0.1). The ladder diagram consists of two networks. The following are the critical events in this timer example:

• The first network initially during the first scan, I0.1, is TRUE because the STOP PB is wired high. Also, the START PB is FALSE because the switch is not yet pressed.

• Once the START PB is pressed, I0.0 becomes TRUE; this, in turn, makes Q0.0 TRUE, and MOTOR1 runs. The next scan, Q0.0, will latch around the START PB and maintain Network 1 TRUE status.

Network 1:

Network 2:

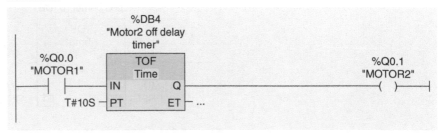

Figure 3.7 OFF-DELAY ladder-logic diagram.

- As shown in the second network, Q0.0 is the input condition for the TOF instruction, which is now TRUE.
- The timer (TOF) will set the output Q0.1 ON, causing MOTOR2 to turn ON. So far the two motors are running after pressing the START PB.
- Once the STOP PB is pressed, MOTOR1 loses power and goes OFF. This will cause the TOF timer to start timing.
- After the prespecified delay of 10 seconds is over, the timer output Q0.1 goes OFF. This will cause MOTOR2 to stop. Thus MOTOR2 stops 10 seconds after MOTOR1 is OFF.
- The timer preset value is cleared when TOF is TRUE.
- Pressing the STOP PB at any time while the two motors are running stops MOTOR1, and 10 seconds later MOTOR2 stops.

The network in Fig. 3.8 shows a simple process-control implementation using three OFF-DELAY timer (TOF) instructions. This network assumes a limit-switch tag name LS1 and three outputs: MOTOR1 Q1.0, MOTOR2 Q1.1, and MOTOR3 Q1.2. Timer preset values of 10, 20, and 30 seconds are used to sequence the operation of the three motors. When LS1 is TRUE, all three motors start running. If LS1 changes from high to low (OFF state), MOTOR1 turns off after 10 seconds, MOTOR2 turns off after 20 seconds, and MOTOR3 turns off after 30 seconds. Once LS1 returns back to the TRUE state, the accumulated values of all three timers are cleared.

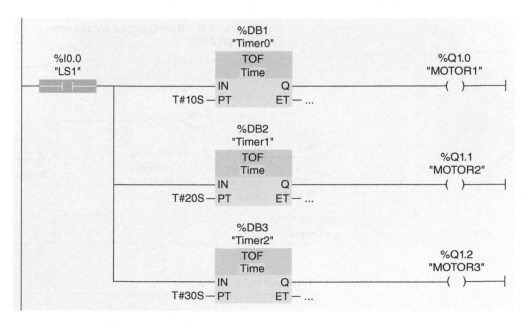

Figure 3.8 A simple process-control implementation using three OFF-DELAY timers.

3.1.3   *Retentive/Time-Accumulator Timers (TONRs)*

The time-accumulator timer (TONR) works exactly the same as the ON-DELAY timer except that the accumulated value is retained while the timer instruction is inactive. To reset the accumulated value to zero, a positive pulse-input instruction is required at the reset input (R). Figure 3.9 shows the timer, and Fig. 3.10 illustrates the timer timing diagram. Please refer to the book's website for an *interactive simulator* illustrating the operation of this timer.

Figure 3.11 shows a ladder-logic diagram for the TONR timer instruction. The TONR ladder-logic diagram shown has an AUTO/MANUAL selector switch, and the timer preset value is set to 1 hour. Once the selector switch is placed in the AUTO mode, the network turns TRUE. The processor scans the ladder, and the following events are observed:

* The timer starts timing once the selector switch is placed in the AUTO mode. If the selector switch is placed in the MANUAL mode, the timer stops timing, and the timer accumulated value is retained.

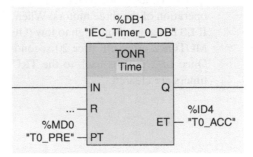

Figure 3.9   Retentive/time-accumulator timer (TONR).

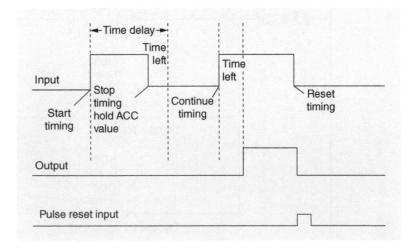

Figure 3.10   TONR delay timing diagram.

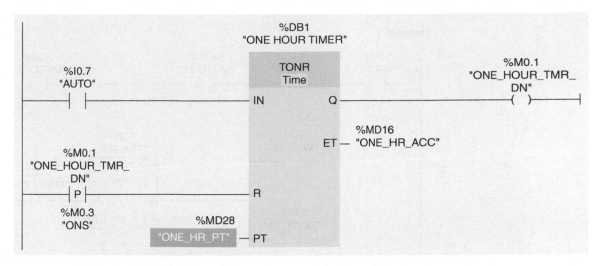

Figure 3.11 TONR delay timing diagram.

- Once the switch is placed back in the AUTO mode, the timer starts timing from where it left off.
- Once the timer accumulated value ET (ONE_HR_ACC) equals the preset value PT (ONE_HR_PT), the timer output (ONE_HOUR_DN) is set to TRUE. The accumulated-value ET (ONE_HR_ACC) is reset to zero at any time by initiating the one scan pulse applied to the timer reset input (R).
- The TONR in this example continuously recycles on an hourly basis.

3.1.4 *Implemented Timer Applications*

Example 3.1

This application assumes four motors represented by four pilot lights on the trainer unit. A push-button START switch is used to start a sequence of motor operations that can be terminated at any time by pressing a push-button STOP switch. The sequence starts MOTOR1, then MOTOR2, then MOTOR3, and then MOTOR4. The same sequence will repeat until the process is stopped. Each selected motor will run for a simulated abbreviated period of 5 seconds while the other motors are idle. This timer project was named by our students as the "Merry-Go-Round Project."

 Figure 3.12 shows the implemented application Network 1. When the START PB is pressed, Network 1 turns TRUE, and the PLC processor performs the following steps:

- MOTOR1 coil is set; associated contact from MOTOR1 latches around the START PB and maintains input to Timer 0 and MOTOR1. The timer and MOTOR1 will stop once MOTOR2 turns ON.

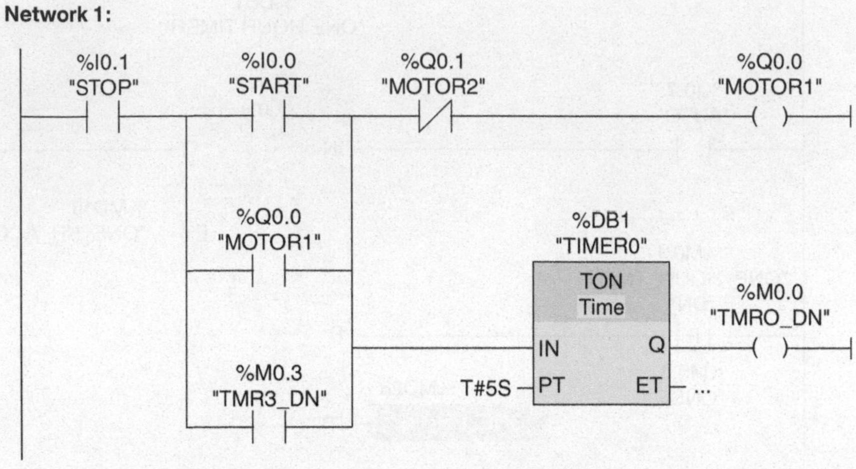

Network 1:

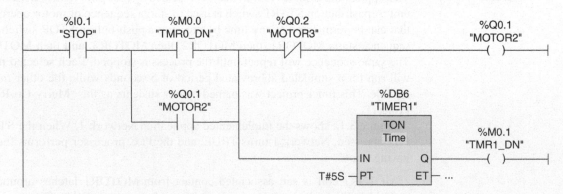

Figure 3.12 Merry-Go-Round Network 1.

- TON Timer 0 starts timing, 5 seconds later, as defined by the preset time (PT), %M0.0 output tag name TMR0_DN is set.
- MOTOR4 timer's done bit (TMR3_DN) is used to latch the START PB in order for MOTOR1 to restart after MOTOR4 stops.
- This sequence continues as long as the STOP PB remains inactive. STOP PB, when activated, stops all outputs and associated motors.

Figure 3.13 shows Network 2. TMR0_DN sets MOTOR2 and Timer 1. MOTOR2 contact latches around TMR0_DN to keep the input to Timer 1 and MOTOR2 TRUE while MOTOR3 is OFF. TON Timer 1 starts timing; 5 seconds later, it sets

Network 2:

Figure 3.13 Merry-Go-Round Network 2.

Network 3:

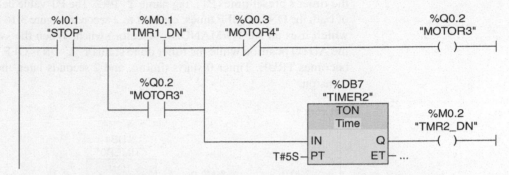

Figure 3.14 Merry-Go-Round Network 3.

its output %M0.1 with tag name TMR1_DN. The STOP PB, when activated, stops all outputs and associated motors. MOTOR2 stops once MOTOR3 runs.

Figure 3.14 shows Network 3. TMR1_DN sets MOTOR3. Contact from MOTOR3 latches around TMR1_DN to keep the input to Timer 2, and MOTOR3 status TRUE while MOTOR4 is OFF. TON Timer 2 starts timing; 5 seconds later, %M0.2 with tag name TMR2_DN output is set. The STOP PB, when activated, stops all outputs.

Figure 3.15 shows Network 4. TMR2_DN sets MOTOR4. Contact from MOTOR4 latches around TMR2_DN to keep the input to the Timer 3 and MOTOR4 status TRUE while MOTOR1 is OFF. TON Timer 3 starts timing; 5 seconds later, TMR3_DN (%M0.3) output is set. The TMR3_DN (%M0.3) latches around the START PB in Network 1 to restart MOTOR1 and repeat the cycle. The STOP PB, when activated, stops all outputs and associated motors.

Network 4:

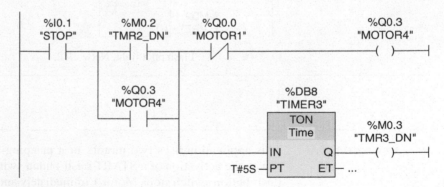

Figure 3.15 Merry-Go-Round Network 4.

Example 3.2

This application flashes a pilot light (PL) ON/OFF using a duty cycle indicated by the timer's preset time (PT), tag name T_PRE. The PT value defines the duration of both the ON and OFF times, each set to 2 seconds. Figure 3.16 shows Network 1, which uses an AUTO/MANUAL selector switch. When the switch is placed in the AUTO position while the timer done status T_1_DN is OFF, Network 1 status becomes TRUE. Timer 0 starts timing, and 2 seconds later, the pilot light (PL) turns on.

Network 1:

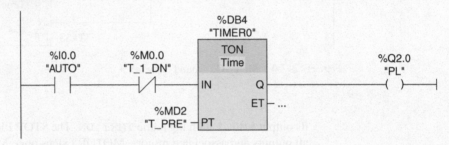

Figure 3.16 Flash pilot light, Network 1.

Figure 3.17 shows Network 2. PL is used as the input to TON Timer 1. Once PL turns ON, Timer 1 starts timing causing Network 2 output (T_1_DN) to turn ON after a 2-second delay. This, in turn, will cause Timer 0 in Network 1 to reset and restart timing. This logic will cause PL to flash, switching status ON/OFF every 2 seconds.

Network 2:

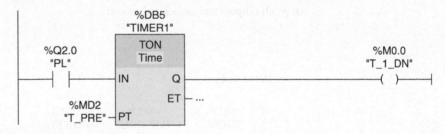

Figure 3.17 Flash pilot light, Network 2.

Example 3.3

This application uses two motors in a pumping-station facility. Both motors run on the activation of a START push-button switch. The activation of a STOP push-button switch stops Motor 1 immediately and then Motor 2 after 10 hours. Figure 3.18 shows a single network realizing this requirements using an OFF-DELAY timer.

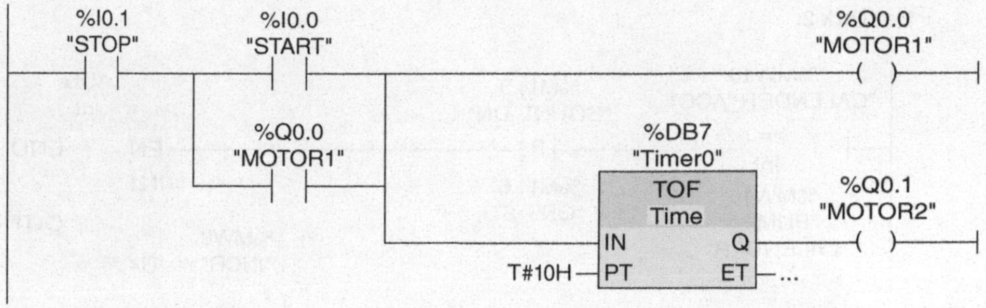

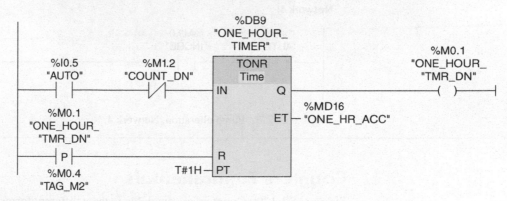

Figure 3.18

Example 3.4　　Two immersed pumps are each driven by a constant-speed motor. One pump is located in the east wet well and the other in the west wet well. The two wells are connected through controlled piping. Each motor provides a discrete input signal indicating whether the motor is running or not. Motors can be started by activating the AUTO/MAN selector switch on the local panel to the AUTO mode. The implemented ladder logic acts to alternate operation between the east and west pumps according to the user calendar. Figure 3.19 shows Network 1, which recycles the time-accumulator (TONR) timer on an hourly basis. The COUNT_DN is a one-shot flag generated through counter logic, which will be discussed in the next two sections. This flag is TRUE when the calendar accumulated value is greater than or equal to the user calendar time.

Figure 3.19　Pump alteration, Network 1.

Figure 3.20 shows Network 2, which increments the register tag name (INCR) when the calendar accumulated value is greater than or equal to the user calendar time. The positive edge instruction is used to prevent the INCR register from overflowing. The ADD instruction is detailed in Chap. 4.

Network 2:

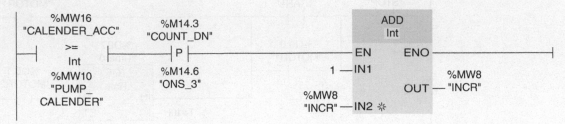

Figure 3.20 Pump alteration, Network 2.

Figure 3.21 shows Network 3, which toggles the east pump on the status of the least significant bit (M9.0) of the increment register (INCR) when the user calendar expires. (Notice that the processor swaps the two memory bytes.)

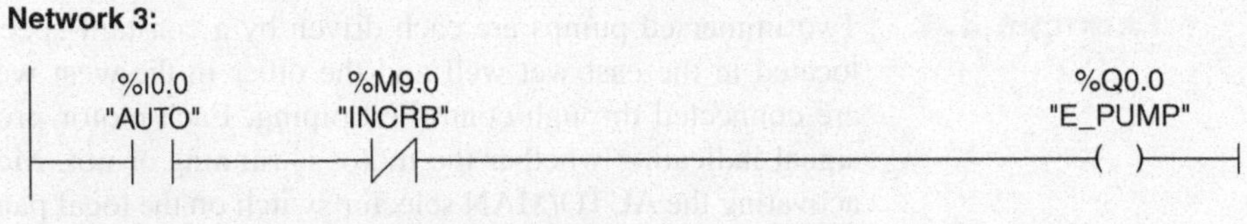

Network 3:

```
         %I0.0              %M9.0                                    %Q0.0
         "AUTO"             "INCRB"                                  "E_PUMP"
     ─────┤ ├────────────────┤/├──────────────────────────────────────( )──────
```

Figure 3.21 Pump alteration, Network 3.

Figure 3.22 shows Network 4, which toggles the west pump on the status of the least significant bit (M9.0) of the increment register (INCR) when the user calendar expires.

Network 4:

```
         %I0.0              %M9.0                                    %Q0.1
         "AUTO"             "INCRB"                                  "W_PUMP"
     ─────┤ ├────────────────┤ ├───────────────────────────────────────( )──────
```

Figure 3.22 Pump alteration, Network 4.

3.2   Counters Fundamentals

Siemens S7-1200 Counters are available in three different forms: count up (CTU), count down (CTD), and count up and count down counter (CTUD). Table 3.3 shows the basic parameters for the CTU, CTD, and CTUD. Operation of the three types of counters will be detailed in the next three subsections. An implemented counter application will be analyzed in Sec. 3.2.4. The concepts covered in this section follows the Siemens notations but can be applied to other PLC brands with minor modification.

Table 3.3 CTU, CTD, and CTD Parameters

Parameter	Data Type	Description
CU, CD	BOOL	Count up or count down, by one count
R(CTU, CTUD)	BOOL	Reset count value to zero
LOAD (CTD, CTUD)	BOOL	Load control for preset value
PV	SINT, INT, DINT, USINT, UINT, UDINT	Preset count value
Q, QU	BOOL	True if CV >= PV
QD	BOOL	True if CV <= 0
CV	SINT, INT, DINT, USINT, UINT, UDINT	Current count value

3.2.1   *Count Up Counters (CTU)*

The count up counter's main function is to increment the current value each time the input to the counter transitions from 0 to 1. If the current count value (CV) is equal to the preset value (PV), the output Q is set. When reset input (R) is TRUE, the accumulated value resets to 0. Counter preset input can be defined as a tag name or a constant value. The counter block is shown in Fig. 3.23 with appropriate tag names assigned. These tags are displayed between double quotes in the figure. All counter-required variables are displayed using the standard system labels, which start with the percent (%) character. Counter preset input can be defined as a tag name or a constant value. Figure 3.24 provides the timing diagram for this timer.

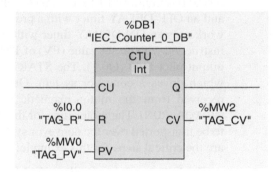

Figure 3.23 Counter block.

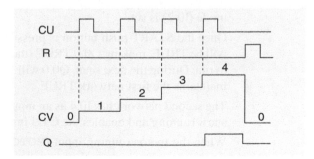

Figure 3.24 CTU timing diagram with PV = 4.

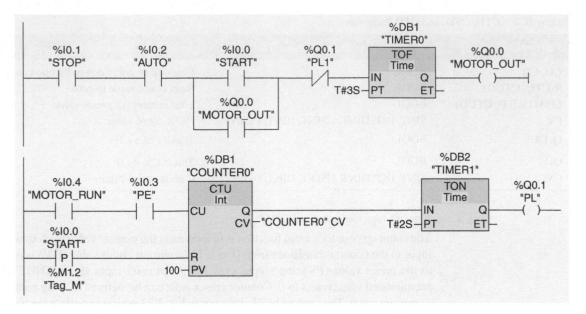

Figure 3.25 CTU ladder-logic diagram.

Figure 3.25 shows two networks to illustrate the CTU instruction. The first network assumes a normally open START push button (I0.0), a normally closed STOP push button (I0.1), an AUTO selector switch (I0.2), motor output (Q0.0), and an OFF-DELAY timer with a preset time (PT) of 3 seconds. The second network uses an ON-DELAY timer with a preset time (PT) of 2 seconds, count up instruction with preset value (PV) of 100, positive edge instruction (M1.2), and an output pilot light (Q0.1). The START push button runs the motor immediately, which drives a conveyor system. Once the conveyor system runs, an input is received from the motor magnetic starter indicating successful system start (MOTOR_RUN). The motor runs for the duration to allow a preset number of parts to be transported over the conveyor system, which is set to 100 in our case. Below are the critical steps for this example:

- The first network initially is FALSE during the first scan because the STOP push button (I0.1) is wired high. Also, the START push button is FALSE because this switch is not pressed (normally open push button), and the AUTO mode (I0.2) is set.

- Once the START push button is pressed, I0.0 becomes TRUE, Timer 0 output will be TRUE, making Q0.0 TRUE (the output to Motor 1) and causing Motor 1 to run. During the next scan, Q0.0 will latch around the START push button and maintains the first network TRUE.

- The second network has I0.4 as an input, which becomes TRUE once the motor starts running and enables the CTU instruction.

- When the motor running input (MOTOR_RUN) is set, indicating that the conveyor is moving, a photoelectric cell will issue a positive narrow pulse as a part crosses its beam, which causes the CTU current value (CV) to increment.

- Once the counter preset value (PV) equals the current value (CV), the counter output Q is set to TRUE and starts Timer 1. After 2 seconds of delay, the pilot light (PL) will be set, indicating the end of the cycle. The 2 seconds allow the last part to cross the counting station.

- Once the pilot light is set, the counter count value (CV) resets to 0 through the positive-edge counter instruction rest input. The TOF timer in the first network will lose power, and Timer 0 will start timing. Three seconds later, MOTOR_OUT will be FALSE, and motor will stop running. The 3 seconds allow the last part to arrive at the end of the conveyor line.

- A solenoid, which is not shown, stops parts from entering the counting station while the conveyor runs during the 5-second delay. Notice that only one timer is needed for this implementation. Two different timers were used for the demonstration of TON and TOF instructions.

3.2.2 Count Down Counters (CTD)

The count down counter's main function is to decrement the current value (CV) each time the input to the counter transitions from 0 to 1. If the current value is equal to or less than 0, the counter output Q is set. The value at the CV is set to the value of the PV parameter when the signal state at the LD input changes to 1. As long as the LOAD input has signal state 1, the signal state at the CTD input has no effect on the instruction. Figure 3.26 shows the CTD counter block, and Fig. 3.27 provides the timing diagram.

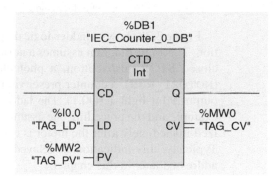

Figure 3.26 CTD counter block.

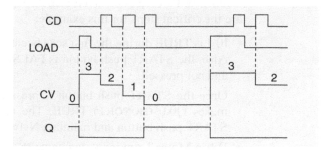

Figure 3.27 CTD timing diagram with PV = 3.

Network 1:

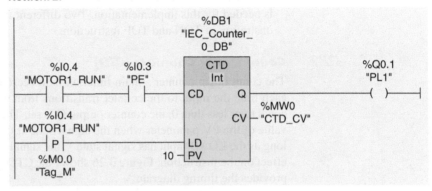

Network 2:

Figure 3.28 CTD ladder-logic diagram.

Figure 3.28 shows a ladder-logic diagram example for the CTD counter instruction. This logic diagram assumes a normally open START push button, a normally closed STOP button, a photoelectric cell (PE), Motor 1 running input (MOTOR1_RUN), a counter preset value (PV) of 10, output Motor 1 (Q0.0), and output Pilot light 1 (Q0.1). The ladder diagram consists of two networks. This example and the preceding one assume that the motor running input signal is realized immediately after the motor is commanded to start and before the next scan. Typically, this indication is delayed for several scan times. This issue will be addressed later in this chapter.

The preceding example uses a motor to derive a conveyor system. The system requires that the conveyor system shuts down once a preset value for rejected parts is reached. Part rejection is initiated through a photoelectric inspection cell. Below are the critical steps for this example:

- I0.1 is TRUE during the first scan because the STOP push button is wired high. Also, the START push button is FALSE because this switch is normally open and not pressed.

- Once the START push button is pressed, I0.0 becomes TRUE; this, in turn, makes Q0.0 (MOTOR1) TRUE. The next scan, Q0.0, will latch around the START push button and maintain Network 1 TRUE status.

- When Motor 1 run input, tag name MOTOR_RUN, is set, the PE will generate a positive pulse for each rejected part.

- The counter (CTD) will decrement the CV, tag name `CTD_CV`, every time PE transitions from 0 to 1.
- Once the CV is 0, Pilot light 1 will turn ON, indicating that 10 parts are rejected. This also will stop the motor and the conveyor system.
- The same sequence can be restarted through the START push button. The counter preset value is maintained at the initial assignment.

It is obvious that the same application discussed in this section can be implemented using the up counter instead of the down counter.

3.2.3 *Count Up and Down Counters (CTUD)*

The main function of the count up and down instruction is to count up or down the value at the current count value, tag name `COUNT_CV`. If the signal state at the input count up (CU) tag named `PE1` transitions from 0 to 1, the current count value with tag name `COUNT_CV` is incremented. If the signal state at the input count down (CD) tag named `PE2` transitions from 0 to 1, the current count value is decremented. When the signal at the LOAD input (LD) tag named `LOAD` changes from 0 to 1, the count value (CV) is set to the value of the PV parameter. As long as the LOAD input has signal state 1, the signal state at the CU and CD inputs has no effect on the instruction. The count value is set to zero when the signal state at the R input tag named `RESET` changes from 0 to 1. As long as the R input has signal state 1, a change in the signal state of the CU, CD, and LOAD inputs has no effect on the count up and down instruction. Figure 3.29 shows the

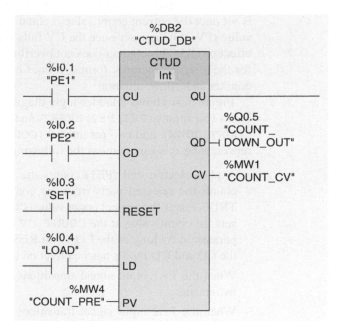

Figure 3.29 Count up and down block.

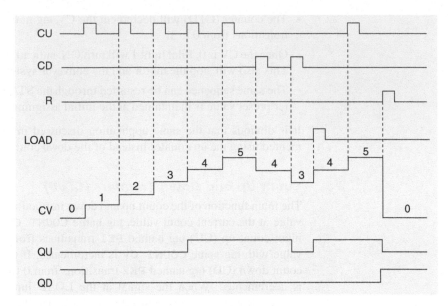

Figure 3.30 CTUD timing diagram with PV = 4.

CTUD instruction block and Fig. 3.30 provide the timing diagram for the up and down counter for a PV value stored in a tag named `COUNT_PRE`. The output QD tag named `COUNT_DOWN_OUT` is used to indicate that the current count value is less than or equal to 0. It is set to 1 when the counter CV is less or equal to 0. A counter load does not affect the status of the QD output. The counter QU output is set once the current count value is equal to or larger than the preassigned preset value (PV). QU is reset once the CV falls below the PV. Load and reset have no effect on QU. This counter does not overflow. Once it reaches the maximum value for the assigned memory format, it does not increment. QD is used to indicate a counter underflow situation.

Figure 3.31 shows a ladder-logic diagram for the CTUD counter instruction. It has four inputs (`PE1`, `PE2`, `RESET`, and `LOAD`), two outputs (`COUNT_UP` and `COUNT_DOWN`), and two parameters (`COUN_CV` and `COUNT_PV`). The network representing this function has the following critical events:

- A photoelectric cell (PE1) counts the parts in a certain conveyor line. PE2 counts the rejected parts from the conveyor line. The RESET signal, when TRUE, resets the current count value (`COUNT_CV`) to 0. A TRUE LOAD signal sets the count value at the `COUNT_CV` output to the value of the `COUNT_PV` parameter. As long as the LOAD or RESET input signal is TRUE, the signal at the CU and CD inputs has no effect on the instruction.

- When the PE1 input signal transitions from 0 to 1, the current count value increments.

- When the PE2 input signal transitions from 0 to 1, the current count value decrements.

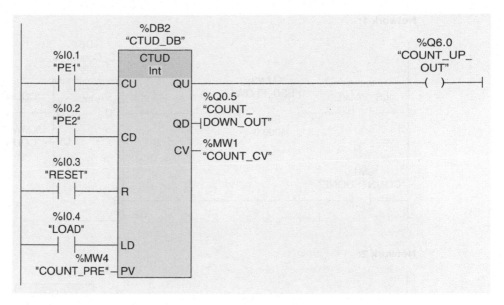

Figure 3.31 Ladder-logic diagram for the CTUD instruction.

- The current count value represents the number of good parts processed by the photoelectric cells on the conveyor line.
- If the current count value is greater than or equal to the value of the PV parameter, the QU output is set.
- If the current count value is less than or equal to zero, the QD output is set.

3.2.4 *Implemented Counter Applications*

This feed-flow application allows 32,000 gal of liquid to flow into an empty reactor tank (Fig. 3.32). It uses a limit switch with tag name **LS_VALVE** mounted on a valve. This switch closes while the liquid feed valve is open. The counter PV represents thousands of gallons and is set to 32. Thus each count increment is equivalant to 1000 gal of flow. **FEED_FLOW** is memory register containing the accumulated flow amount in gallons. The following are the critical events for this application:

- Network1 limit switch input (**LS_VALVE**) is set when the valve is open. Once **FEED_FLOW** is greater than or equal to 1000, the counter increments the count current value (**COUNT_CV**). When **COUNT_CV** is greater than or equal to 32, which is equivalent to 32,000 gal, the counter output Q is set. The counter current count value resets to 0 during the next program scan.
- In Network 2, when the valve is open and **FEED_FLOW** is greater than or equal to 1000, its value is reduced by 1000.

Network 1:

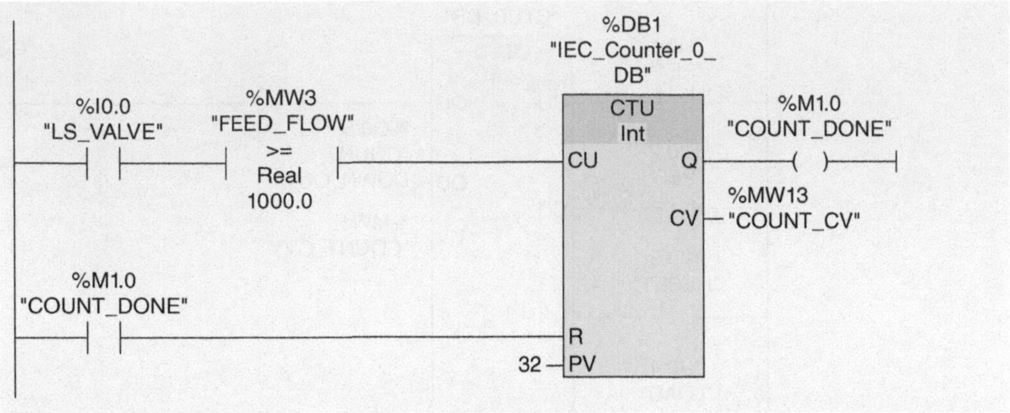

Network 2:

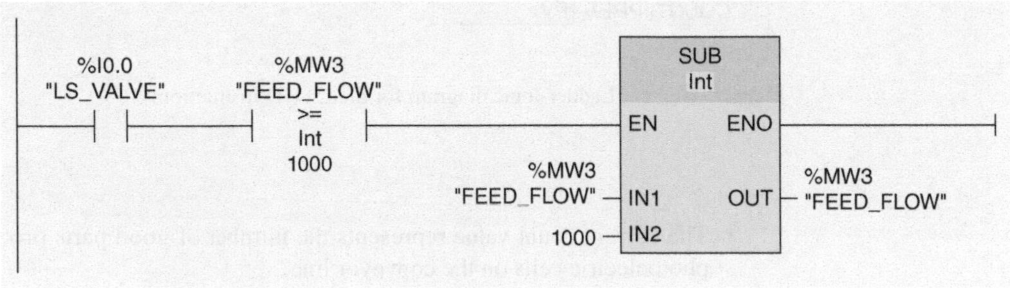

Figure 3.32 Feed flow control.

3.3 Special Timing Instructions

Two commonly used special timing instructions will be discussed in this section: the pulse timer (generate pulse timer) and the positive-edge/negative-edge instructions. An implemented ladder-logic program also will show the application of such timers in real-time industrial-control situations.

3.3.1 Pulse Generation/Pulse Timer (TP)

A pulse timer (TP) generates a pulse with a preset width time. If the rung input is TRUE, the output Q is set TRUE for the period identified by the preset value PT, tag name T_PRE. Figure 3.33 shows the TP instruction block, and Fig. 3.34 provides the TP timing diagram.

Figure 3.35 shows the ladder-logic diagram for the TP timer instruction. This diagram assumes a normally open input M27.6 pulse signal, tag name VG1 _AUTO_START; a timer preset TP, tag name VG1_PRE, with a value of 15 seconds; and output Q0.0, tag name VG1_RAISE. When the START pulse (VG1_AUTO_START) is set, the network is TRUE, and the TP timer output Q is set for 15 seconds, causing vertical gate Motor 1 to run for 15 seconds, which raises the vertical gate.

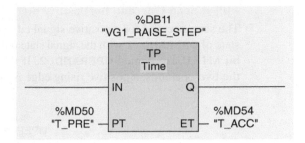

Figure 3.33 PT block.

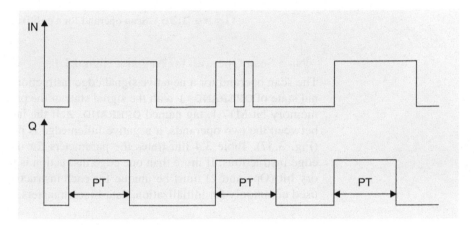

Figure 3.34 TP timing diagram.

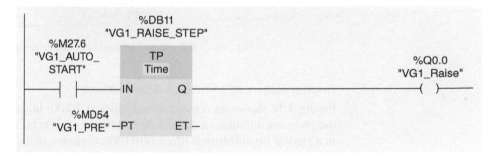

Figure 3.35 TP ladder-logic diagram.

3.3.2 *One-Shot Operations*

The one-shot instruction (scan operand for positive or negative signal edge in Siemens PLCs) is designed to detect a positive or a negative signal edge. The two types are detailed below. The names *one shot* and *positive edge/negative edge* are used interchangeably in the literature.

--|P|--: Scan Operand for Positive Signal Edge

The scan operand for a positive signal edge instruction compares the current signal state of OPERAND_1 with the signal state of the previous scan stored in edge memory bit M17.0, tag named OPERAND_2. If the instruction detects a change between the two operands, a positive rising edge is produced for only one scan (Fig. 3.36).

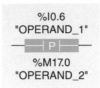

Figure 3.36 Scan operand for a positive signal edge.

--|N|--: Scan Operand for Negative Signal Edge

The scan operand for a negative signal edge instruction compares the current signal state of OPERAND_1 with the signal state of the previous scan stored in edge memory bit M17.1, tag named OPERAND_2. If the instruction detects a change between the two operands, a negative fallen edge is produced for only one scan (Fig. 3.37). Table 3.4 illustrates the parameters for the positive/negative signal edge instructions. If more than one edge instruction is used, the associated memory bit (Operand 2) must be unique for each instruction. These instructions are used commonly for initialization, data-block transfers, and event detection.

Figure 3.37 Scan operand for a negative signal edge.

3.3.3 *Implemented One-Shot Application*

Figure 3.38 shows an organizational block (OB100) initialization. On power up, the program initializes a set of values. This action is achieved by placing the code in a startup organizational block (OB100), as shown in the figure, without the use

Table 3.4 Scan Operand for Positive/Negative Signal Edge Instructions

Parameter	Declaration	Data Type	Memory Area	Description
<Operand1>	Input	BOOL	I, Q, M, D, L	Signal to be scanned
<Operand2>	Memory bit	BOOL	I, Q, M, D, L	Edge memory bit in which the signal state of the previous scan is saved

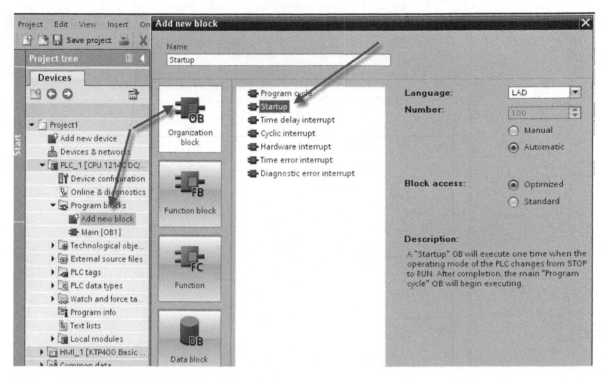

Figure 3.38 Startup OB initiation.

of the ONS/positive-edge trigger instruction, as was done in Chap. 2. All the code included in a startup OB executes one time when the operating mode of the CPU changes from STOP to RUN.

The startup program consists of one or more startup OBs (OB numbers 100 or ≥123). The startup program is executed one time when the PLC transitions from STOP mode to RUN mode. After complete execution of the startup OBs, the process image of the inputs is read in, and the cyclic program is started. Figure 3.38 shows the configuration of startup OB100. The following are the steps needed to create a startup OB:

- Click "Add new block."
- Click "Organizational block."
- Click "Startup." Automatically, the number 100 will be assigned to the Startup OB.

The MOV instruction shown in Fig. 3.39 was implemented in Chap. 2 (Example 2.3) using the scan operand for a positive signal edge, which is a one-shot (OS) logic used to reset the timers' preset values. In this example, the MOVE instruction is placed in the OB100 to clear all timer preset values when the PLC processor switches from STOP to RUN.

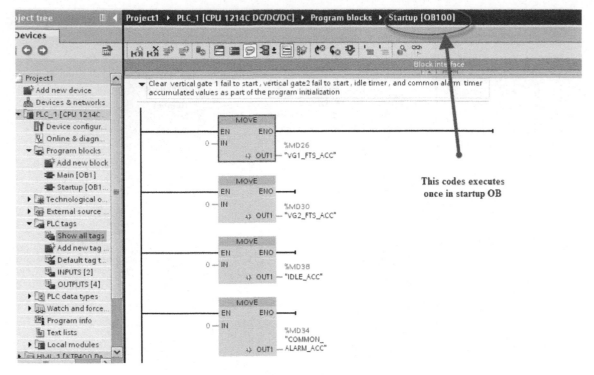

Figure 3.39 Startup OB trigger ladder-logic diagram initialization.

3.3.4 *Implemented Counter Applications*

The PLC counters and timers instructions covered earlier in this chapter function in a similar way. The timer instructions will continually increment the assigned accumulated value at a rate determined by the time base when the contacts that power the instruction are enabled. On the other hand, the counter requires a complete enable-contact transition from open to close each time it increments/decrements the accumulated value. This means that the contacts must return to their open state before they can transition for a second time. The counter does not care how long the contacts stay closed once they transition; it only looks for the transition. The transition must take place at a rate less than the PLC program scan rate; otherwise, some transitions can be lost.

Example 3.5   Figure 3.40 shows a tank flow process. This application uses counters in the implementation of tank process control. It assumes two solenoid valves, FILL (SV1) and DRAIN (SV2). The tank level sensor is simulated by a count up and down (CTUD) counter. If the tank level is greater than or equal to 10 m, drain the tank by activating SV2, and if tank level is less than or equal to 1 m, fill the tank by activating SV1.

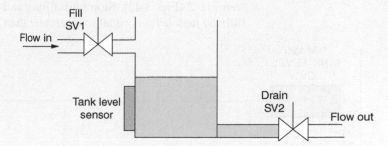

Figure 3.40 Tank level and flow control.

The Siemens PLC S7 implementation detailed next uses a processor built-in flash bit to simulate the tank filling and draining actions. It assumes a 1-m level increase during filling per flash and a 1-m decrease in level during draining per flash. Three ladder networks are used to implement this part of the specification. The ladder networks are listed below with brief details:

Network 1 (Fig. 3.41). An up/down counter (CTUD) controls the liquid level in the tank, which is given the name **TANK_LEVEL**. The PLC **%M100.5** bit flashes once every second, which simulates a 1-m change in tank level. Tank control is active when the AUTO/MANUAL selector switch is set to the AUTO position. The STOP push button, which is wired high, will reset the counter when activated.

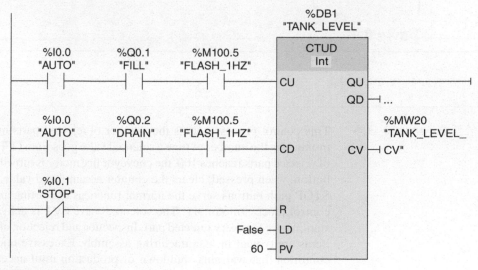

Figure 3.41

Network 2 (Fig. 3.42). Stop tank filling and start draining when the tank is full; the tank level is equal to or greater than 10 m.

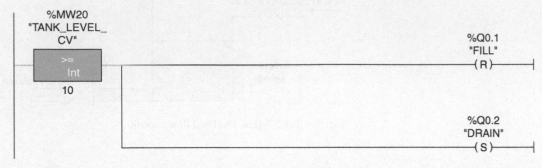

Figure 3.42

Network 3 (Fig. 3.43). Start tank filling and stop draining when the tank level is high; the tank level is less than or equal to 1 m.

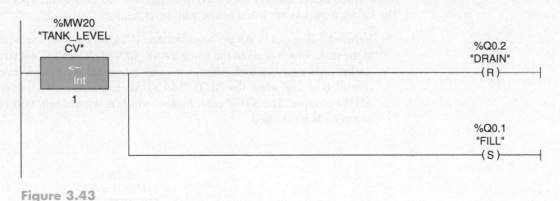

Figure 3.43

Example 3.6

This control process counts the number of rejected parts on a conveyor line by monitoring the number of time a solenoid valve goes from OFF to ON. If the number of rejected parts reaches 100, the conveyor line motor is turned OFF. A RESET push button, when pressed, clears the counter accumulated value. The START and the STOP push buttons serve the normal functions of starting and stopping the PLC-control process/motor M1. The solenoid valve SV1 is activated at an inspection station once for every rejected part. Inspection and rejection of assembled/produced items are typical in manufacturing assembly. Excessive rejection is an alarming condition that warrants shutdown of production until necessary corrections are carried out by the operator. The two network ladder-logic diagrams in Figs. 3.44 and 3.45 implement the required control specification.

Network 1:

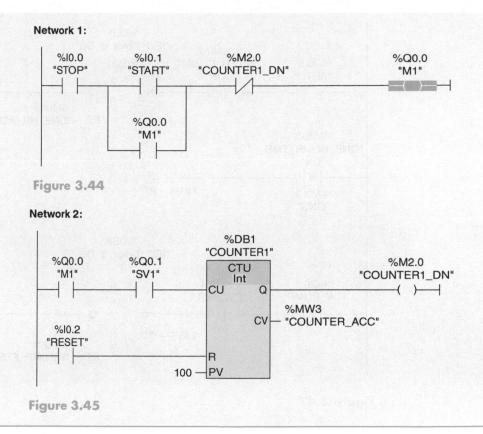

Figure 3.44

Network 2:

Figure 3.45

Homework Problems and Laboratory Projects

Problems

3.1 Explain the difference between ON-DELAY timer (TON) and retentive timer (TONR) instructions.

3.2 Complete the timing diagram for the ON-DELAY (TON) timer in Fig. 3.46. The input signal is the timer enabling input, whereas the output is derived by the timer done bit. Indicate when the timer accumulated value starts and stops timing.

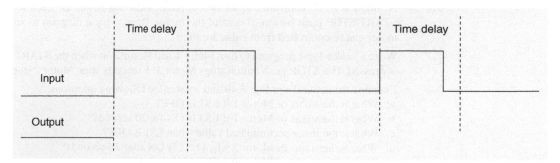

Figure 3.46

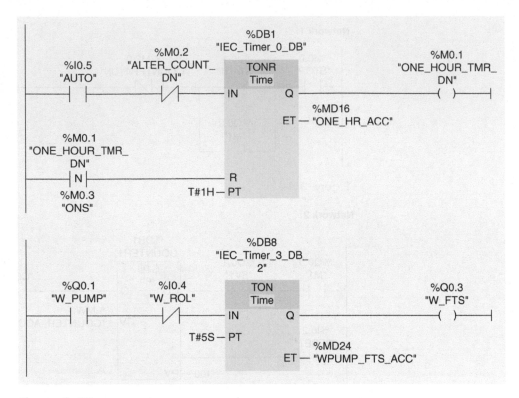

Figure 3.47

3.3   The ladder network shown in Fig. 3.47 was programmed to recycle a timer on an hourly basis. When the program was tested, it did not work correctly. Explain why.

3.4   Examine the networks shown in Fig. 3.48 and answer the following questions:
      a. What is the status of Motor 1 if LS1 is OFF?
      b. What is the status of Motor 1 if LS1 is ON for 30 seconds?
      c. What is the value of the timer accumulated value when LS1 is OFF?
      d. What is the status of Motor 2 when LS1 is ON for 10 seconds?
      e. What is the status of SOL1 and SOL2 if LS1 is OFF?

3.5   Write a ladder-logic program to turn ON Motor 1 when the START push button is pressed. Ten seconds later, Motor 2 turns ON. The STOP push button stops both motors.

3.6   A motor is to be controlled from two different locations A and B. Each location has a START/STOP push button to control the motor. Draw a logic diagram to show how the motor can be controlled from either location.

3.7   Write a ladder-logic program to turn Motor 1 and Motor 2 on when the START push button is pressed. The STOP push button stops Motor 1; 5 seconds later, Motor 2 stops.

3.8   Examine the networks in Fig. 3.49 and answer the following questions:
      a. What is the status of Motor 1 if LS1 is OFF?
      b. What is the status of Motor 1 if LS1 is ON for 20 seconds?
      c. What is the timer accumulated value when LS1 is OFF?
      d. What is the status of Motor 2 when LS1 is ON after 10 seconds?
      e. What is the status of SOL1 and SOL2 if LS1 is OFF?

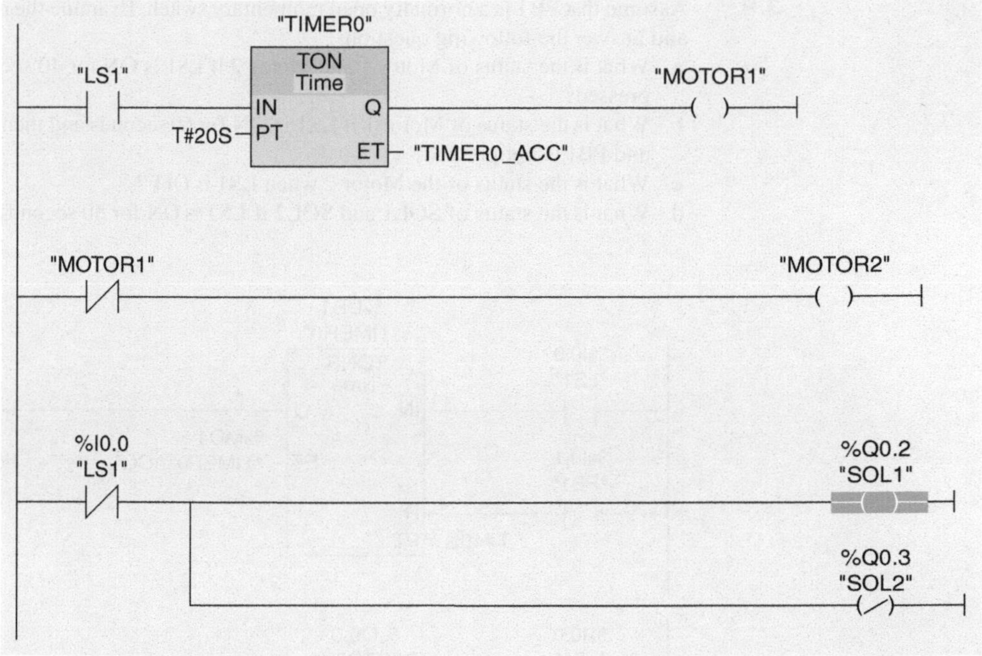

Figure 3.48

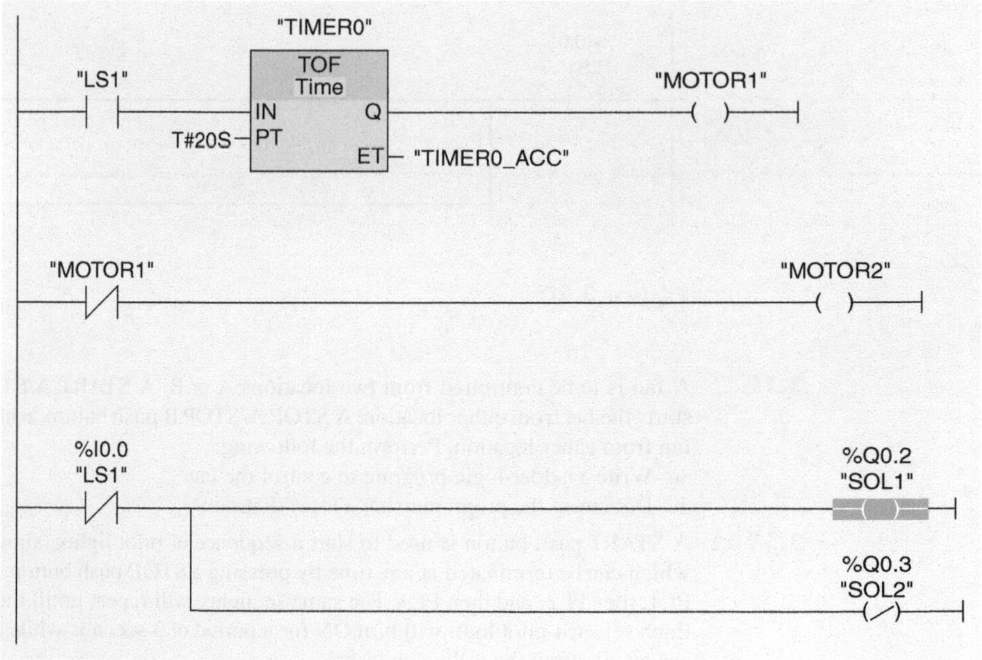

Figure 3.49

3.9   Assume that PB1 is a normally open momentary switch. Examine the networks in Fig. 3.50 and answer the following questions:
    a. What is the status of Motor 1 and Motor 2 if LS1 is ON for 40 seconds and PB1 is not pressed?
    b. What is the status of Motor 1 if LS1 is ON for 60 seconds and then OFF for 40 seconds and PB1 is not pressed?
    c. What is the status of the Motor 2 when LS1 is OFF?
    d. What is the status of SOL1 and SOL2 if LS1 is ON for 50 seconds?

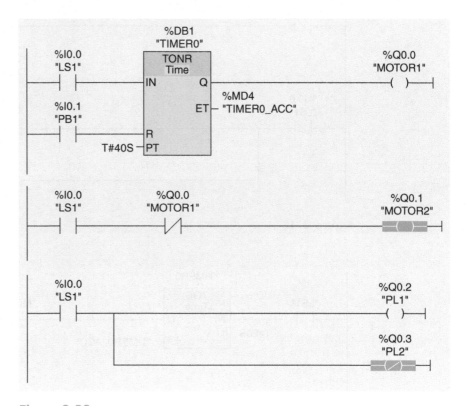

Figure 3.50

3.10   A fan is to be controlled from two locations A or B. A START A/START B push button starts the fan from either location. A STOP A/STOP B push button, when pressed, stops the fan from either location. Perform the following:
    a. Write a ladder-logic program to control the fan.
    b. Document the program using a logic diagram.

3.11   A START push button is used to start a sequence of pilot lights (simulating real motors), which can be terminated at any time by pressing a STOP push button. The sequence starts PL1, then PL2, and then PL3. The same sequence will repeat untill the process is stopped. Each selected pilot light will turn ON for a period of 3 seconds while the other pilot lights are off. Perform the following tasks:
    a. Write a ladder-logic program to sequence the pilot lights as specified.
    b. Modify the ladder logic to stop the pilot lights if the sequence is repeated five times.

3.12 Write a ladder-logic diagram using the TONR timer instruction under the following conditions:
  a. The timer starts timing if the AUTO/MANUAL selector switch is placed in the AUTO position. If the AUTO/MANUAL selector switch is placed in MANUAL, the timer stops timing, and the timer accumulated value retains its current count.
  b. If the switch is placed back to AUTO, the timer starts timing from where it left off. When the timer accumulated value equals the preset value, the timer output is set to TRUE, and the timer accumulated value is reset to zero.

3.13 Draw the Q output for the up counter shown in Fig. 3.51 assuming that Q is initially low.

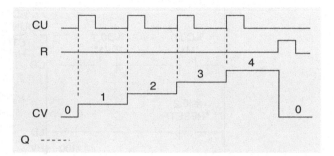

Figure 3.51

3.14 Study the networks in Fig. 3.52 for the count up (CTU) timer, and answer the following questions:
  a. Under what condition will M1 turn ON?
  b. Under what condition will PL1 turn ON?
  c. What is the value of the counter accumulated value (**COUNTER1_ACC**) when I10.2 is ON?

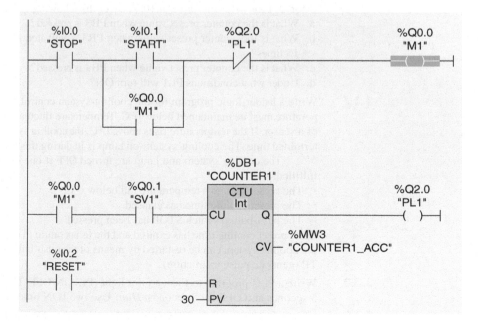

Figure 3.52

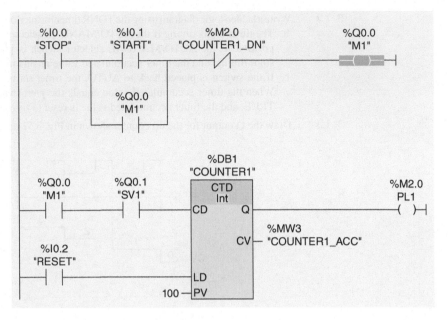

Figure 3.53

3.15   Study the networks in Fig. 3.53 for the down counter (CTD), and answer the following questions:
   a. What condition(s) should be met for Motor 1 (M1) to run?
   b. What condition(s) should be met for Pilot light 1 (PL1) to turn ON?

3.16   Study the network in Fig. 3.54 assuming that the counter preset value with tag name COUNT_PRE holds the value of 10. Answer the following questions:
   a. What is the counter preset value when PB3 is pressed?
   b. What is the counter preset value when PB3 is activated 12 times and PB2 is activated 15 times?
   c. What is the counter preset value when PB4 is pressed?
   d. Under what conditions PL1 will turn ON?

3.17   Write a ladder-logic program for the cooling-system control shown in Fig. 3.55. The temperature must be maintained below 0°C. Temperature fluctuations are monitored by means of a sensor. If the temperature rises above 0°C, the cooling system switches on for a predetermined time. The cooling-system-on lamp is lit during this time.
        The cooling system and lamp are turned OFF if one of the following conditions is fulfilled:
   • The sensor reports a temperature fall below 0°C.
   • The preset cooling time has elapsed.
   • The push-button switch STOP has been pressed.
   If the preset cooling time has expired and the temperature in the cold room is still too high, the cooling system can be restarted by means of the push-button switch RESET. *Hint:* Use TP (generate pulse instruction).

3.18   Write a PLC program to flash a pilot light (PL) ON/OFF. The pilot light (PL) is ON for 5 seconds and OFF for 3 seconds. *Hint:* Use two TON timers.

3.19   Write a PLC program to allow the operator to run the conveyor line for luggage transportation by pushing a START push button. The START push button is a normally open momen-

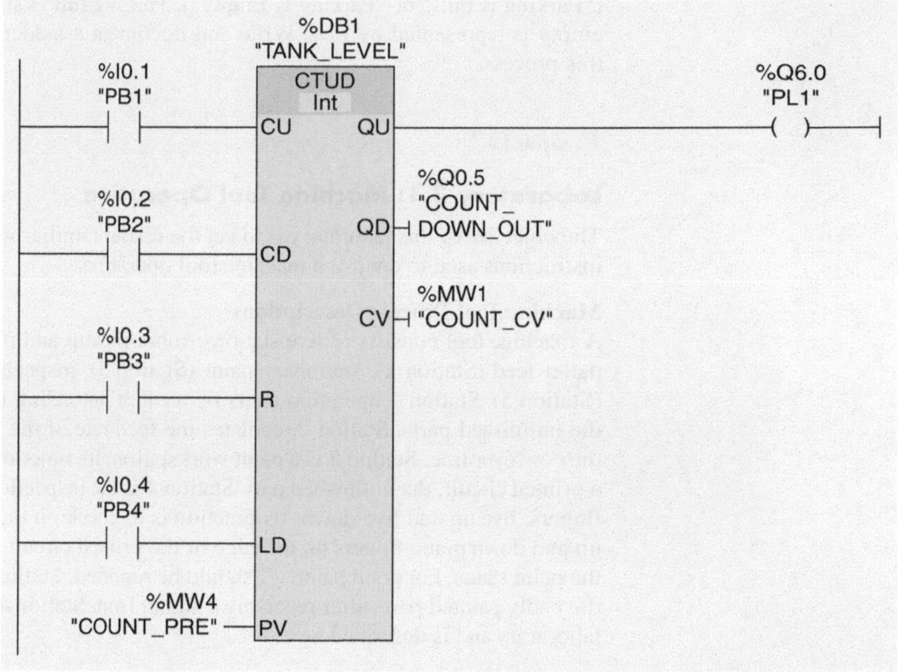

%DB1
"TANK_LEVEL"

%I0.1
"PB1"
CTUD
Int

%Q6.0
"PL1"

CU   QU   ( )

%I0.2
"PB2"

%Q0.5
"COUNT_
DOWN_OUT"

QD

CD

%MW1
CV "COUNT_CV"

%I0.3
"PB3"

R

%I0.4
"PB4"

LD

%MW4
"COUNT_PRE"   PV

Figure 3.54

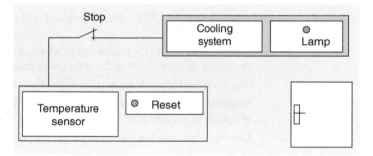

Figure 3.55

tary switch, and the STOP push button is normally closed and wired high. The following is a brief process description:

a. When the operator pushes START, flash a pilot light (PL) every 2 seconds to warn people that the conveyor belt is about to start.
b. After 80 seconds, start the conveyor motor, and turn off the pilot light.
c. The operator stops the conveyor line by pushing the STOP push button.

3.20 Modify Problem 19 to count the number of bags on the conveyor line, and stop the motor after 100 bags have passed. Add sensors as needed.

3.21 A parking lot has two momentary action sensors to count the number of cars entering or exiting the garage. One sensor is placed at the entrance, and the other is placed at the exit. Two messages should be displayed to customers indicating the status of the parking lot

("Parking is Full" or "Parking is Empty"). Parking full is simulated by PL1, and parking empty is represented by PL2. Write and document a ladder-logic program to implement this process.

Projects

Laboratory 3.1: Machine Tool Operation

The objective of this laboratory is to get the reader familiar with basic timing and counting instructions used to control a machine tool operation.

Machine Tool Process Descriptions

A machine tool consists of five stations: robot pickup and placement of parts (Station 1), pallet feed (Station 2), stationary paint (Station 3), inspection (Station 4), and rejection (Station 5). Station 1 operation starts by a robot unloading the finished parts and placing the unfinished parts. Station 2 regulates the feed rate of the pallets that carry the parts on the conveyor line. Station 3 is a paint workstation. Its function is to spray-paint the edge of a printed circuit, the unfinished part. Station 4 is the inspection station, which has 10 piano fingers, five up and five down. Its function is to check on the sprayed paint by closing the up and down piano fingers on the edge of the printed circuit and reporting 1 or 0 based on the paint status. For good paint, a 1 should be reported. Station 5's main function is to reject the badly painted parts after receiving a signal from Station 4. Station 3 is the focus of this laboratory and is described next.

Station 3 (Spray Paint) Operations

A part in place limit switch (LS1) is located at the station to indicate that a part exists at the paint work area.

- A solenoid valve (SV1), when energized, will raise the pallet to the level of the spray-paint chamber.
- A limit switch (LS2) is placed in the upper location, and when closed, it indicates that the part is up at the level of the spray-paint chamber.
- The spraying time is 5 seconds.
- When the spray-paint cycle is done, the solenoid valve (SV1) should deenergize to lower the pallet to the conveyor-belt level.
- Conveyor-belt and feed-rate control are not part of this laboratory.

Laboratory Requirements

Write a ladder-logic program for Station 3 (paint workstation) to raise the pallet that carries the printed circuit, maintain it in the paint chamber for 5 seconds, and lower the painted part to return the conveyor belt.
a. Assign the I/O addresses.
b. Assign the bit addresses.
c. Assume and use any sensors required to control the spray-paint process.
d. Apply the concept of what-if scenarios to check on and resolve any unusual conditions.
e. Check out the program and document the ladder-logic program using logic diagrams.

Laboratory 3.2: Conveyor System Control

The objective of this laboratory is to familiarize users with the basic timing and counting instructions used in conveyor-system control. A part is moving on a conveyor line crossing a photoelectric cell. The photoelectric cell function is primarily to count parts. The conveyor line stops after 100 parts are counted.

Design Specifications

Design and implement a documented ladder-logic program to satisfy the following sequence of operations:

a. The START push button starts the conveyor motor after a 5-second delay. The motor must start only if the AUTO/MANUAL switch is placed in the AUTO position. The photoelectric cell should count only when the motor is running.

b. When the count reaches 100, stop the conveyor after a 2-second delay.

c. The ON pilot light indicates the end of the sequence.

d. The STOP switch, when activated, takes the system to initial conditions.

e. The operator can restart the same sequence by activating the START switch.

NOTE Use SS1 for the photoelectric cell input signal, SS2 for the motor running input indication signal, PL1 for the motor starting output signal, and PL2 for the pilot light output signal.

Lab Requirements

a. Assign the system inputs and outputs.

b. Enter the networks shown in Fig. 3.56.

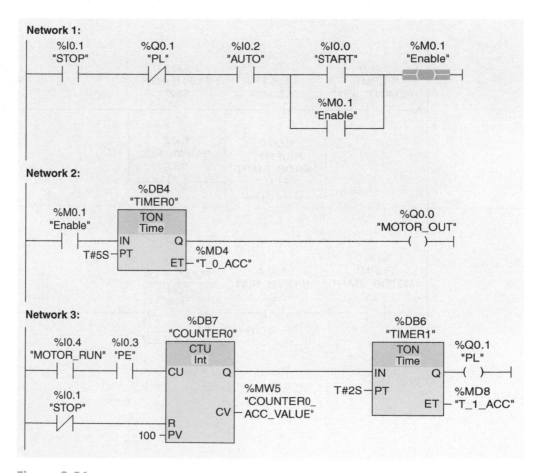

Figure 3.56

   c. Add comments to each network.

   d. Download the program and go online.

   e. Simulate the program using the training unit's I/O or the Siemens simulator, and verify that the program is running according to the process description.

   f. Modify this laboratory such that the part count is retained if the operator activates a new switch HALT while the motor is running. The HALT switch stops the motor and halts the counter operation. Reactivation of the HALT switch will start the motor and resume the part count.

Laboratory 3.3: Pump Fail-to-Start Alarm

The objective of this laboratory is to get users familiar with the methods used to trigger an alarm if a motor fails to start using the timer TON instruction. For each of the following pump start/pump stop digital outputs, search the program listed in Fig. 3.57 for the following:

   a. List the conditions for successful pump start/pump stop.

   b. Change the timer preset value for fail to start, fail to stop, start, and stop operation.

   c. Use large values of preset time, and monitor the program operation in the PLC.

   d. Simulate and monitor the pump fail alarms.

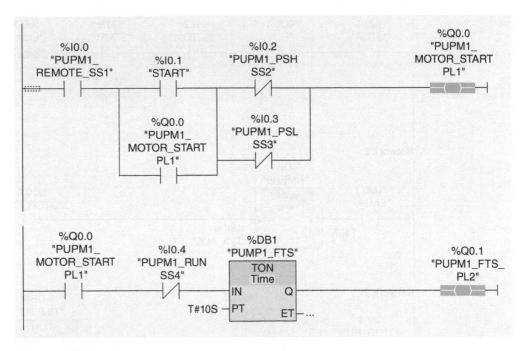

Figure 3.57

4

Math, Move, and Comparison Instructions

This chapter details three classes of ladder instructions: mathematical, comparison, and move. Numbering systems and representation in the S7-1200 PLC system are briefly covered.

Chapter Objectives

▶ Understand math instructions.

▶ Understand move and comparison instructions.

▶ Be able to implement industrial control applications.

▶ Be familiar with the S7-1200 PLC development software.

This chapter details three classes of programmable logic controller (PLC) ladder instructions: mathematical, comparison, and move instructions. It complements the coverage of logic and input-output (I/O) instructions in Chap. 2, and it also follows the detailed discussions of timer and counter instructions in Chap. 3. The coverage here is not intended to be comprehensive and repetitive, but it lays the ground work for readers to be able independently to study, comprehend, and use other instructions in any application. The coverage of mathematical instructions is preceded by a short review of numbering systems and numbers representation in the S7-1200 PLC system.

4.1   Math Instructions

Mathematical operations are an essential part of digital process control and automation. These operations are performed over input values from analog measurements or user interfaces and execute control strategies that lead to the desired outputs for regulation of the control process and associated user-interface displays. Analog I/O scaling and validation are typical examples of mathematical instruction use.

4.1.1   Numbering Systems

The commonly used decimal number system is based on module 10 representation. The system accommodates both integers and real numbers. Real numbers have an integer part and a fraction part separated by a decimal point. The integer-part digits have increasing weight as you move to the left of the decimal point, whereas the weights decrease for the fraction digits as you move to the right of the decimal point. Consider the decimal number 9623.154, which is equivalent to the following sum:

$$9623.154 = 3 \times 10^0 + 2 \times 10^1 + 6 \times 10^2 + 9 \times 10^3 + 1 \times 10^{-1} + 5 \times 10^{-2} + 4 \times 10^{-3}$$

A normalized scientific representation is frequently used for real numbers as follows: $9623.154 = 0.9623154 \times 10^4$; the fraction part (0.9623154) is called the *mantissa*, whereas the power of ten (4) is called the *exponent*. This system assumes infinite range and unlimited precession, which neither can be implemented in limited-resource systems such as digital computers nor is needed in practical real-time control applications.

PLCs, as well as digital computers, use the binary number system, or module 2 arithmetic. Numbers, integers or real, are represented in fixed-size memory words. A memory word is typically 1, 2, 4, or 8 bytes long. The size is either assigned to the default standard or can be altered through available user declaration. We will assume a 1-byte (8-bit) representation to demonstrate this binary system for unsigned and signed integers. Octal (base 8) and hexadecimal (base 16) representations, which are also part of our discussion, are just derivatives of the binary system conveniently used in documentation. The following is a summary of integer presentation:

- *Unsigned-integer presentation* (range from 0 to 255): $(97)_{10} = (0110\ 0001)_2$ and $(1110\ 0001)_2 = (225)_{10}$.
- *Signed-integer presentation* (range from –128 to 127): $(97)_{10} = (0110\ 0001)_2$ and $(-97)_{10} = $ two's complement of $(0110\ 0001)_2 = (1001\ 1110)_2 + 1 = (1001\ 1111)_2$.

Normalized real numbers are decomposed into two components: mantissa and exponent. Exponents use the signed-integer representation; mantissas use the binary-fraction presentation, and a single bit is used to indicate the real-number sign. Four bytes is the default for real-number presentations: 24 bits for the mantissa, 7 bits for the exponent, and 1 bit for the whole-number sign. More accuracy in the representation can be achieved through declaration: an additional 4 bytes are added to the mantissa.

4.1.2   *S7-1200 Data and Number Representation*

Variables used in the S7-1200 can be declared as real, integer, or Boolean. The smallest memory unit is a bit, which can accommodate a Boolean variable as a discrete I/O. Eight consecutive bits form a *memory byte*, whereas 16 consecutive bits (2 bytes) constitute a *memory word*, which is often used to represent integers. Four consecutive bytes form a *double word*, which is used to accommodate real and some integer variables. Typical Boolean variables include discrete I/O, comparison flags, computational flags, and other intermediate-logic variables. Each such variable occupies one memory bit and carries the value 0 (FALSE/OFF) or 1 (TRUE/ON). Each integer uses one-memory word (2 bytes) or double-memory word (4 bytes). Each real number uses double-memory words (4 bytes) or four-words (8 bytes). The S7-1200 processor does not use memory swapping; the high byte of a 2-byte integer is stored first at the lower memory address, and then the low byte is stored at the next memory address. Table 4.1 shows available forms of variable representation in the Siemens S7-1200 PLC system. Several other data types are available, including characters, strings, arrays, and other element-specific parameters/tags.

Table 4.1 documents common symbol types and allowed addresses. The types shown are divided into five categories: I/O signals, marker memory, peripheral I/O, timers and counters, and data blocks. Each type is assigned unique mnemonic identifier and a data type. The address range defines the start address and the upper address for each type. These ranges are extremely large, but only limited subranges are used in most practical applications. The S7-1200 processor and associated I/O interfaces are limited and do not require such large available address ranges. The size of the PLC program in terms of the number of networks, the number of I/O points used, and configured memory addresses defines the length of the PLC scan cycle time. This scan cycle time is associated with the process real-time requirements. During each scan cycle, the CPU executes the entire ladder-logic program and performs all hardware diagnostics. The longer the ladder program scans time, the less frequently the process-control system updates. In some cases, the application dynamics are slow, and scan time is not important. In other applications with fast process dynamics, scan time is critical and needs to be minimized. Also, systems with redundant PLC hardware will

Table 4.1 Symbol Table of Allowed Addresses and Data Types

Mnemonics	Description	Data Type	Address Range
I/O Signals			
I	Input bit	BOOL	0.0–65535.7
IB	Input byte	BYTE, CHAR	0–65535
IW	Input word	WORD, INT, S5TIME, DATE	0–65534
ID	Input double word	DWORD, DINT, REAL, TOD, TIME	0–65532
Q	Output bit	BOOL	0.0–65535.7
QB	Output byte	BYTE, CHAR	0–65535
QW	Output word	WORD, INT, S5TIME, DATE	0–65534
QD	Output double word	DWORD, DINT, REAL, TOD, TIME	0–65532
Marker Memory			
M	Memory bit	BOOL	0.0–65535.7
MB	Memory byte	BYTE, CHAR	0–65535
MW	Memory word	WORD, INT, S5TIME, DATE	0–65534
MD	Memory double word	DWORD, DINT, REAL, TOD, TIME	0–65532
Peripheral I/O			
PIB	Peripheral input byte	BYTE, CHAR	0–65535
PIW	Peripheral input word	WORD, INT, S5TIME, DATE	0–65534
PID	Peripheral input double word	DWORD, DINT, REAL, TOD, TIME	0–65532
PQB	Peripheral output byte	BYTE, CHAR	0–65535
PQW	Peripheral output word	WORD, INT, S5TIME, DATE	0–65534
PQD	Peripheral output double word	DWORD, DINT, REAL, TOD, TIME	0–65532
Timers and Counters			
T	Timer	TIMER	0–65535
C	Counter	COUNTER	0–65535
Data Blocks			
DB	Data block	DB, FB, SFB, UDT	1–65535

have greater scan times as a result of the housekeeping and duplication overhead. PLCs come in different sizes and processor speeds to accommodate different application requirements.

The following is an example illustrating the requirement of matching operands in arithmetic and move instructions both in type and in size.

Example 4.1   **Data Formatting**   Two signed integers are represented in different formats, the first in a 2-byte word and the second in a 4-byte word. Before we can add up the two integers, the first signed integer needs to be converted to a 4-byte signed integer. The conversion is straightforward: If the integer is positive, we just add 16 insignificant zeroes to the integer's original representation (hex 0000 0013). Notice that the most significant bit is zero, indicating a positive integer. What if the integer is negative? In this case we need to represent this negative integer in 32 bits, using the two's complement format. Let us assume that the value is −19, which is represented originally as the two's complement of 19 (hex FFED). Representing −19 in 32 bits would produce the hex value FFFF FFED. Notice that the most significant bit is 1, signifying a negative integer.

4.1.3   *Common Math Instructions*

This section covers the seven most commonly used mathematical instructions: addition, subtraction, multiplication, division, increment, decrement, and general equation calculation. Each instruction is covered individually as a block with input enable condition, output enable signal, and I/O tags. Later in this chapter we will discuss an implemented application using these mathematical instructions.

Conversion instructions are typically available with most microprocessors as part of the instruction set. They are also available in PLCs and can accommodate a wide variety of format changes, which includes conversions on integers, real numbers, and strings. Care should be taken when using such instructions in user programs.

ADD Instruction (Fig. 4.1)

If **TAG_IN** is TRUE, the ADD instruction is executed. The value of **TAG_VALUE1** is added to the value of **TAG_VALUE2**. The result of the addition is stored in the value of **TAG_RESULT**.

SUB Instruction (Fig. 4.2)

If **TAG_IN** is TRUE, the SUB instruction is executed. The value of **TAG_VALUE2** is subtracted from the value of **TAG_VALUE1**. The result of the subtraction is stored in the value of **TAG_RESULT**.

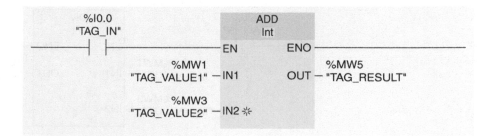

Figure 4.1   ADD instruction block.

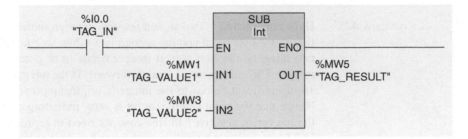

Figure 4.2 SUB instruction block.

MUL Instruction (Fig. 4.3)

If **TAG_IN** is TRUE, the MUL instruction is executed. The value of **TAG_VALUE1** is multiplied by the value of **TAG_VALUE2**. The result of the multiplication is stored in the value of **TAG_RESULT**.

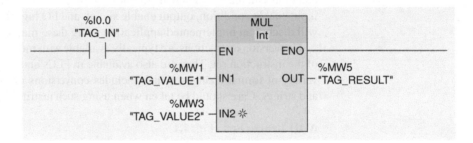

Figure 4.3 MUL instruction block.

DIV Instruction (Fig. 4.4)

If **TAG_IN** is TRUE, the DIV instruction is executed. The value of **TAG_VALUE1** is divided by the value of **TAG_VALUE2**. The result of the division is stored in the value of **TAG_RESULT**.

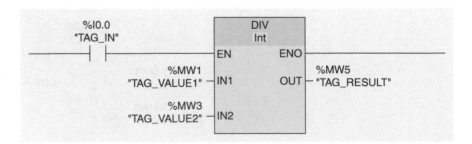

Figure 4.4 DIV instruction block.

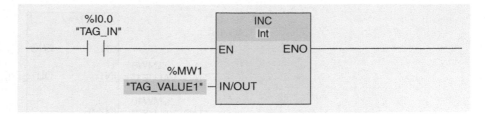

Figure 4.5 INC instruction block.

INC Instruction (Fig. 4.5)

If **TAG_IN** transitions from low to high, the INC instruction is executed. The value of **TAG_VALUE1** is incremented by one. If the **TAG_IN** is maintained TRUE, **TAG_VALUE1** keeps incrementing until it overflows.

DEC Instruction (Fig. 4.6)

If **TAG_IN** transitions from low to high, the DEC instruction is executed. The value of **TAG_VALUE1** is decremented. If the **TAG_IN** is maintained TRUE, **TAG_VALUE1** keeps decrementing until it underflows.

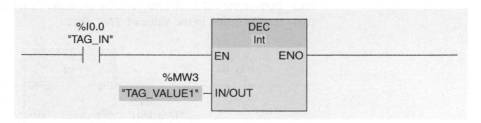

Figure 4.6 DEC instruction block.

MIN Instruction (Fig. 4.7)

If **TAG_IN** is TRUE, the MIN instruction is executed. The value of **TAG_VALUE1** is compared with the value of **TAG_VALUE2**, and the minimum of the two values is stored in the value of **TAG_RESULT**.

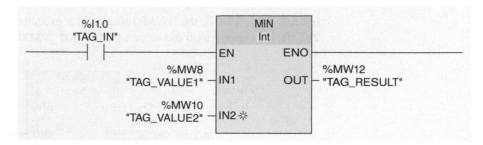

Figure 4.7 MIN instruction block.

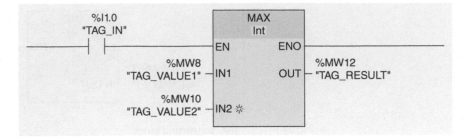

Figure 4.8 MAX instruction block.

MAX Instruction (Fig. 4.8)

If **TAG_IN** is TRUE, the MAX instruction is executed. The value of **TAG_VALUE1** is compared with the value of **TAG_VALUE2**, and the maximum of the two values is stored in the value of **TAG_RESULT**.

LIMIT Instruction (Fig. 4.9)

If **TAG_IN** is TRUE, the LIMIT instruction is executed. The value of **TAG_VALUE** is compared with the values of **TAG_MIN** and **TAG_MAX**. If the value of **TAG_VALUE** is less than the low limit, the value of **TAG_MIN** is stored in the value of **TAG_OUT**. If the value of **TAG_VALUE** is greater than the high limit, the value of **TAG_MAX** is stored in the value of **TAG_OUT**.

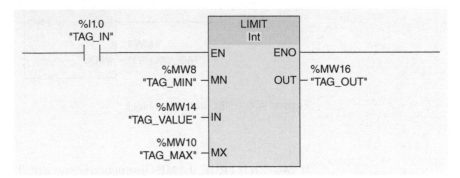

Figure 4.9 LIMIT instruction block.

SWAP Instruction (Fig. 4.10)

If **TAG_IN** is TRUE, the SWAP instruction is executed. The value of **TAG_IN_VALUE** is swapped with the value of **TAG_OUT_VALUE**, as shown in Fig. 4.11.

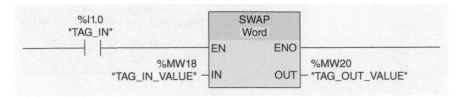

Figure 4.10 SWAP instruction block.

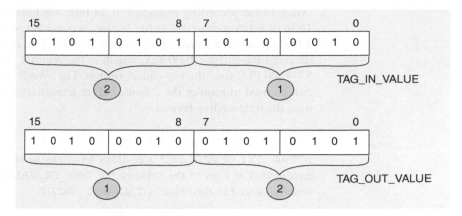

Figure 4.11   Byte swap operation.

It is commonly known that a 16-bit integer occupies 2 consecutive bytes of memory. Many people do not realize, however, that there is no standard order in which computers and instruments store these 2 bytes in memory or transmit them. As a result, computers that use one storage method cannot transmit integer data to devices that use another without first programmatically swapping each pair of bytes.

Although all 16-bit integers are stored in memory as two 8-bit bytes, these 2 bytes can be stored in two different orders. For example, consider the decimal number 16, which in hexadecimal notation is written 0010 hex. The following 2 bytes make up this number:

00_{16} Most significant byte

10_{16} Least significant byte

The significance we have assigned to each byte is generally acceptable. However, the order in which these bytes are stored in memory varies from processor to processor.

Computers based on the Intel 80×86 family of processors store the most significant byte in the higher address and the least significant byte in the lower address. Thus, on an Intel PC, if the integer 16 (0010 hex) were stored at memory location 2000, the 10 hex (the least significant byte) would be stored in byte 2000 and the 00 hex (the most significant byte) in location 2001. This form of integer storage has been dubbed *little-endian* because the least significant byte is stored in the low end of the memory word.

Computers based on the Motorola 68000 family of processors, however, store integers in the opposite order. That is, in the preceding example, the most significant byte, 00 hex, would be stored at the lower address, 2000, whereas the least significant byte, 10 hex, would be stored at address 2001. As you might expect, this form of storage has been named *big-endian* because the most significant byte is in the lower address.

If two computers exchange integer data, each must know the order in which the data will be sent and received; otherwise, the bytes might be interpreted in reverse order. In the preceding examples, if an Intel machine transfers the integer value 16 in its native storage mode to a Motorola computer, which, in turn, interprets the 2 bytes in its own native storage mode, the Motorola computer would think it had received the integer 1000 hex, which is the decimal integer 4096. The Siemens S7-1200 PLC uses the big-endian format. The SWAP instruction, covered earlier, can be used to reorder the 2 bytes before transmission to another processor that uses the little-endian format.

NEG Instruction (Fig. 4.12)

If **TAG_IN1** or **TAG_IN2** transitions from low to high, the NEG instruction is executed. The sign of the value in the **TAG_IN_VALUE** input changes, and the result is stored in the value of **TAG_OUT_VALUE**.

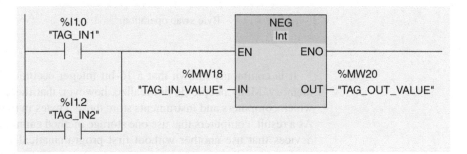

Figure 4.12 NEG instruction block.

SCALE_X Instruction (Fig. 4.13)

If **TAG_IN1** transitions from low to high, the SCALE_X instruction is executed. The value in the **TAG_IN_VALUE** input is scaled to the range of values defined by the values in the **TAG_MIN** and **TAG_MAX** inputs. The result is stored in the **TAG_RESULT** output.

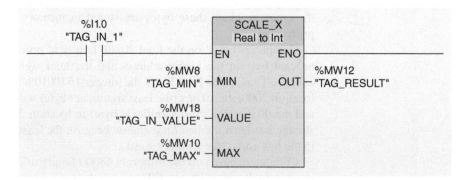

Figure 4.13 SCALE_X instruction block.

The SCALE instruction scales the value at input **TAG_IN_VALUE** by mapping it to a specified value range (MIN to MAX). When the SCALE instruction is executed, the floating-point value at the VALUE input (0.5) is scaled to the value

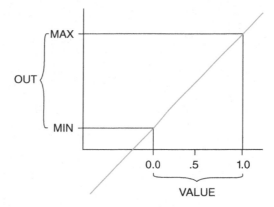

Figure 4.14 Scaling values.

range (10 to 30) that was defined by the MIN and MAX parameters. The result of the scaling (20) is an integer, which is stored in the OUT output. Figure 4.14 provides an example of how values can be scaled. Table 4.2 lists the parameters.

Table 4.2 SCALE Instruction Parameters

Parameter	Tag Name	Value
MIN	TAG_MIN	10
MAX	TAG_MAX	30
VALUE	TAG_IN_VALUE	0.5
OUT	TAG_RESULT	20

NORM_X (Normalize) Instruction (Fig. 4.15)

If **TAG_IN** is TRUE, the NORM_X instruction is executed. The instruction normalizes the value of **TAG_VALUE** by mapping it to a linear scale. The MIN and MAX parameters are used to define the limits of a value range that is applied to the scale. The result at the **TAG_RESULT** output is calculated and stored as a real number. Figure 4.16 shows how values can be normalized.

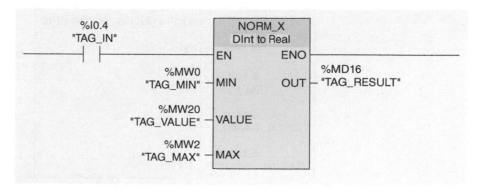

Figure 4.15 NORM_X instruction block.

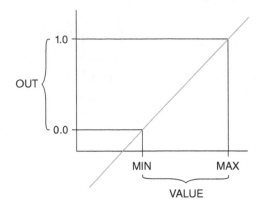

Figure 4.16 Normalized values.

CALCULATE EQUATION Instruction

The CALCULATE instruction allows for multiple inputs (IN1, IN2, IN3, etc.) that can be assigned as the independent variables of an equation including constants and produce the value for the dependent variable (OUT) as stated in the equation. The CALCULATE EQUATION instruction is used to calculate the following equation, which converts an analog input digital count "Digital Input" to a temperature value in degrees Fahrenheit "Degree F":

$$\text{Degree F} = (\text{MAX} - \text{MIN})/\text{Digital Range} \times \text{Digital Input} + \text{MIN}$$

Figure 4.17 shows the CALCULATE instruction for the temperature-conversion problem. Table 4.3 documents the parameters for the CALCULATE instruction.

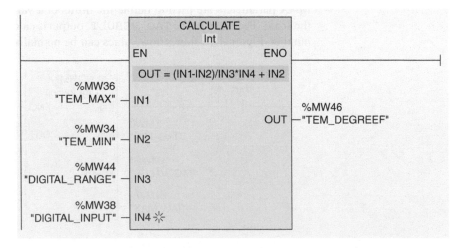

Figure 4.17 CALCULATE instruction for the temperature-conversion problem.

Table 4.3 CALCULATE Instruction Parameters

Parameters	Tag Name	Value
IN1	TEM_MAX	400
IN2	TEM_MIN	50
IN3	DIGITAL_RANGE	4095
IN4	DIGITAL_INPUT	1000
OUT	TEMP_DEGREEF	135

4.1.4 *MOVE and TRANSFER Instructions*

The MOVE instruction (Fig. 4.18) is used to copy a data element stored at a specified memory address to a new address location. The TRANSFER instruction (block move) is used to move a block of data from a memory location to another in one step using declared arrays. Also, mathematical instructions can be done over arrays. Array declarations are not covered here but can be reviewed using the Siemens programming manuals online.

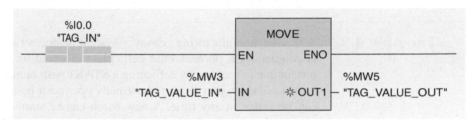

Figure 4.18 MOVE instruction block.

If **TAG_IN** is TRUE, the MOVE instruction is executed. The value of **TAG_VALUE_IN** is copied to the value of **TAG_VALUE_OUT**.

MOVE BLOCK Instruction (Fig. 4.19)

If **TAG_IN** transitions from low to high, the MOVE BLOCK instruction is executed. The instruction selects three elements from the array #A_ARRAY (#A_ARRAY [2...4] tag, as specified by the number 3 in COUNT) and copies their contents into the array **B_ARRAY** (**B_ARRAY** [1...3]).

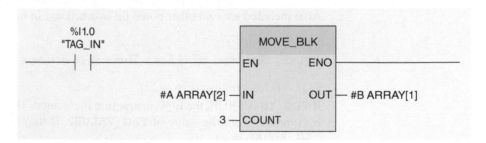

Figure 4.19 MOVE BLOCK instruction block.

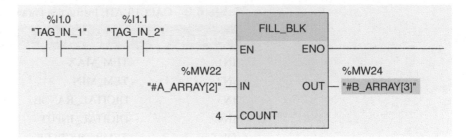

Figure 4.20 FILL BLOCK instruction block.

FILL BLOCK Instruction (Fig. 4.20)

If **TAG_IN1** and **TAG_IN2** transition from low to high, the FILL BLOCK instruction is executed. The instruction copies the second element (**#A_ARRAY [2]**) of the array (**#A_ARRAY**) four times (as specified by the number 4 in COUNT) to the array **B_ARRAY** starting from the third element **B_ARRAY [3]** and ending at **B_ARRAY [6]**.

Example 4.2

A flexible manufacturing conveyor system produces two products. Each product is counted by a photoelectric cell (PE1) mounted on the conveyor line. Batch production is enabled by activating a START push button. Each batch processing is initiated through a dedicated normally open push-button switch. Only one batch can be active at any time. A new batch can be started after completion of the ongoing batch or by forcing a termination through the STOP push button. Product 1 is produced in the amount of 2000 total parts and product 2 in the amount of 5000 total parts. A pilot light PL1 is ON after a batch is completed. The process begins by pressing the push button for the selected batch and can be terminated at any time by pressing the STOP push button. A RESET push button is used to reset the counter accumulated value at any time. Figure 4.21 provides a listing of four ladder networks that achieve the required control.

4.2 Comparison Instructions

The most commonly used comparison instructions are covered in this section. Also included are two other powerful instructions: in range and out of range.

4.2.1 Equal, Greater, and Less Than Instructions

EQUAL Instruction (Fig. 4.22)

If **TAG_IN** is TRUE, the EQU instruction is executed. The value of **TAG_VALUE1** is compared with the value of **TAG_VALUE2**. If they are equal, then the output **TAG_EQUAL** is set.

Network 1:

```
%I0.4        %I0.0                                                      %M10.0
"STOP"       "START"                                                    "ENABLE_BATCH"
──┤├──────────┤├───────┬─────────────────────────────────────────────────( )──────────

            %M10.0     │
            "ENABLE_BATCH"
         ──────┤├──────┘
```

Network 2:

```
%M10.0          %I0.1          %M10.2            MOVE              %M10.1
"ENABLE_BATCH"  "PB1"          "ACTIVE2"                          "ACTIVE1"
──┤├──────┬──────┤├──────────────┤/├─────────┤EN          ENO├──────( )──────

          │    %M10.1                   2000 ─┤IN
          │    "ACTIVE1"                               %MW0
          └──────┤├─────┘                        ※ OUT1├─"COUNTER0_PRESET"
```

Network 3:

```
%M10.0          %I0.2          %M10.1            MOVE              %M10.2
"ENABLE_BATCH"  "PB2"          "ACTIVE1"                          "ACTIVE2"
──┤├──────┬──────┤├──────────────┤/├─────────┤EN          ENO├──────( )──────

          │    %M10.2                   5000 ─┤IN
          │    "ACTIVE2"                               %MW0
          └──────┤├─────┘                        ※ OUT1├─"COUNTER0_PRESET"
```

Network 4:

```
                                          %DB1
                                         "COUNTER0"
%M10.0          %I10.5                      CTU                   %Q0.0
"ENABLE_BATCH"  "PE1"                        Int                  "END_BATCH_PL1"
──┤├──────────────┤├──────────────────────┤CU          Q├──────────( )──────

                                                        CV├─ ....
   %I0.6
   "RESET"
──────┤├────────────────────────────────┤R

                              %MW0
                       "COUNTER0_PRESET" ─┤PV
```

Figure 4.21 The four ladder networks that achieve the required control.

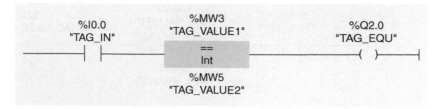

Figure 4.22 EQUAL instruction block.

NOT EQUAL Instruction (Fig. 4.23)

If `TAG_IN` is TRUE, the NOT EQUAL instruction is executed. The value of `TAG_VALUE1` is compared with the value of `TAG_VALUE2`. If they are not equal, the output `TAG_NOT_EQU` is set.

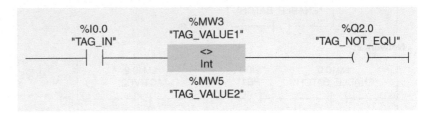

Figure 4.23   NOT EQUAL instruction block.

GREATER THAN OR EQUAL Instruction (Fig. 4.24)

If `TAG_IN` is TRUE, the GREATER THAN OR EQUAL instruction is executed. The value of `TAG_VALUE1` is compared with the value of `TAG_VALUE2`, and if `TAG_VALUE1` is greater than or equal to `TAG_VALUE2`, the output `TAG_GREATER_OR_EQU` is set.

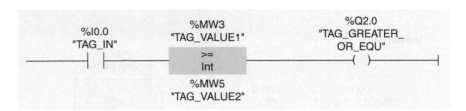

Figure 4.24   GREATER THAN OR EQUAL instruction block.

LESS THAN OR EQUAL Instruction (Fig. 4.25)

If `TAG_IN` is TRUE, the LESS THAN OR EQUAL instruction is executed. The value of `TAG_VALUE1` is compared with the value of `TAG_VALUE2`, and if

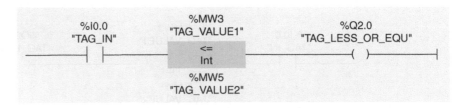

Figure 4.25   LESS THAN OR EQUAL instruction block.

TAG_VALUE1 is less than or equal to TAG_VALUE2, the output TAG_LESS_
OR_EQU is set.

GREATER THAN Instruction (Fig. 4.26)

If TAG_IN is TRUE, the GREATER THAN instruction is executed. The value of
TAG_VALUE1 is compared with the value of TAG_VALUE2, and if TAG_
VALUE1 is greater than TAG_VALUE2, the output TAG_GREATER is set.

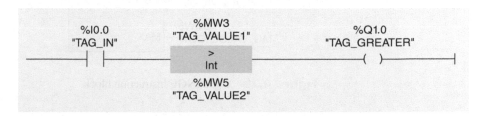

Figure 4.26 GREATER THAN instruction block.

LESS THAN Instruction (Fig. 4.27)

If TAG_IN is TRUE, the LESS THAN instruction is executed. The value of TAG_
VALUE1 is compared with the value of TAG_VALUE2, and if TAG_VALUE1 is
less than TAG_VALUE2, the output TAG_LESS is set.

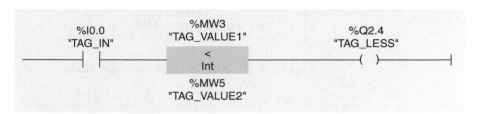

Figure 4.27 LESS THAN instruction block.

4.2.2 *IN RANGE and OUT RANGE Instructions*

IN RANGE Instruction (Fig. 4.28)

If TAG_IN is TRUE, the IN RANGE instruction is executed. If the value TAG_
VALUE is within the value range specified by the current values of TAG_MIN and
TAG_MAX, that is, MIN <= VALUE or VALUE <= MAX, then output value of
TA_IN_RANGE is set.

OUT RANGE Instruction (Fig. 4.29)

If TAG_IN is TRUE, the OUT RANGE instruction is executed. If the value of
TAG_VALUE is outside the value range specified by the current values of TAG_
MIN and TAG_MAX, that is, VALUE < MIN or VAL > MAX, then the output
value TAG_OUT_RANGE is set.

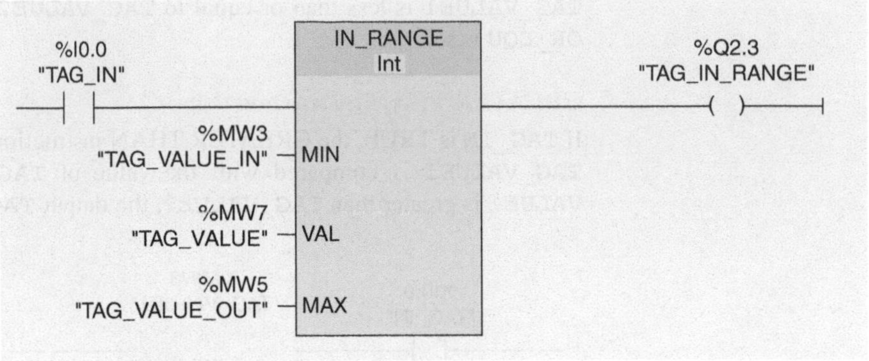

Figure 4.28 IN RANGE instruction block.

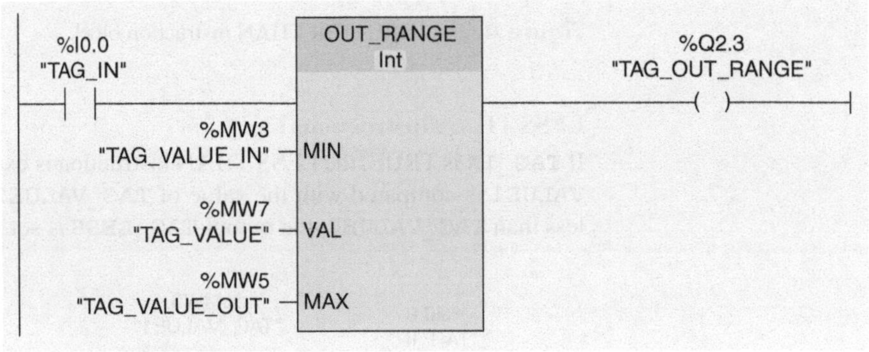

Figure 4.29 OUT RANGE instruction block.

Example 4.3

A cascaded-tanks reactor is a typical chemical process requiring accurate timing to allow for the desired reactions among mixed materials in a recipe production. In this reactor, three tanks are cascaded through a series of solenoid valves: SV1, SV2, and SV3. Materials for the selected recipe are poured into tank 1. START push-button activation will open tank 1 SV1. Reactions take place in tank 2 for 7 hours before the material goes to tank 3 for an additional 8 hours of reaction. The final delivery of finished material through tank 3 consumes the final 5 hours. Figure 4.30 shows the four networks used to implement the required process control.

Example 4.4

This example demonstrates the operation of the ON-DLAY timer. The timer timing bit (TT in Allen Bradley notation) is generated using a comparison instruction. The timer done bit (DN) and timer accumulated value (ACC) are also shown (Fig. 4.31). The following are the key timer operation details:

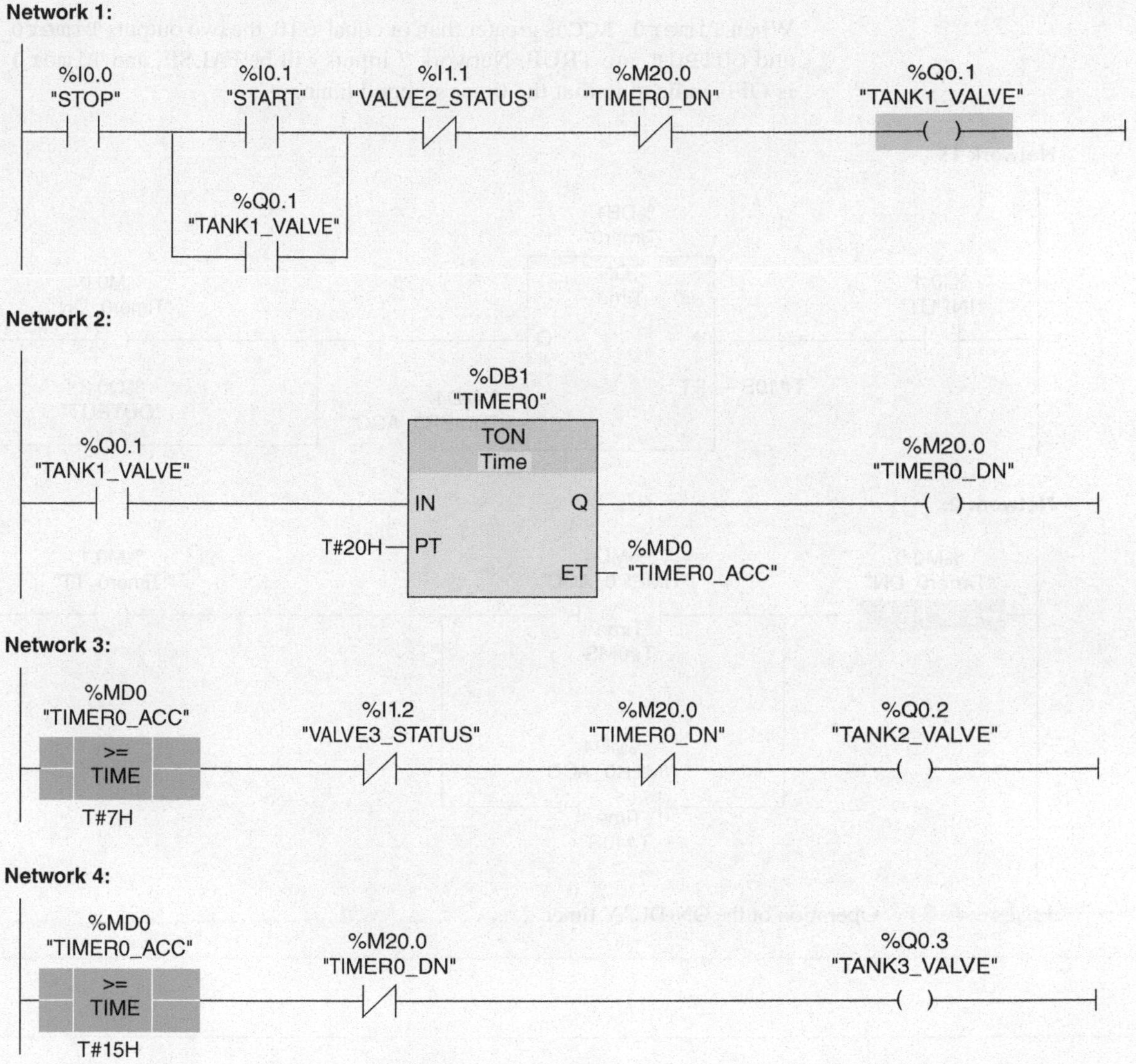

Network 1:

Network 2:

Network 3:

Network 4:

Figure 4.30   The four networks used to implement the required process control.

- *Network 1:* When INPUT (I0.1) is TRUE, Timer 0 starts trimming. Output Timer0_DN and OUTPUT are FALSE as long as Timer0_ACC did not reach 10.
- *Network 2:* When Timer0_DN is FALSE, compare instructions are executed. Timer0_TT is set, indicating that the timer is timing.

- When `Timer0_ACC` is greater than or equal to 10, the two outputs `Timer0_DN` and `OUTPUT` are TRUE. Network 2 inputs will be FALSE, and `Timer0_TT` is OFF, indicating that the timer stopped timing.

Network 1:

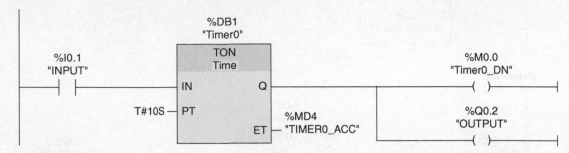

Network 2:

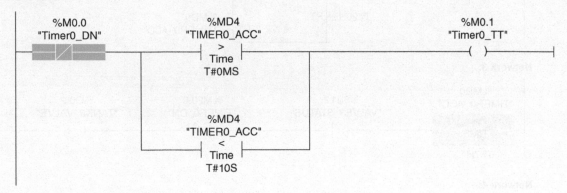

Figure 4.31 Operation of the ON-DLAY timer.

Example 4.5   This example demonstrates the operation of the DOWN COUNTER and the comparison instruction. A solenoid valve SV1 is activated to allow passage of the first 40 items counted by the photoelectric cell. Figure 4.32 shows the two networks used to implement the required operation.

- When the CTD instruction executes the first time, the count value of the CV parameter will be set to the value of the PV parameter (100).
- Network 1: When the photoelectric cell PE1 (I0.0) transitions from FALSE to TRUE, current count value (`Counter0_ACC`) decreases by 1. Output `Counter0_DN` is set to 1 when current count value is 0.
- Network 2: When current count value is greater than 60, the > instruction is TRUE, and solenoid valve SV1 is energized. When COUNT reaches 0, `Counter0_DN` is ON, reloading the counter accumulated value with 100.

Network 1:

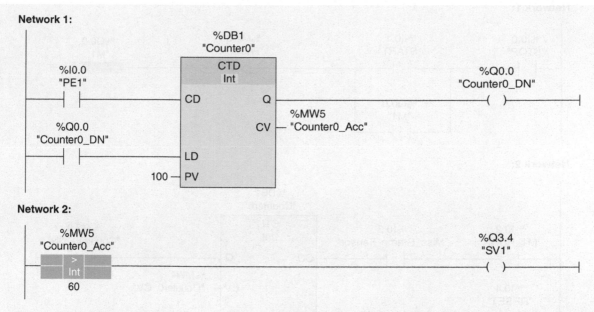

Network 2:

Figure 4.32 The two networks used to implement the required operation in Example 4.5.

Example 4.6 An inspection station is designed to detect 10 missing stamps in a 2-hour time window. Once the detection is TRUE, the motor (M1) should stop running, and the pilot light will turn ON to indicate that a fault has occurred. To restart the motor, the operator should fix the fault and push the RESET normally closed push-button switch. The four networks in Fig. 4.33 implement the desired control.

4.3 Implemented Industrial Application

This section includes a few implemented tasks that are common requirements in most process-control implementations. It is followed by three small industrial-control module applications illustrating instructions and concepts covered so far in this book. All implementations are done using the Siemens S7-1200 PLC hardware and software. The concepts and techniques used are applicable to any other PLC environment implementation.

4.3.1 *Common Process-Control Tasks*

Example 4.7 calculates the average downstream water level in an irrigation canal. Two sensors placed at two different locations downstream from a vertical gate system provide online measurments of the water level. Sensor data validation is not part of this example and must be carried out prior to this calculation. Redundant measurements typically are used to provide sensory information for the same controlled variable, especially in cases where measurements are not accurate and reliable at all times. Sensory measurement fusion requires an understanding of both the control process and the instrumentation but should be taken into consideration during the design and implementation phases.

Network 1:

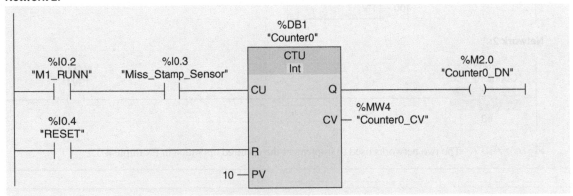

Network 2:

Network 3:

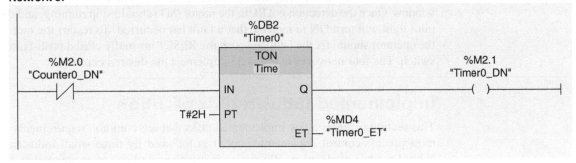

Network 4:

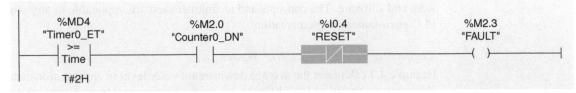

Figure 4.33 Inspection station ladder implementation.

Example 4.7   This application assumes two downstream water-level tag names (`DS1_LEVEL` and `DS2_LEVEL`). The network in Fig. 4.34 detrmines the average downstream water level using ADD and DIV instructions. The ADD instruction adds downstram level 1 (`DS1_LEVEL`) to downstream level 2 (`DS2_LEVEL`) and places the result in `DS_SUM`. Then the DIV instruction (`DS_SUM`) divides by 2 and places the result in `DS_AVE`.

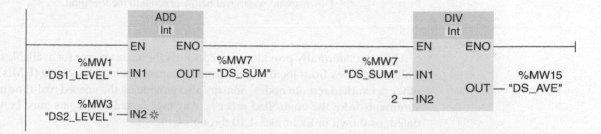

Figure 4.34   Downstream water-level average.

Example 4.8 illustrates the use of comparison instructions in making actuator decisions based on the process-control requirements. Regulation of the downstream water level is governed by both the desired water level set point and its defined dead band. Action is not required while the downstream water level stays within the dead band. This is an important condition to prevent excessive and wasteful activations of control.

Example 4.8   The network in Fig. 4.35 determines whether the downstream average level is greater than or equal to the downstream high treshold. The GREATER THAN OR EQUAL instruction compares `DS_AVE` with `DS_HIGH_LEVEL`; if TRUE, power flows to the output coil, and `DS_HLEVEL` is set.

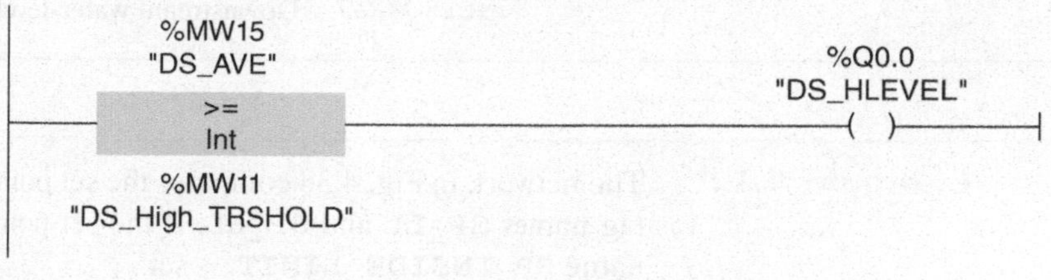

```
            %MW15
            "DS_AVE"                                    %Q0.0
                                                     "DS_HLEVEL"
              >=                                        ─( )─
              Int
            %MW11
         "DS_High_TRSHOLD"
```

Figure 4.35   Downstream water level above or equal to the high limit.

Example 4.9   The network in Fig. 4.36 determines whether the downstream average level is less than or equal to the downstream low treshold. The LESS THAN OR EQUAL instruction compares `DS_AVE` with `DS_LOW_LEVEL`; if TRUE, power flows to the output coil, and `DS_LLEVEL` is set.

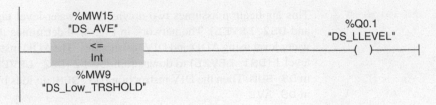

Figure 4.36 Downstream waterlevel below or equal to the low limit.

Operators normally provide required values for the set points for available controlled varaibles from local control panels, human-machine interfaces (HMIs), or other networked remote nodes. Sensors also provide all the needed real-time measurements from the controlled process. Any user or sensory inputs must be validated, as shown in Examples 4.10 through 4.14.

Example 4.10   The network in Fig. 4.37 compares the set point to minimum and maximum limits, tag names `SP_LL` and `SP_HL`. If the set point is outside the limit, the output, tag name `SP_OUTSIDE_LIMIT`, is set.

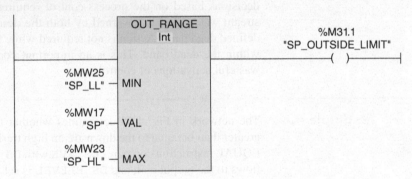

Figure 4.37 Downstream water-level set point outside the limits.

Example 4.11   The network in Fig. 4.38 compares the set point to minimum and maximum limits, tag names `SP_LL` and `SP_HL`. If the set point is inside the limit, the output, tag name `SP_INSIDE_LIMIT`, is set.

Example 4.12   The networks in Fig. 4.39 show sensory-input-signal validation. The input signal with tag `INPUT_SIGNAL` is a 12-bit-resolution analog-to-digital (A/D) count ranging from 0 to 4095. The count is set to 0 if it comes negative and to 4095 if it is received at a value above the maximum count (4095).

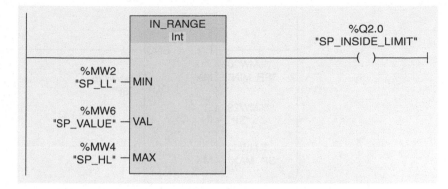

Figure 4.38 Downstream water-level set-point-limits validation.

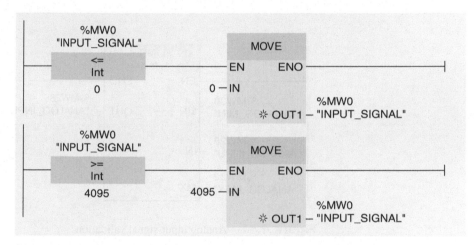

Figure 4.39 Input sensory-signal validation.

Example 4.13 The networks in the preceding example show sensory-input-signal validation using two networks, two MOVE instructions, and two comparison instructions. This example shows a more efficient way to accomplish the same task by using one network and a LIMIT instruction. The set point received from the operator is compared with the set-point minmum and maximum values. If the set point is less than the minimum value, the minimum value will be moved to the set point, and if the set point is greater than the maximum value, the maximum value will be moved to the set point (Fig. 4.40).

Example 4.14 This example shows another sensory-input-signal validation for a raw 12-bit analog input signal using the LIMIT instruction. The analog input signal for 12-bit resolution should be between 0 and 4095. If the analog input signal is less than the minimum value (0), the minimum value will be moved to the analog input signal, and if it is greater than the maximum value (4095), the maximum value will be moved to the analog input signal (Fig. 4.41).

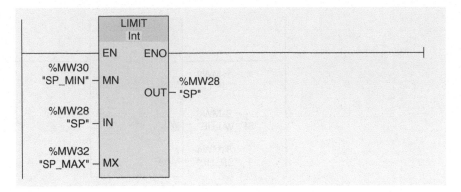

Figure 4.40 Set-point validation.

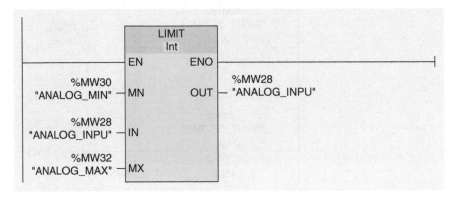

Figure 4.41 Analog input-signal validation.

Example 4.15 This example shows the instruction used for scaling a raw analog input signal as a digital count between 0 and 4095 to engineering units, which is required by the test-cell operator to monitor the temprature. An oven has a range of tempratures between 50 and 400°F. The oven temprature is measured using a 12-bit-resolution analog input module. The CALCULATE instruction in Fig. 4.42 will be used to scale the digital counts to the engineering units (°F) to be displayed on the human-machine interface (HMI) for the operator. HMI configuration, interfacing, and operation will be covered in Chap. 5.

The preceding scaling can be done using the CALCULATE instruction or by using the NORMALIZE and SCALE instructions, as shown in Figs. 4.43 and 4.44.

4.3.2 **_Small Industrial Process-Control Application_**

This section presents three small industrial-control applications aimed at further demonstrating the concepts covered through implementation of small modules of real industrial process-control applications.

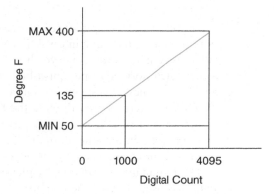

Figure 4.42 Scaling of digital count to engineering units.

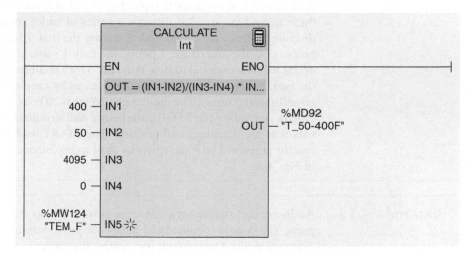

Figure 4.43 Scaling using the CALCULATE instruction.

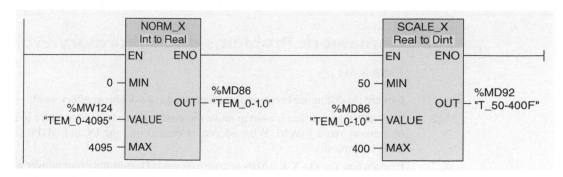

Figure 4.44 Scaling using the NORMALIZE and SCALE instructions.

Example 4.16   A flexible manufacturing conveyor-system module produces three different products but only one at any time interval. Each product is counted by a photoelectric cell (PE1) mounted on the conveyor line. Each batch processing is initiated through a dedicated normally open push-button switch. Only one batch can be active at any time. A new batch can be started after completion of the ongoing batch or by forcing a termination through the STOP push button. Product 1 is produced in the amount of 2000 total parts, product 2 in the amount of 2000 total parts, and product 3 in the amount of 5000 total parts. A pilot light (PL1) is ON after a batch is completed. The process begins by pressing the START push button and can be terminated at any time by pressing the STOP push button. A RESET push button is used to reset the counter accumulated value. Figure 4.45 is the ladder implementation, which uses five networks.

Example 4.17   A cascaded-tanks reactor is typical in chemical process control. In this example, three tanks are cascaded through a series of outlet solenoid valves. Material is discharged from tank 1 to tank 2 during the first 7 hours after the start of the process. At the end of the 7 hours, the tank 1 valve closes, and the tank 2 valve opens to allow material to flow into tank 3 for 8 additional hours. In the final stage, the tank 2 valve closes, and tank 3 is cleared by keeping its outlet valve open for an additional 5 hours. The entire reaction takes 20 hours, after which all valves are closed. Activating the STOP push button will terminate the process and return all valves to the default closed position. The START push button initiates the entire reactor process. The four networks used to implement this application are shown in Fig. 4.46.

Example 4.18   An inspection station on a conveyor system detects missing stamps on produced parts. The system is considered faulty once three missing stamps are detected in a 2-hour window. Once a fault is detected, the conveyor motor should stop running, and a pilot light turns ON to indicate the faulty condition. In order to restart the motor, the operator must fix the fault prior to pushing the START push button. Figure 4.47 shows the six networks used to implement this control task.

Homework Problems and Laboratory Projects

Problems

4.1   Explain the difference between a memory word and a double-memory word.

4.2   The MOVE instruction is used to move the content of memory word MW4 to the content of memory word MW10. What address is entered into the IN and OUT1 fields of the MOVE instruction?

4.3   Explain how the MAX and MIN instructions work. Demonstrate your answer with a documented and tested ladder network.

4.4   How many compare instructions are used in the S7-1200? List and explain three instructions.

4.5   Explain the function of each of the two networks in Fig. 4.48.

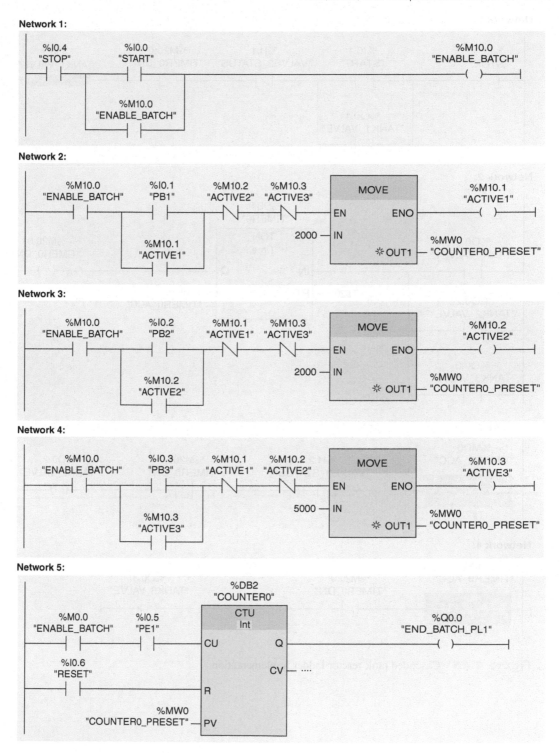

Figure 4.45 Conveyor system ladder implementation.

Network 1:

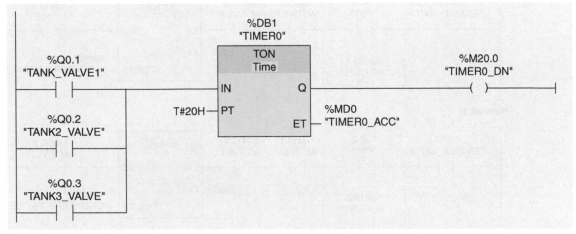

Network 2:

Network 3:

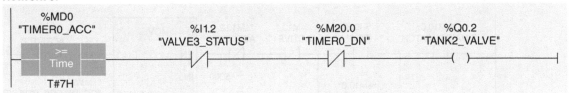

Network 4:

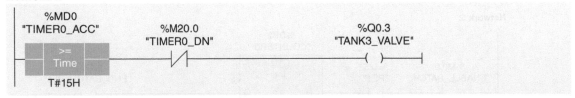

Figure 4.46 Cascaded tank reactor ladder implementation.

Network 1:

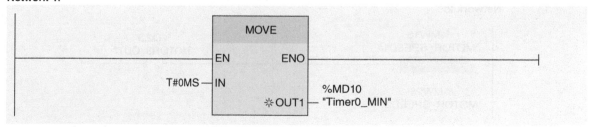

Network 2:

Network 3:

Network 4:

Figure 4.47 Missing stamp ladder implementation.

Network 5:

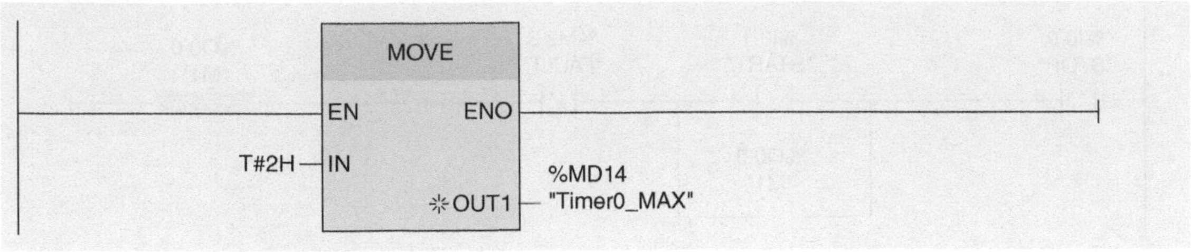

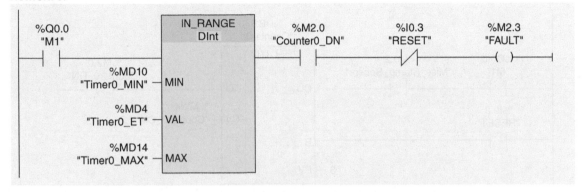

Network 6:

Figure 4.47 (*Continued*)

Network 1:

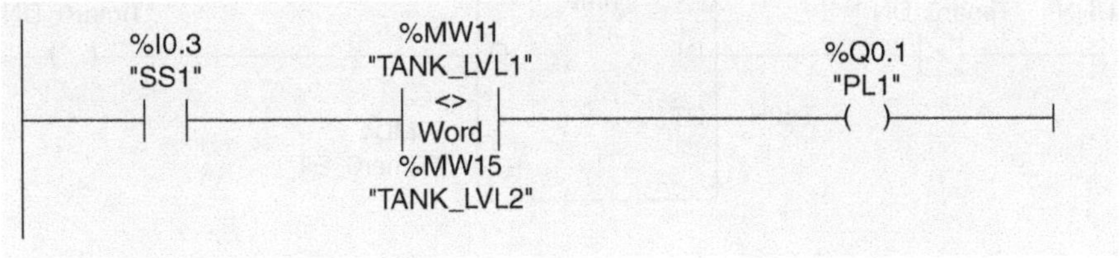

Network 2:

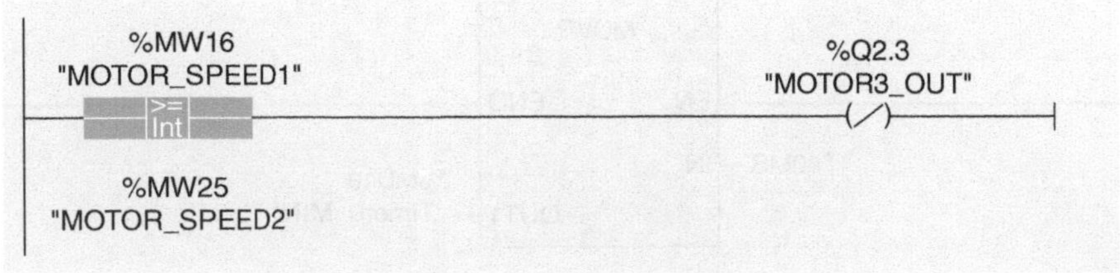

Figure 4.48 Networks for homework problem 4.5.

4.6   Answer the following questions for the network shown in Fig. 4.49, assuming that PB1 is a normally open push button and `TAG_1` (MW20) has value of 0.
a. What is the value of `TAG_1` (MW20) if PB1 is pushed three times?
b. What is the value of `TAG_1` (MW20) if PB1 is replaced by maintained switch SS1, and the switch is closed?

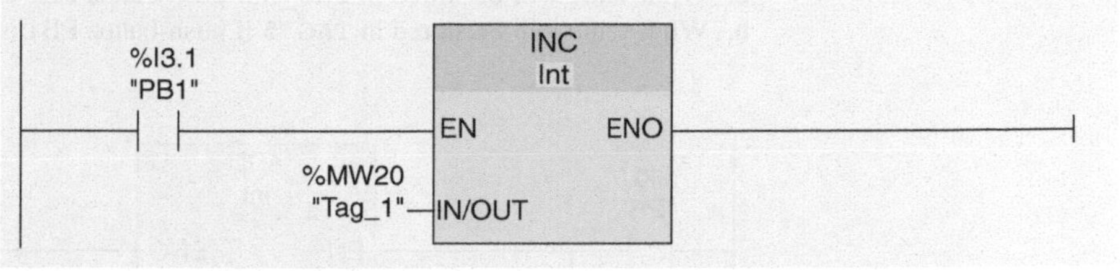

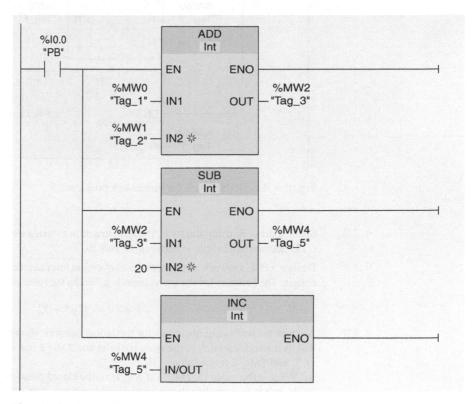

Figure 4.49   Network for homework problem 4.6.

4.7   Write a ladder-logic program to turn ON memory bit tag name `SP_OUTSIDE_LIMIT` if a set-point value is outside the minimum and maximum limits.

4.8   In reference to the math instructions shown in Fig. 4.50, assume the `TAG_1` and `TAG_2` have the value of 5 and 10, respectively. Answer the following questions:
a. What value will be stored in `TAG_5` if push-button PB is pushed one time?
b. What value will be stored in `TAG_5` if push-button PB is pushed two times?

Figure 4.50   Network for homework problem 4.8.

c. This network executes repeatedly during the time the PB is pressed. Modify the network to execute once for every activation of the push button regardless of the push action duration.

4.9   In reference to the math instructions in the network shown in Fig. 4.51, assume that `TAG_1` and `TAG_2` have the values of 5 and 10, respectively. Answer the following questions:

a. What value will be stored in `TAG_5` if push-button PB1 is pushed one time?

b. What value will be stored in `TAG_5` if push-button PB1 is pushed two times?

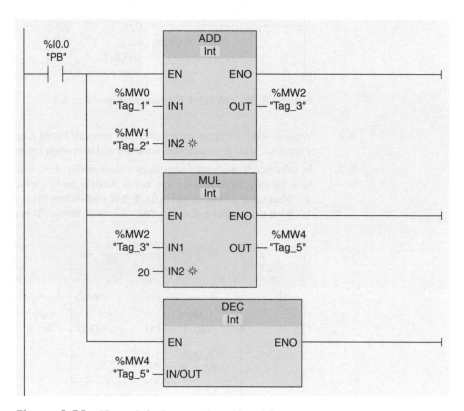

Figure 4.51   Network for homework problem 4.9.

4.10   Show a network using the IN RANGE instruction to turn a memory bit M0.0 if the counter CTU accumulated value is between 10 and 20.

4.11   Design a PLC network to perform the conversion from degrees centigrade to degrees Fahrenheit. The formula for the conversion is given by the following equation:

$$°F = (9 \times °C)/5 + 32$$

4.12   Answer the following questions for the ladder network shown in Fig. 4.52, assuming that SS1 is a selector switch in the open position and `TAG_1` has value of 10, `TAG_2` has value of 3, and `TAG_3` has a value of 0.

a. What is the value of `TAG_3` if SS1 is in the closed position?

b. How can you modify the network to add only one time without replacing SS1?

c. If SS1 is closed and the program is running, which memory word(s) will change?

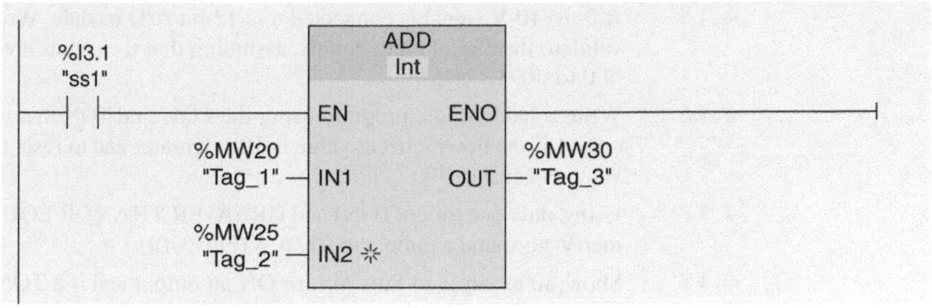

Figure 4.52   Network for homework problem 4.12.

4.13   Study the two networks in Fig. 4.53 and answer the followig questions:
a. Under what condition `Counter0_DN` output will be true?
b. Under what condition SV1 output will be true?

Network 1:

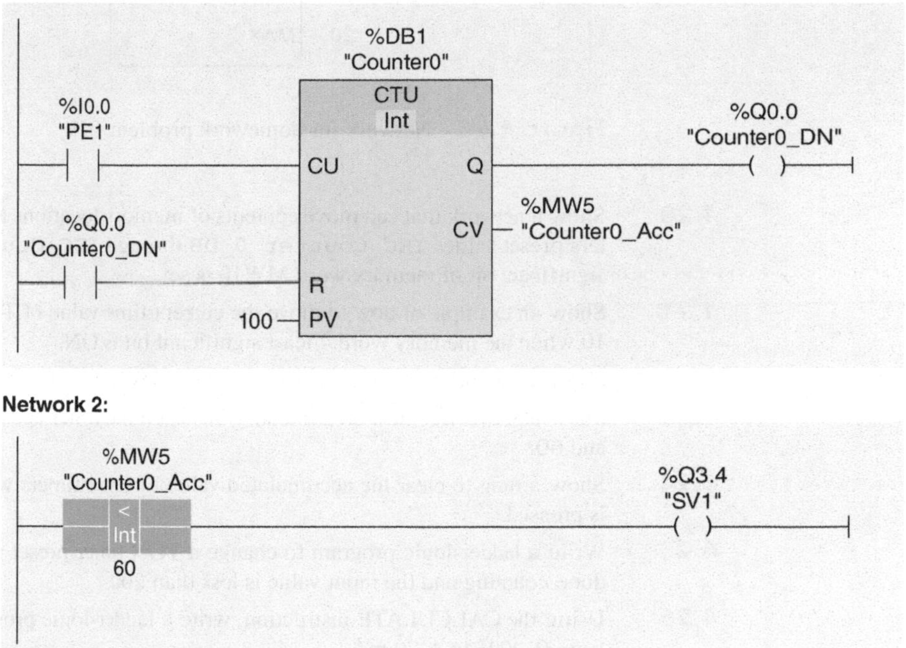

Network 2:

Figure 4.53   Networks for homework problem 4.13.

4.14   Write a ladder-logic program to calculate the average of two analog signals (input 1 and input 2). Use IW64 and IW66 for the analog inputs. MD40 is allocated to the average value.

4.15   A 0- to 10-V signal is connected to a 12-bit A/D module. Write the network(s) required to validate the digital input counts, assuming that the counts are already coming in the range of 0 to 4095.

4.16   Write a ladder-logic program using the EQU and TON instructions to set a motor 1 (M1) output if the timer current value is 10 or greater and to reset the motor if the timer current value is equal to 30.

4.17   Using only one timer (TON) and GREATER THAN OR EQUAL instructions, program the merry-go-round example in Chap. 3 (Fig. 3.12).

4.18   Show an example of how to turn ON an output coil if a TON timer accumulated value is greater than or equal to 15.

4.19   What is the value of tag value in the ladder network shown in Fig. 4.54 to turn ON M1?

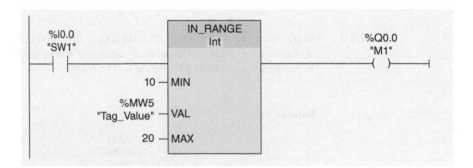

Figure 4.54   Network for homework problem 4.19.

4.20   Show a network that can move contents of memory locations MW4 through MW6 to counters preset values `IEC_Counter_0_DB` through `IEC_Counter_2_DB` when the least significant bit of memory word MW10 is set.

4.21   Show an example of how to move the current time value of TON timer 0 to memory word 10 when the memory word 4 least significant bit is ON.

4.22   Show a network to clear memory locations MW10 through MW20 when input I0.2 is set.

4.23   Show a network to turn on a memory bit when the timer accumulated value is between 40 and 60.

4.24   Show a rung to clear the accumulated value of three timers when the START push button is pressed.

4.25   Write a ladder-logic program to change a TON timer preset value to 19 once the timer is done counting and the input value is less than 200.

4.26   Using the CALCULATE instruction, write a ladder-logic program to scale an input signal from 0–4095 to 4–20 mA.

Projects

Laboratory 4.1: Tank Alarm

This laboratory covers math/comparison instructions and their use in S7-1200 ladder programming to produce a high-tank-level alarm.

Tank Alarm Process Description

A tank (Tank No_1) high-level alarm will turn ON once the tank level exceeds the prespecified high threshold level (120). The alarm remains ON until the level goes below the low threshold level (100). This logic is called the *Schmitt trigger* in digital logic design applications. The scaled analog input signal for the tank is received in memory location MW20, and analog programming will be discussed in a later chapter. The ladder-logic program is given in Fig. 4.55.

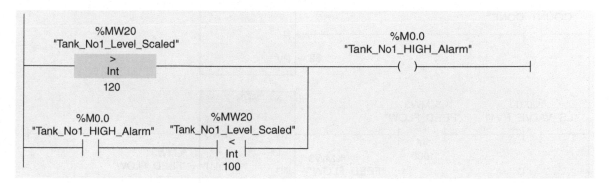

Figure 4.55   Tank alarm ladder network.

Laboratory Requirements

- Change the tank high-level alarm address to match the pilot light (PL1) address on the training unit.
- Simulate the alarm, and monitor PL1.
- Add the logic needed to flash the PL1 ON and OFF using 2 seconds for ON duration and 2 seconds for OFF duration.
- Document and debug your program.

Laboratory 4.2: Feed-Flow Digester Control

In this laboratory, you will learn how to use math and comparison instructions as well as counters in a real industrial application.

Laboratory Specification

An up-counter (Counter 0) is used to count the feed-flow amount into a digester in thousands of gallons. A total of 32,000 gal needs to be discharged into the digester. A valve PV1 is calibrated to provide an accurate measure of the flow amount into the digester. Every time valve PV1 opens the **FEED_FLOW** increments based on the flow from 0 to 1000 gal. Once it reaches 1000 gal, it resets, as shown in Network 2. The counter accumulated value is incremented by one for every 1000 gal, as shown in Network 1. **LS_VALVE_PV1** is the tag for the valve limit switch, which goes on while the valve is open. High-speed counters are typically used to implement the amount of flow in industrial automation applications, including chemical process control.

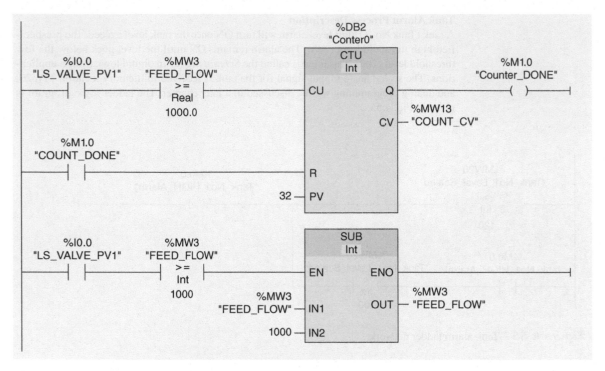

Figure 4.56 Ladder logic for the feed-flow digester.

Laboratory Requirements

- Modify the counter operation to simulate the feed flow into the digester. Use toggle switch SS1 to enable the counter. Change the 1000-gal value to 10 and the total amount from 32,000 to 60 gal.
- Toggle switch SS1, and monitor the counter operation.
- When the feed flow exceeds 50 gal, turn a pilot light (PL1) ON.

Laboratory 4.3: Tank Fill/Drain Control

A cylinderical tank has an area of 10 m². This application maintains a liquid volume of 10 to 50 m³ inside the tank at all times, assuming that the tank is equipped with two solenoid valves: fill (SV1) and drain (SV2), as shown in Fig. 4.57. The tank level sensor is simulated by a count up (CTU) and a count down (CTD) counter. If the tank level is greater than or equal to 50 m, drain the tank by activating SV2. If the tank level is less than or equal to 10 m, fill the tank by activating SV1. Report the volume values in a tag labeled `Tank_Volume`.

Laboratory Requirements

1. Write a documented ladder-logic program for the preceding process.
2. Download the program, and perform the checkout.
3. Doccument the program using a logic diagram.

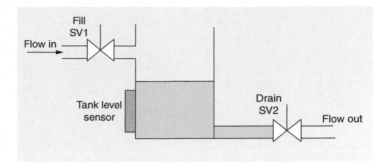

Figure 4.57 Tank fill/drain process.

Laboratory 4.4: Merry-Go-Round Using One Timer

The objective of this laboratory is to become familiar with basic timing, positive edge trigger, and equal instructions used to control a series of running motors simulated by four pilot lights. The same requirement was implemented in Example 3.1 using four timers. This laboratory requires implementation using only one timer.

Process Descriptions

This laboratory assumes four motors represented by four pilot lights on the trainer unit. A push-button START switch is used to start a sequence of pilot light operation that can be terminated at any time by pressing a push-button STOP switch. The sequence starts with PL1, then PL2, then PL3, and then PL4. The same sequence will repeat until the process is stopped. Each selected pilot light will turn ON for a period of 5 seconds while the other pilot lights are idle.

Laboratory Requirements
a. Write a documented ladder-logic program using only one ON-DELAY timer.
b. Assign the I/O addresses.
c. Assign the bit addresses.
d. Provide the laboratory checkout.

5

Device Configuration and the Human-Machine Interface

This chapter will introduce and detail HMI fundamentals as typically used in PLC process control. It also will detail the steps needed to configure the PLC and associated HMI devices.

Chapter Objectives

▶ Be able to implement the PLC/HMI configuration.

▶ Be able to perform HMI programming/communication.

▶ Be able to implement HMI remote control and monitoring in industrial automation.

Programmable logic controllers (PLCs) are the evolution of hardwired analog control systems. PLCs use a wide array of networked human-machine interfaces (HMIs) to allow users to interact with controlled processes through visual and user-friendly tools. HMIs are used for reporting/status display purposes, but they can provide user command from any desired location or through the Internet anywhere in the world. The HMI infrastructure is both flexible and expandable. Its sole critical function is to allow for process-control continuous quality improvements and thus continuous enhancements in the final products produced. This chapter introduces and details the fundamentals of HMIs as typically used in PLC process control. In addition to allowing system monitoring and command from convenient locations at any time, HMIs also provide an excellent instrument for overall system performance improvements. Some of the world biggest chemical companies provide such access to any company control system from any place at any time with adequate authentication control. Access is obtained using a PC/laptop/smart phone or other intelligent Internet-networked devices.

5.1 Device and PLC/HMI Configuration

This section is intended to help users in configuring S7-1200 hardware modules. It also details the steps needed to configure a PLC and associated HMI device. The approach used in our coverage is one of many options available under the Siemens S7-1200 basic PLC software. The reader is encouraged to learn these concepts through hands-on and familiarity with the software and its effective online help facilities.

5.1.1 Siemens S7-1200 Hardware Arrangement

To configure a PLC, a CPU and additional modules are needed, as shown in Fig. 5.1:

1. Communications module (CM): up to three, inserted in slots 101, 102, and 103
2. CPU: slot 1
3. Ethernet port of CPU
4. Signal board (SB): up to 1, inserted in the CPU
5. Signal modules (SMs) for digital or analog input-output (I/O): up to 8, inserted in slots 2 through 9 (CPU 1214C allows 8, CPU 1212C allows 2, and CPU 1211C does not allow any.)

5.1.2 PLC/HMI Configuration

Follow these steps to configure PLC/HMI devices for your new project, which is named *Project 1*:

- Allow device configuration by inserting a CPU into the new project.
- Select Create new project, and type "Project 1"; then click Create, as shown in Fig. 5.2.
- In the Portal view, select PLC programming, and click on "Write PLC program," as shown in Fig. 5.3.
- Under Project view, click on "Configure a device" as shown in Fig. 5.4.

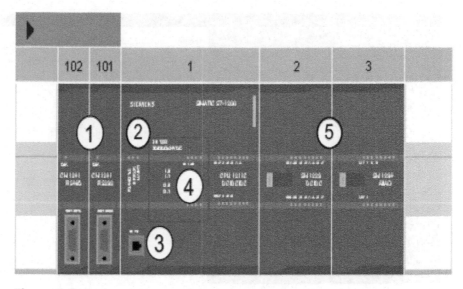

Figure 5.1

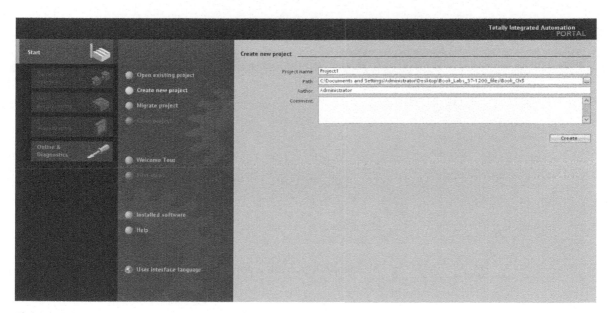

Figure 5.2

- Click on "Add new device," as shown in Fig. 5.5. You will be prompted to add a new device, PLC, or HMI.
- Under Add new device, choose PLC, and then select the CPU type; then click "The CPU Series" as shown in Fig. 5.6. Make sure to select the correct firmware version, for example, V2.0, V2.1, V2.2, and so on.
- Click on "Add new device," as shown in Fig. 5.7.

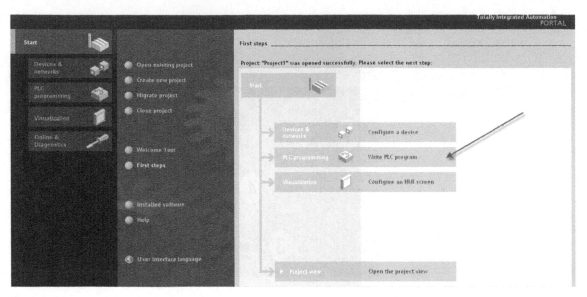

Figure 5.3

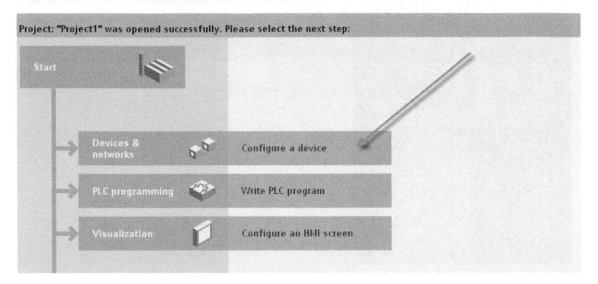

Figure 5.4

- Under Device name, select the HMI type, and then click on "The HMI series," as shown in Fig. 5.8. Make sure that the correct HMI model is selected.
- Under Browse, select PLC1, and then click "Finish," as shown in Fig. 5.9.
- Click on "Add new device" and then Devices & networks, as shown in Fig. 5.10 to view the final configuration.

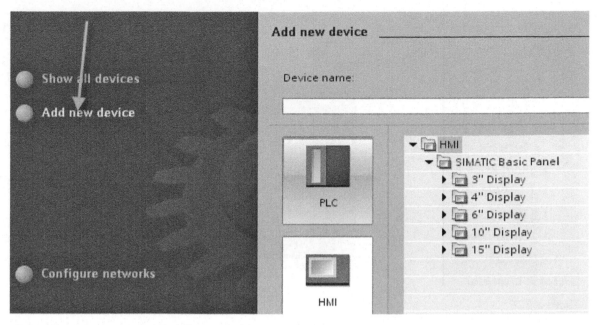

Figure 5.5

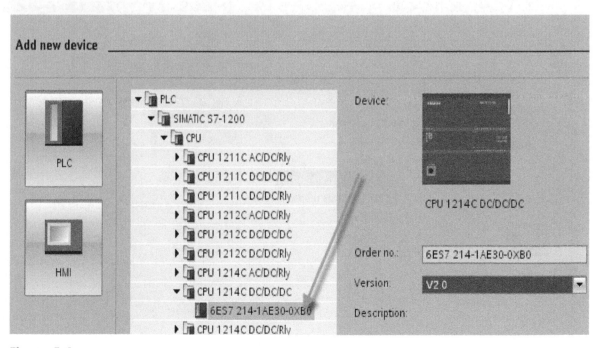

Figure 5.6

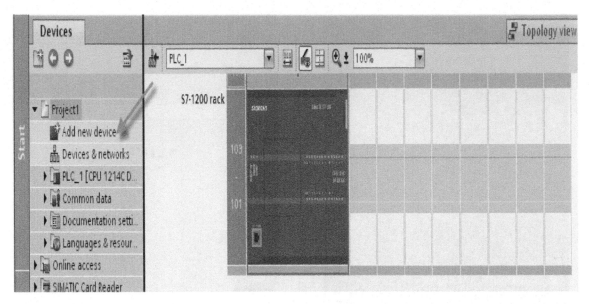

Figure 5.7

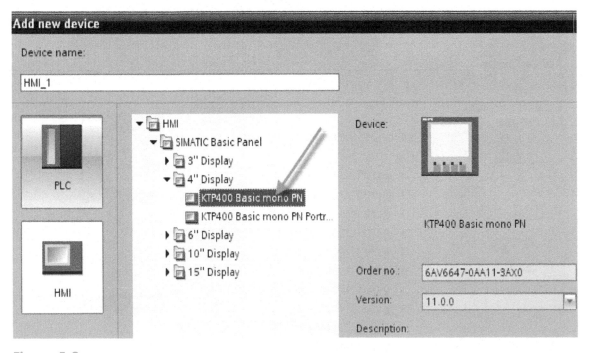

Figure 5.8

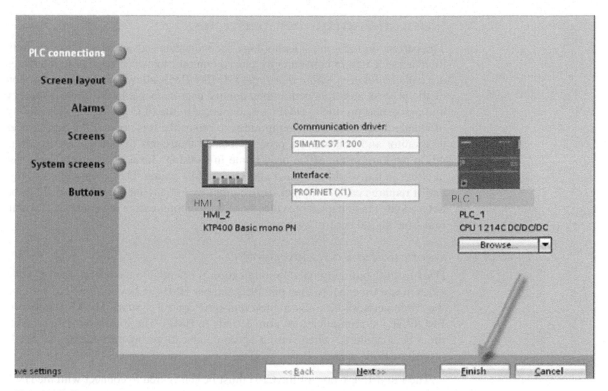

Figure 5.9

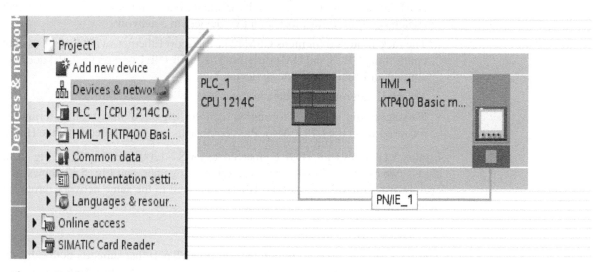

Figure 5.10

5.2 Human-Machine Interfacing

The current digital control technology accommodates advanced and user-friendly interfaces located in conveniently placed control rooms/centers. These interfaces are called *human-machine interfaces* (HMIs). HMIs allow for visual observation of the process and thus continuous quality improvements in the control strategy and associated products. HMIs are connected to the PLC through a wide variety of standard networks and communication protocols. It is possible for an engineer to monitor and command a process located thousands of miles away in China while working in the office or at home in Houston, Texas. The largest computer network in the world is owned and operated by a large U.S. chemical company, and it requires each computer device employed at any facility in the world to be networked. This chapter provides brief but comprehensive coverage of HMIs in real-time digital control systems.

5.2.1 Communication Fundamentals

HMI-to-PLC bidirectional communication is essential to the operation of the two sides in real time. More than one PLC and one HMI can be configured to work on the same network as a distributed real-time control system. The CPU supports PROFINET communications connections to HMIs. The following requirements must be considered when setting up communications between CPUs and HMIs (configuration/setup):

- The PROFINET port of the CPU must be configured to connect with the HMI. The HMI must be set up and configured.
- The HMI configuration information is part of the CPU project and can be configured and downloaded from within the project.
- No Ethernet switch is required for one-to-one communication; an Ethernet switch *is* required for more than two devices in a network.
- The rack-mounted Siemens CSM1277 four-port Ethernet switch can be used to connect the CPUs and HMI devices. The PROFINET port on the CPU does not contain an Ethernet switching device.
- The HMI can read/write data to the CPU.
- Messages and status information can be triggered based on information retrieved from the CPU. Also, PLC commands can be initiated from the HMIs.

Figure 5.11 shows an HMI connected to an S7-1200 Siemens PLC, Fig. 5.12 shows two S7-1200 PLCs connected directly together, and Fig. 5.13 shows several devices (three PLCs and an HMI) connected to the same network through a network switch.

5.2.2 PROFINET and Ethernet Protocol

The S7-1200 CPU has an integrated PROFINET port that supports both Ethernet and direct connections. Direct communication is used when using a programming device, HMI, or another CPU that can still connect to a single CPU. Network communication is used when two or more devices are connecting (e.g., CPUs, HMIs, programming devices, and non-Siemens devices). Figure 5.14 shows a

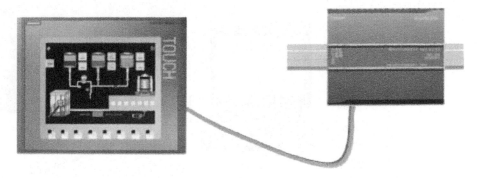

Figure 5.11 HMI and S7-1200 PLC connection.

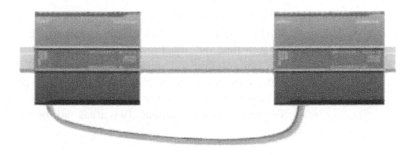

Figure 5.12 S7-1200 PLC–to–S7-1200 PLC connection.

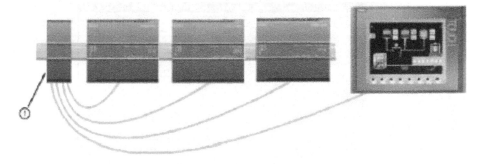

Figure 5.13 Network connection (more than two devices connected).

programming terminal connected to an S7-1200 PLC. In a direct connection, it is possible to connect a programming device to an S7-1200 CPU, an HMI to an S7-1200 CPU, and an S7-1200 CPU to another S7-1200 CPU. Multiple PLCs/ HMIs can be configured and connected to the same network, each with a unique network address identifier. Use of a PROFINET Ethernet communication/ network interface will be assumed in this chapter.

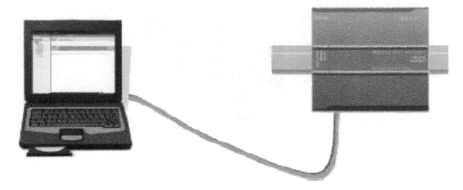

Figure 5.14 Programming terminals and S7-1200 PLC connection.

5.2.3 *HMI Programming*

HMI programming will be illustrated through the use of a series of examples showing real-time industrial project implementation. These examples will demonstrate the configuration and implementation of such user visual interactive interfaces and at the same time introduce common industrial specification implementations.

Figure 5.15 shows the HMI main development page, as documented in the Siemens technical manual. Two areas will be used repeatedly in the examples in

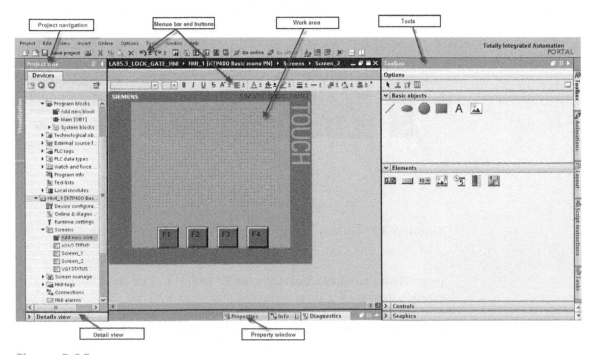

Figure 5.15

this section: the Tools and Property windows. In the Tools window, a selection of objects is provided that can be inserted in pictures, such as graphic objects and operating elements. In addition, the Tools window includes libraries with preassembled objects and collections of picture blocks. Objects are moved to the work area with drag and drop.

In the Property window, you can edit the properties of objects, for example, the colors of picture objects. The Property window is available only in certain editors. In the Property window, the properties of the selected object are displayed according to categories. As soon as you exit an input field, the value change becomes effective. If you enter an invalid value, its background is colored. Via the Quick Information window, you are then provided with information regarding the valid value range. In addition, animations and events of the selected object are configured in the Property window; in the figure, for example, a display change occurs when a button is released.

Example 5.1   This example explains HMI circle and text field programming. This application was disscused in Chap. 3 as the "Merry-Go-Round." It assumes four motors represented by four pilot lights on the trainer unit. A push-button START switch is used to start a sequence of motor operation, which can be terminated at any time by pressing a push-button STOP switch. The sequence starts Motor 1, then Motor 2, then Motor 3, and then Motor 4. The same sequence will repeat until the process is stopped. Each selected motor will run for a period of 5 seconds while the other motors are idle. Figure 5.16 shows the status of the four motors during operation. The HMI screen shot in the figure was taken during the time when Motor 1 was running. The other three motors are shown as not running, as expected.

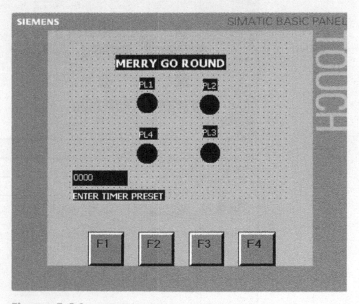

Figure 5.16   Merry-Go-Round HMI page.

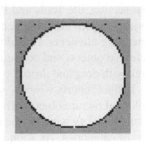

Figure 5.17 The circle object.

Circle (Fig. 5.17)

The circle object is a closed object that can be filled with a color or pattern. In the Inspector window, you can customize the settings for the object position, geometry, style, frame, and color. You can adapt the following properties in particular:

- *Circle configuration.* From the Tools window, drag and drop the circle.
- *Circle appearance.* As shown in Fig. 5.18, right-click on the circle, and then choose Properties; from Properties, choose the background and border color.

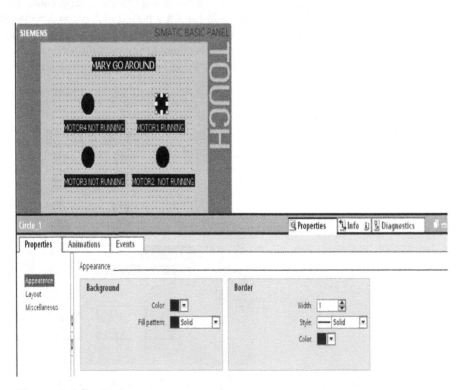

Figure 5.18 Circle appearance.

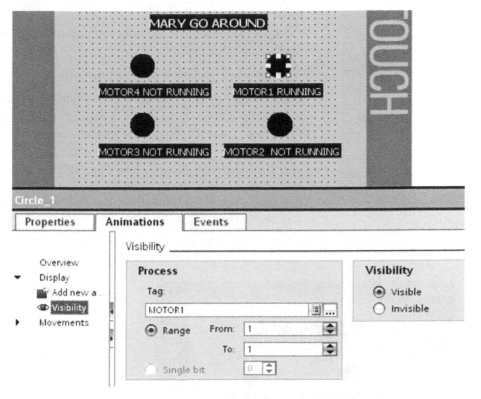

Figure 5.19   Circle visibility.

- *Circle visibility.* As shown in Fig. 5.19, right-click on the circle, and then choose Animations; from Animations, click Display and Add new animation, and then choose visibility. Then enter the PLC tag and the range, which make the circle visible.

Text Field (Fig. 5.20)

In the Inspector window, you can customize the settings for the object position, geometry, style, frame, and color. Specify the text for the text field in the Inspector window. You can adapt the following properties in particular:

- In the Inspector window, select Properties → Properties → General.

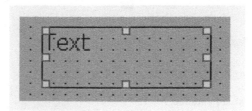

Figure 5.20   The text field.

- Enter your text.
- For texts over several lines long, you can set a line break by pressing the key combination Shift + Enter.

You can adapt the following properties in particular:

- *Text field configuration.* From the Tools window, drag and drop the text.
- *Text field appearance.* As shown in Fig. 5.21, click on the text, and then choose Properties; from the Properties window, choose the background and border color.
- *Text field visibility.* As shown in Fig. 5.22, click on the text field, and then choose Animations. From Animations, click Display and add the new animation; then choose Visibility. Enter the PLC tag and the range, which make the text visible (Fig. 5.23).

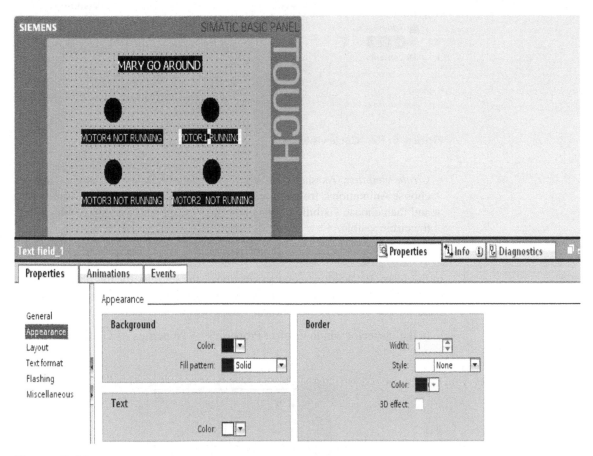

Figure 5.21 Text appearance.

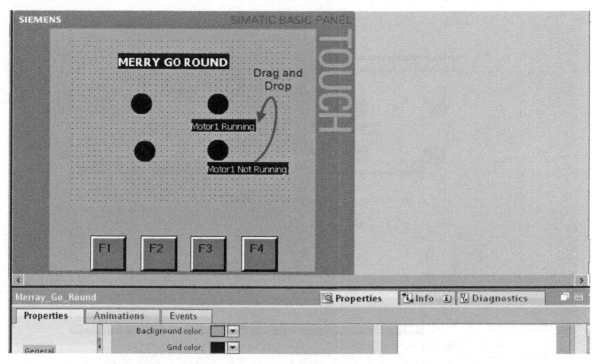

Figure 5.22 Text visibility.

Example 5.2   This example explains HMI input-output (I/O) and Button field programming. This application was disscused in Chap. 3 under Section 3.3.1. It assumes one push button and one motor. A push-button (AUTO START PULSE) switch is used to run the Motor 1 for a preset time defined by the user. Figure 5.24 shows the motor status page. The user enters the preset time after activitation of the AUTO START PULSE push button on the HMI page as shown. The page function keys are also shown and can be configured and assigned to page navigation tasks, as will be explained later.

I/O Field

The I/O field object is used to enter and display a process value. You can adapt the following properties in particular:

- *Mode.* Specifies the response of the object in run time.
- *Display format.* Specifies the display format in the I/O field for input and output.
- *Input.* Values can only be input into the I/O field in run time.
- The response of the I/O field is specified in the Inspector window in Properties → General → Type. The I/O field is used for the output of values only.

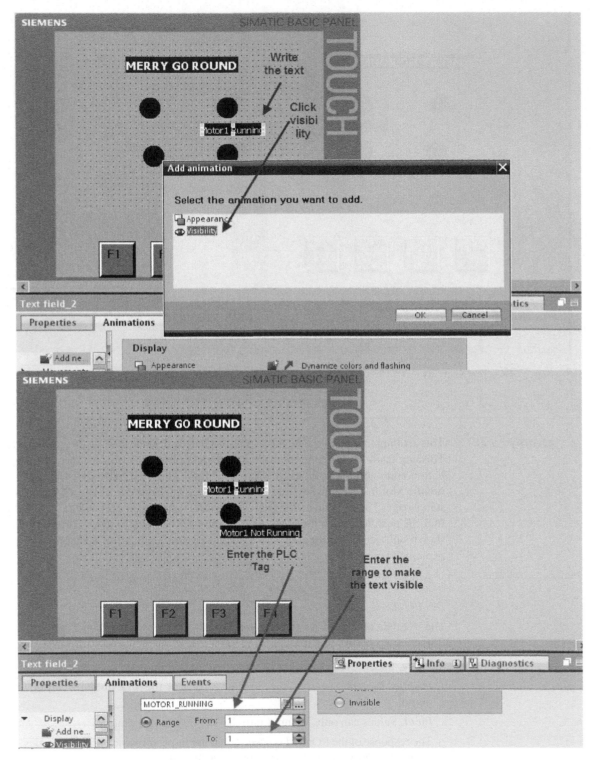

Figure 5.23

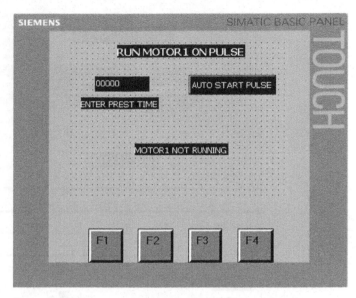

Figure 5.24 Run Motor on Pulse HMI page.

In the Inspector window, you can customize the settings for the object position, geometry, style, frame, and color. You can adapt the following properties in particular:

- *I/O field configuration.* From the Tools window, drag and drop the I/O field.
- *I/O field appearance.* As shown in Fig. 5.25, right-click on the I/O field, and then choose Properties; from the Properties, choose the background and border color.
- *I/O field programming.* As shown in Fig. 5.26, right-click on the I/O field, and then choose Properties, and then General; then enter the PLC tag under Process, and choose Input under Type. Enter the I/O information under Format.

Button Field

The Button object allows the user to configure an object that the operator can use in run time to execute any configurable function. In the Inspector window, you customize the position, shape, style, color, and font of the object. You can adapt the following properties in particular:

- *Mode.* Defines the graphic representation of the object.
- *Text/graphic.* Defines whether the graphic view is static or dynamic.
- *Define hotkey.* Defines a key, or shortcut, that the operator can use to actuate the button.

The button is defined in Properties → General → Mode in the Inspector window.

- *Button field configuration.* From the Tools window, drag and drop the Button field.
- *Button field appearance.* As shown in Fig. 5.27, right-click on the Button field, and then choose Properties; from the Properties, choose the background and border color.

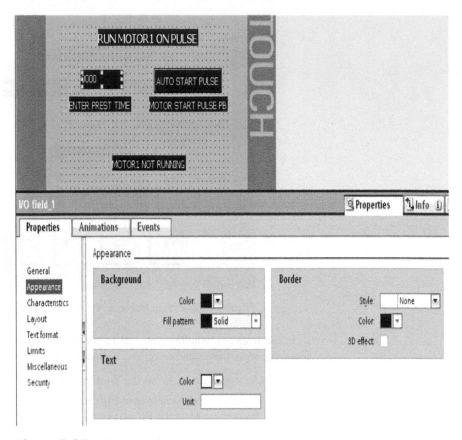

Figure 5.25 I/O appearance.

- *Button field programming.* As shown in Fig. 5.28, set and reset bit is explained as follows.
- *Set bit programming.* Under Events, click Press and SetBit, and then enter the PLC tag to set the bit.
- *Reset bit programming.* Under Events, click Press and ResetBit, and then enter the PLC tag to reset.

Example 5.3   This example explains HMI function key programming. This application uses three pages (FUNCTION KEYS, STATUS, and CONTROL). Pressing any of the two soft keys (F1 and F2) will take the user from the FUNCTION KEYS page to one of the two other pages. Pressing F1 while in either the CONTROL or the STATUS page will switch the display to the FUNCTION KEYS page. Figure 5.29 shows the steps to program the function keys.

- From the Project Tree, create three pages: FUNCTION KEYS, CONTROL, and STATUS as shown in Fig. 5.29(*a*).

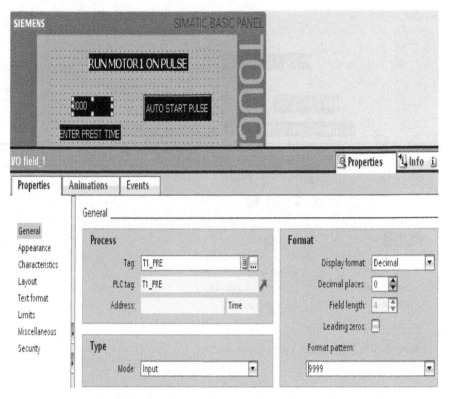

Figure 5.26   I/O configuration.

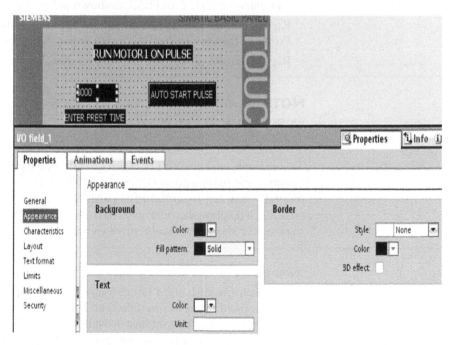

Figure 5.27   Button field appearance.

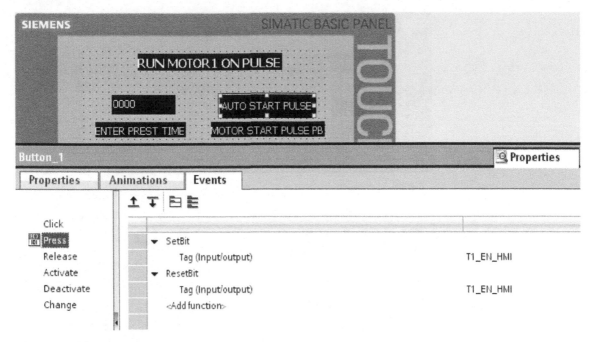

Figure 5.28 I/O preset event programming.

- From the Tools window, drag and drop the text field, configure the apperance, and enter the text "STATUS," as shown in Fig. 5.29(b).
- Right-click on F2, and a red square appears on the "STATUS" text. Under Property, then event select Activate Screen, and choose "STATUS" from the list.
- Repeat the preceding steps to program the CONTROL page (Fig. 5.30).

NOTE Configure the Function Keys page as the start screen befor downloading the program to the PLC and the HMI.

Example 5.4   This example explains SWITCH and OUTPUT field programming. This application assumes one two-position switch (LEFT/RIGHT). In the left position, it displays the value entered from the PLC (VALUE_LOCAL_STATUS), and in the right position, it displays the value entered from the HMI (VALUE_REMOTE_STATUS). Figure 5.31 shows the local/remote switch page.

Switch Field

The Switch object is used to configure a switch that is used to switch between two predefined states in run time. The Button display is defined in Properties → Properties → General → Settings in the Inspector window. The position of the switch indicates the current state. The state is changed in run time by sliding the switch.

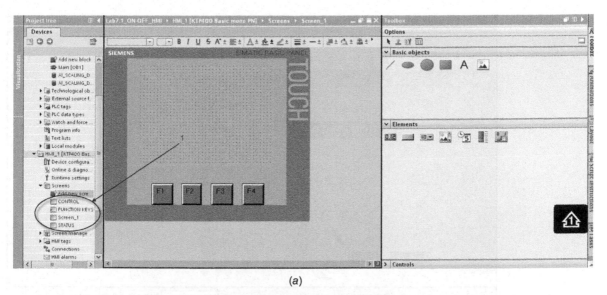

(a)

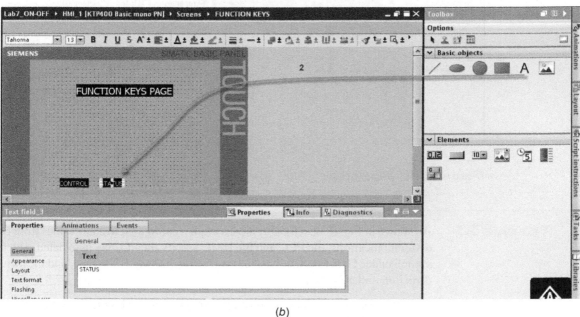

(b)

Figure 5.29 (*a, b*) I/O Function Keys page. (*c*) I/O function key programming.

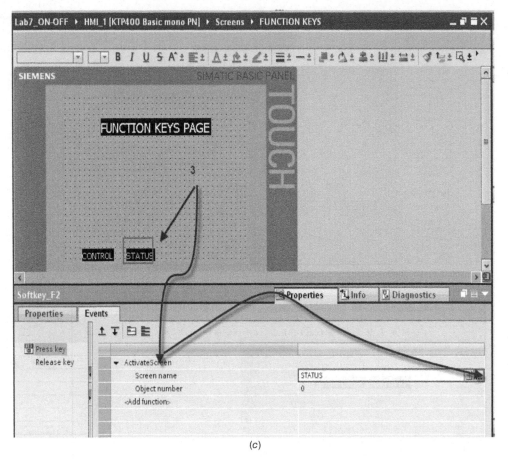

(c)

Figure 5.29 (*Continued*)

Example 5.5

As shown in Fig. 5.32, a local/remote switch is configured to enter values from two different locations, the PLC and the HMI. Specify the PLC tag associated with the switch under General. Specify the PLC tag for the switch, the type, and the switch direction under Graphic.

- *Switch field configuration.* From the Tools window, drag and drop the Switch field.

- *Switch field appearance.* As shown in Fig. 5.32, right-click on the Switch field, and then choose Properties; from Properties, choose the background and border color.

- *Switch field programming.* As shown in Fig. 5.33, right-click on the Switch field, and then choose Properties and then General; then enter the PLC tag under Process and from Type, enter the switch information.

- *Output field programming.* As shown in Fig. 5.34, under General process, enter the PLC tag, and under Type, enter output to monitor the value associated with the PLC tag. Under Format, enter the display format and pattern.

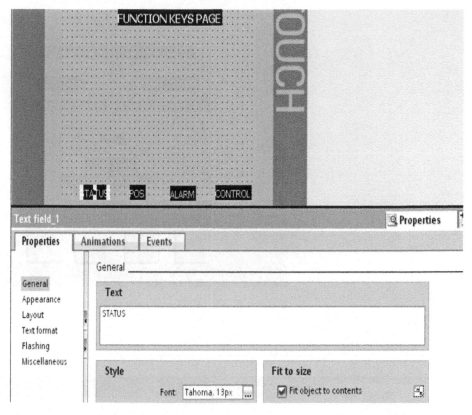

Figure 5.30 Function Keys page status.

Bar Field

The tags are displayed graphically using the Bar object. The bar graph can be labeled with a scale of values. The color change is represented in Properties $\rightarrow$ Appearance in the Inspector window.

If a particular limit was reached, the bar changes color segment by segment. With segment-by-segment representation, you visualize which limits are exceeded by the displayed value. If a particular limit was reached, the entire bar changes color.

Displaying limit lines and limit markers are configured and displayed in the HMI page shown in Fig. 5.35.

5.3 Control and Monitoring

Control and monitoring are the core of the HMI's function and use. The entire system, PLCs and HMIs, can be viewed as a distributed control system connected through adequate network and real-time communication facilities. This section illustrates this concept, with emphasis on the role of the HMI in overall process monitoring and control. The discussion will be centered on a real industrial application of pump-station control in a wastewater treatment facility.

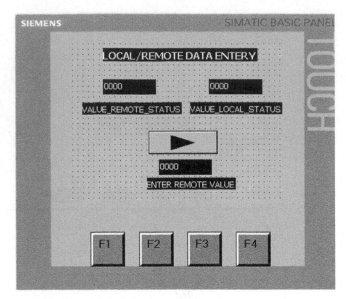

Figure 5.31 I/O local/remote switch.

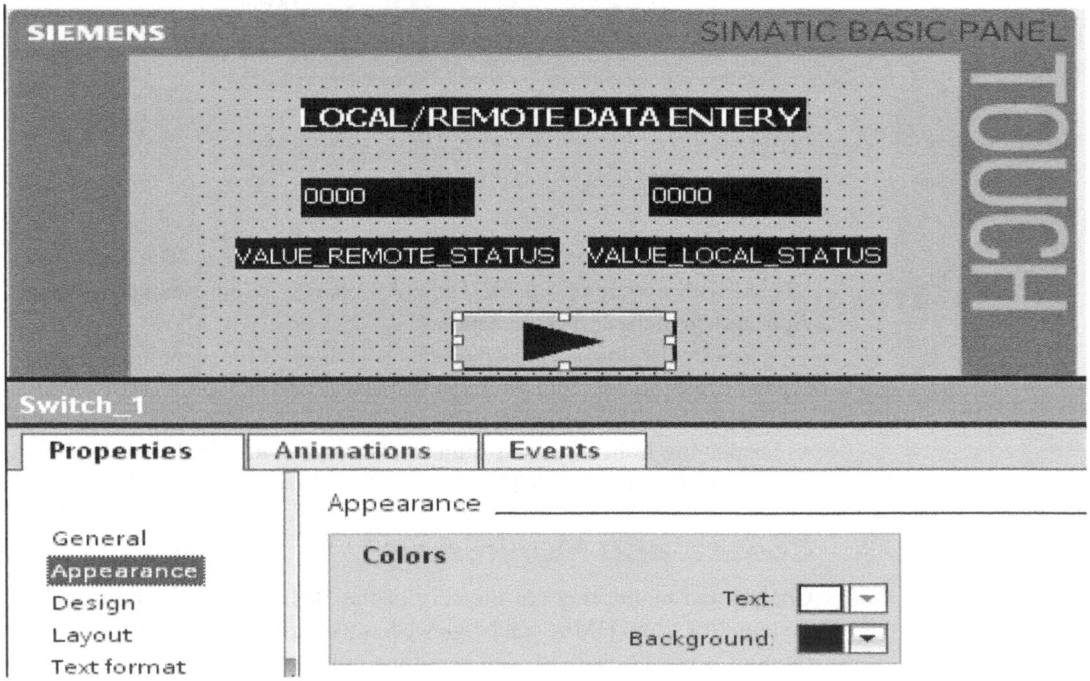

Figure 5.32 I/O local/remote switch appearances.

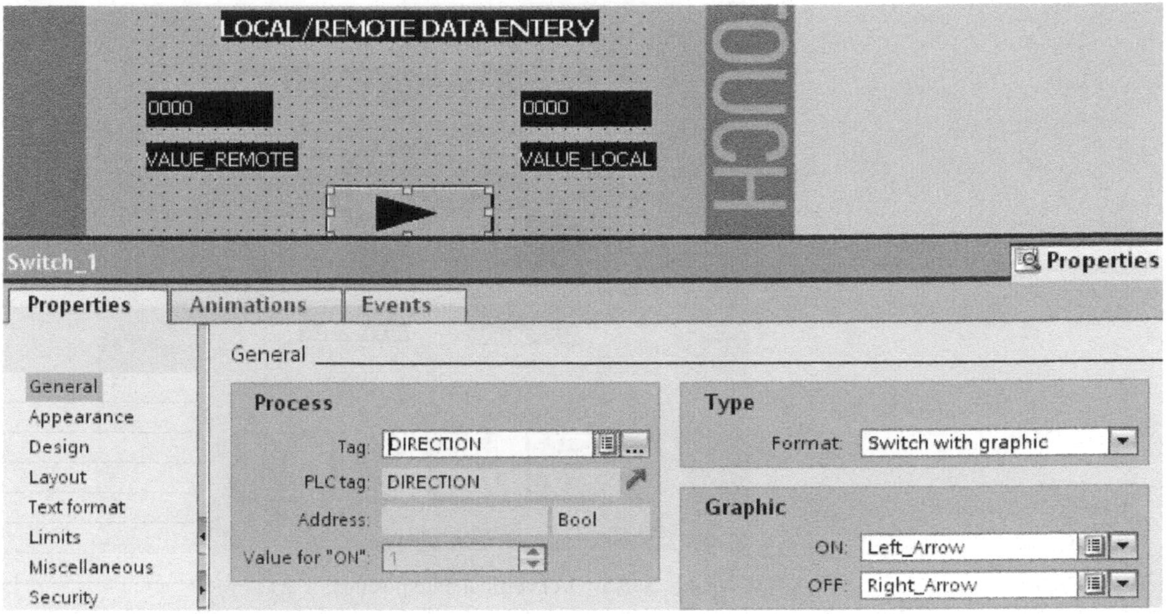

Figure 5.33   I/O local/remote switch field programming.

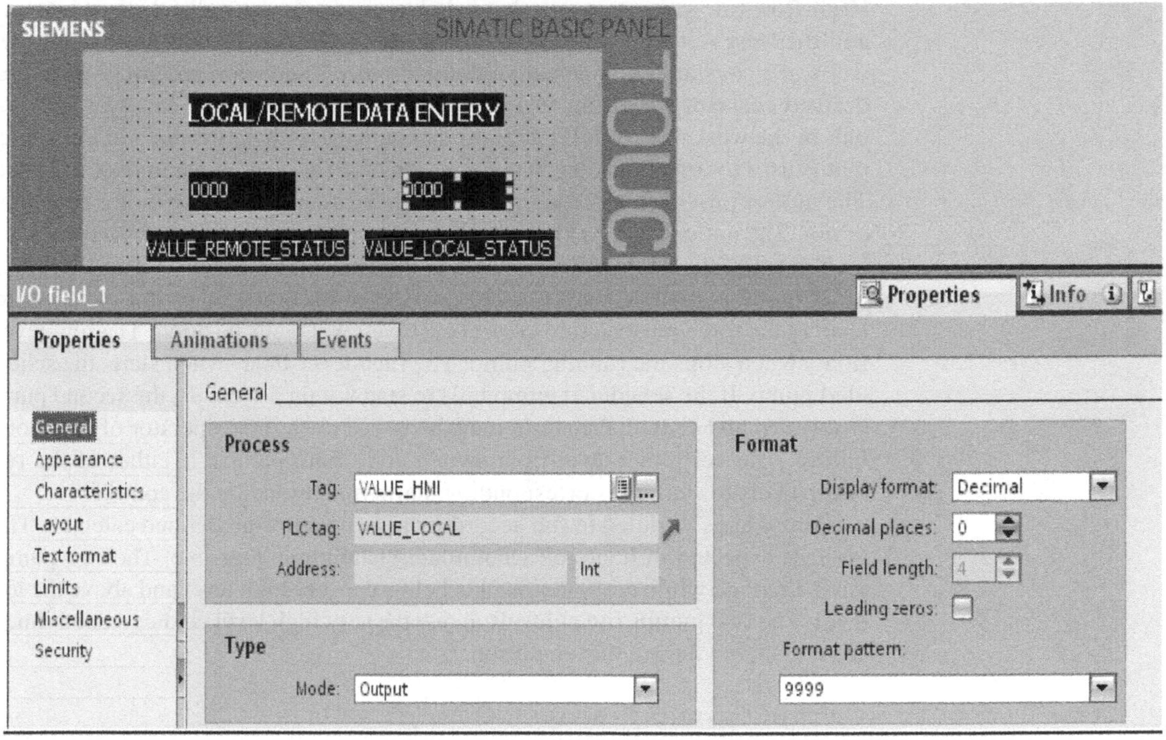

Figure 5.34   `VALUE_LOCAL_STATUS`.

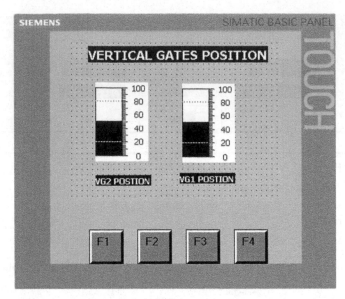

Figure 5.35 I/O vertical gate position.

5.3.1 *Distributed Control System Process Description*

High-flow-rate storm water is channeled to two large wet wells, the east wet well and the west wet well. The water is pumped to the river from the two connected wells at a constant rate using a predefined process sequence control. Two motor-derived constant-speed immersed pumps are used, one in the east wet well and one in the west wet well. Each pump is equipped with an overload alarm switch that is used to trigger any unusual conditions such as overtemperature or overload. The motors provide a discrete input signal indicating whether the motor is running or not. The motors also can be started by activating the push button located on the local panel if the AUTO/MAN switch is in manual mode.

Three float switches are used to provide an accurate indication of the water level at the three prespecified critical east/west wet wells position. The low-level-float switch stops the running pump. The high-level-float switch starts the scheduled pump. If the scheduled pump fails to start within 5 seconds, the second pump is selected and started. An alarm must be issued to alert the operator of any motor failure. The very-high-level-float switch starts both pumps. If either of the two pumps fails to start, the corresponding alarm is activated by the control.

Pumps are scheduled to run according to an operator-predefined calendar. This input is expected in hours of accumulated total pump run time. The two pumps must alternate while the water level is below the very high level and above the low level. The two pumps run at levels above the very high level, and cascaded timers are not altered during this condition.

5.3.2 *Control System Process I/O Map*

Figures 5.36 and 5.37 show the system I/O as it should be documented initially in the project specification, whereas Figs. 5.38 and 5.39 show the I/O as PLC tags in the PLC ladder program.

Tag Name	Address Number	Comments
PL1	Q 0.0	East Pump
PL2	Q0.1	West Pump
PL3	Q0.2	East Pump Fail to Start
PL4	Q0.3	West Pump Fail to Start
PL5	Q0.4	Common Alarm

Figure 5.36 Pump station system outputs.

Design and Execution of Pump Station Ladder Program

The CPU supports the following types of code blocks that allow an efficient structure for the program:

- Organization blocks (OBs) define the structure of the program.
- Functions (FCs) and function blocks (FBs) contain the program code that corresponds to specific tasks or combinations of parameters. Each FC or FB provides a set of input and output parameters for sharing data with the calling block.
- Data blocks (DBs) store data that can be used by the program blocks.

Functions

The network in Fig. 5.40 shows the PLC functions (Pump START/STOP and Pump ALTERNATION, ALARMS, and INITIALIZATION). The following list provides definitions for each of the functions:

- *Pump Start/Stop.* The Pump Start/Stop function will include the logic to START/STOP the pumps based on the indications of the water level of the three float switches at three critical wet-well locations.

Tag Name	Address Number	Comments
SS1	I0.0	OFF Float Switch
SS2	I0.1	ON Float switch
SS3	I0.2	Override Float switch
SS4	I0.3	East Running On Line Contact
SS5	I0.4	West Running On Line Contact
SS6	I0.5	AUTO Switch
SS7	I0.6	ESD Switch
SS8	I0.7	East Overload
SS9	I1.0	West Overload

Figure 5.37 Pump station system inputs.

INPUTS

	Name	Data type	Address
01	OFF_FLOAT	Bool	%I0.0
01	ON_FLOAT	Bool	%I0.1
01	OVERIDE_FLOAT	Bool	%I0.2
01	E_ROL	Bool	%I0.3
01	W_ROL	Bool	%I0.4
01	AUTO	Bool	%I0.5
01	ESD	Bool	%I0.6
01	E_OVERLOAD	Bool	%I0.7
01	W_OVERLOAD	Bool	%I1.0

Figure 5.38 Pump station PLC input tags.

OUTPUTS

	Name	Data type	Address
01	E_PUMP	Bool	%Q0.0
01	W_PUMP	Bool	%Q0.1
01	E_FTS	Bool	%Q0.2
01	W_FTS	Bool	%Q0.3
01	COMMON_ALARM	Bool	%Q0.4

Figure 5.39 Pump station PLC output tags.

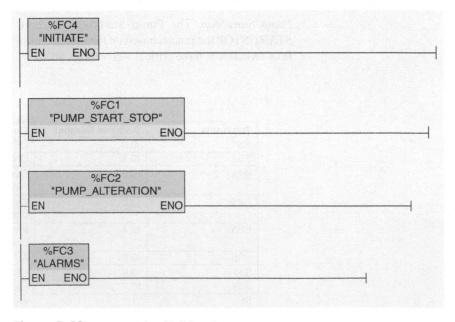

Figure 5.40 Pump station PLC functions.

- *Pump Alternation function.* The Pump Alternation function will include the two pumps' alternation logic according to the available calendar while the water level is below the very high level and above the low level. The alternation calendar is stored as a 16-bit integer in memory location MW4: the law byte in %M5 and the high byte in %M4 (big-endian notation). The alternation calendar is incremented once on the alternation timer done bit. Thus even values (M5.0 is 0) are assigned for one pump, whereas odd values (M5.0 is 1) are assigned to the other pump.

- *Pump Alarm function.* The Pump Alarm function will include two common alarms. One alarm is dedicated to the east wet well and the other to the west well. Each common alarm will be triggered from pump motor fail to start, overload, or emergency shutdown.

- *Initialization function.* The Initialization function will include all the system parameters required to start the process-control project.

- *Initialization Logic function.* The initialization network (Fig. 5.41) consists of one network that clears the accumulated values for the timers, counters, and increment register because the system is placed in AUTO mode.

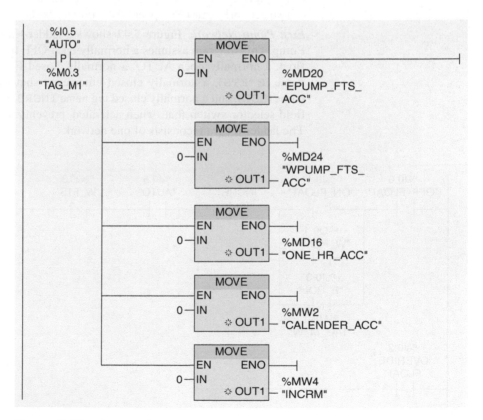

Figure 5.41 Pump station Initialization Network.

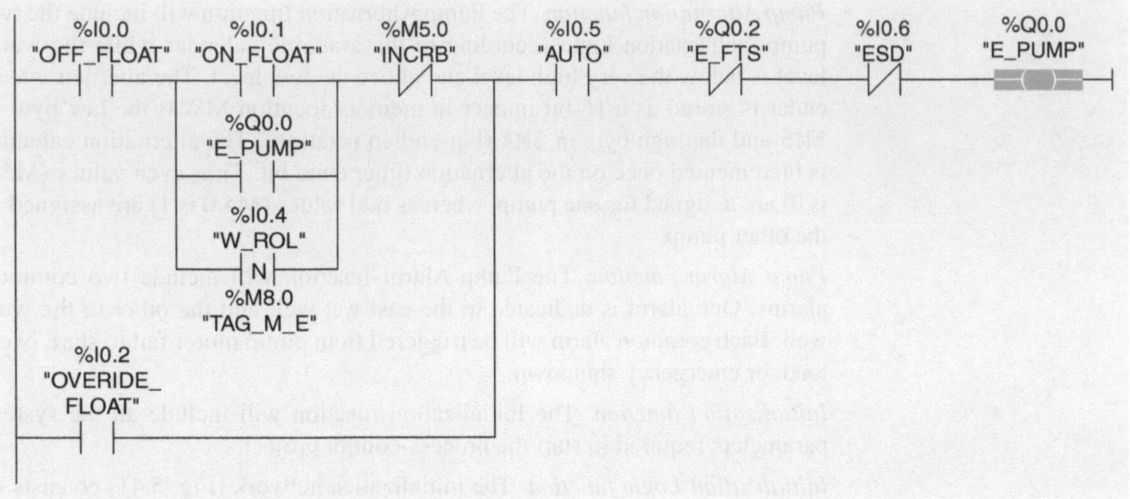

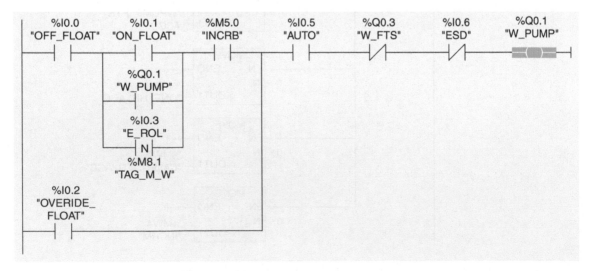

Figure 5.42   East Pump Network.

- *Pump START/STOP Logic function.* The Pump Start/Stop Logic function consists of two networks (East Pump and West Pump), as shown in Figs. 5.42 and 5.43.

- *East Pump Network.* Figure 5.42 shows a ladder-logic diagram for the East Pump. This diagram assumes a normally open OFF float, a normally open ON float, a normally open AUTO, a normally closed East Pump fail-to-start tag name (E_FTS), a normally closed emergency shutdown selector switch tag name (ESD), and a normally closed tag name INCRB. ESD is a normally closed field selector switch that, when activated, prevents the pumps from running. The ladder diagram consists of one network.

Figure 5.43   West Pump Network.

- The network is set initially during the first scan as the water level goes above the low limit (I0.0 is TRUE) and as the water level goes above the high limit (I0.1 is TRUE), also assuming that M5.0 is off (East Pump next to go).
- `INCRB` from the Pump Alternation Add instruction is FALSE before the calendar expires, and the override float switch is FALSE, assuming that the water did not reach the very high limit (also known in industry as high-high limit) or above.
- If the system is placed in AUTO mode (I0.5 is TRUE), if the East Pump did not fail to start (Q0.2 is FALSE), and if no emergency shutdown exists, then the East Pump is selected.
- Assuming all the preceding, the rung condition is TRUE, and Q0.0 will turn ON, causing the East Pump to run.
- If the water level is above the very high limit, it will override the `INCRB`, and the East Pump will run with no variation in accumulated calendar time. In this case, the West Pump will also run until water is reassessed below the low-level set value.
- The negative edge–triggered `W_ROL` becomes TRUE when the West Pump stops; running online goes from ON to OFF. If this condition arises at the alternation time while the water level is below the high level and above the low level, then the selected East Pump must run and override the `ON_FLOAT`.

- *West Pump Network.* Figure 5.43 shows a ladder-logic diagram for the West Pump. This diagram assumes a normally open OFF float, a normally open ON float, a normally open AUTO, a normally closed West Pump fail-to-start tag name (`W_FTS`), a normally closed emergency shutdown tag name (`ESD`), and a normally open `INCRB`. The ladder diagram consists of one network and operates as follows:

  - The network is set initially during the first scan as the water level goes above the low limit (I0.0 is TRUE) and as the water level goes above the high limit (I0.1 is TRUE), also assuming that M5.0 is on (West Pump next to go).
  - `INCRB` from the Pump Alternation Add instruction becomes TRUE once the assigned calendar expires. The override float switch stays FALSE, assuming that water did not reach the very high limit or above.
  - If system is placed in AUTO mode (I0.5 is TRUE), West Pump did not fail to start (Q0.2 is FALSE), and if no emergency shutdown exists, then the West Pump is selected.
  - Assuming all the preceding, the rung condition is TRUE, and Q0.1 will turn ON, causing the West Pump to run.
  - If the water level is above the very high limit, it will override the `INCRB`, and West Pump will run with no variation in accumulated calendar time. In this case, the East Pump also will run. Notice in this case the two pumps will run until the water level is below the `OFF_FLOAT` setting (I0.0 goes OFF).

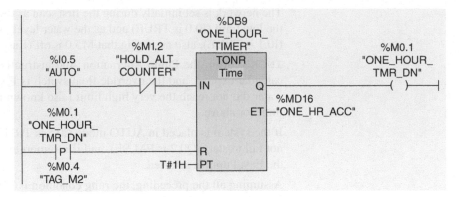

Figure 5.44 Pump station One-Hour-Timer Network.

- The negative edge–triggered **E_ROL** becomes TRUE when the East Pump stops; running online goes from ON to OFF. If this condition arises at the alteration time while the water level is below the high level and above the low level, then the selected West Pump must run and override the **ON_FLOAT**.
- *Pump alternation.* The Pump Alternation Logic function consists of three networks (One-Hour Timer, Counter, and Add instructions), as shown in Figs. 5.42, 5.43, and 5.44.
  - *One-Hour Timer.* The One-Hour Timer (Fig. 5.44) is reset to zero every time 1 hour of accumulated time is reached. The **HOLD_ALT_COUNTER** is a condition causing the timer to halt its count, which will be detailed in Chap. 9.
  - *Counter logic.* The Counter instruction (Fig. 5.45) is used to extend the One-Hour Timer for the Pump Alternation per calendar as defined by the user.

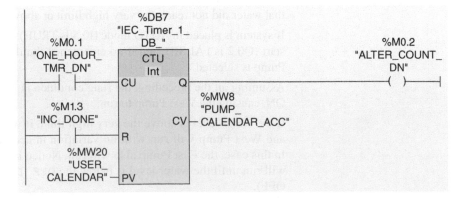

Figure 5.45 Pump station Counter Network.

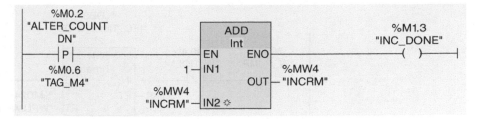

Figure 5.46 Pump station Add Network.

- *Add logic.* The Add instruction (Fig. 5.46) is used to increment register tag name (`INCRM`) to alternate between pumps as the least significant bit (LSB) toggles.
- *Pump ALARM Logic function.* The Pump ALARM Logic function consists of three networks (East Pump Failed to Start, West Pump Failed to Start, and Common Alarm), as shown in Figs. 5.45, 5.46, and 5.47.
- *East Pump Failed-to-Start Network.* Figure 5.47 shows a ladder-logic diagram for the East Pump Failed-to-Start alarm. This diagram assumes a normally open Q0.0 tag name `E_PUMP`, a normally closed contact I0.3 tag name `E_ROL`, an ON-DELAY timer (TON) with a 5-second preset time, and Q0.2 output coil tag name `E_FTS`. The ladder diagram works as follows:
  - The network initially is set during the first scan because the East Pump is ON (Q0.0 is set) and I0.3 is OFF (because it takes at least 5 seconds to receive the pump contact when the pump is started). The power flows to the timer block, and the timer starts timing.
  - If the timer accumulated time equals the timer preset time (5 seconds), Q0.2 will turn ON, indicating that the East Pump failed to start. Under normal conditions, the East Pump running on line (`E_ROL`) will become TRUE before the 5-second timer is done, thus disabling this timer.
- *West Pump Failed-to-Start Network.* Figure 5.48 shows a ladder-logic diagram for the West Pump Failed-to-Start alarm. This diagram assumes a normally open Q0.1 tag name `W_PUMP`; a normally closed contact I0.4 tag name `W_ROL`,

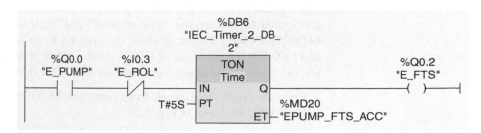

Figure 5.47 East Pump Failed-to-Start Network.

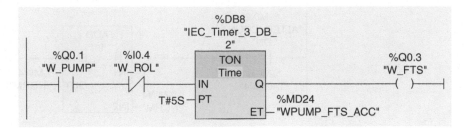

Figure 5.48 West Pump Failed-to-Start Network.

an ON-DELAY timer (TON) with a 5-second preset time, and a Q0.3 output coil tag name `W_FTS`. The ladder diagram works as follows:

- Initially, during the first scan when the West Pump is ON, Q0.1 is set, and I0.4 is OFF (because it takes up to 5 seconds to receive the pump running online contact when the pump runs successfully). The power flows to the timer block, and the timer starts timing.

- If the timer accumulated time equals the timer preset time (5 seconds), Q0.3 will turn ON, indicating that West Pump failed to start. The timer preset value is an experimental value that varies with the type and size of the motor used. Careful selection of this timer preset value is critical to prevent premature motor Failed-to-Start alarming action.

Once a pump fails to start, the associated alarm will be activated along with the common alarm, as shown in Fig. 5.49. The failed pump will not be considered until the operator takes the required action to clear the alarm and indicate to the PLC such action through the user-interface available control panels. This part of the logic is not shown here but will be included in more comprehensive coverage in Chap. 6.

- *Common Alarm Network.* Figure 5.49 shows a ladder-logic network for the Common Alarm. This diagram assumes a normally open Q0.2 tag name `E_FTS` for East Pump Fail to Start or a normally open contact Q0.3 tag name `W_FTS` for West Pump Fail to Start or I0.7 East Overload tag name `E_OVERLOAD` or I1.0 West Overload tag name `W_OVERLOAD`, or I0.6 Emergency Shutdown tag name `ESD` for the common alarm.

5.3.4 *HMI-PLC Application*

The Function key is a physical key on your HMI device, and its function can be configured for the events "Key pressed" and "Release key." Programming the FUNCTION KEYS page, to allow the operator to switch to two pages (STATUS and CONTROL), is shown as follows. The number of function keys available varies, but typically there are four or six on most HMI devices.

- From the project tree, create three pages (FUNCTION KEYS, CONTROL, and STATUS), as shown in Fig. 5.50.

- From the Tools window, drag and drop the text field, configure the apperance, and enter the text (STATUS) (Fig. 5.51).

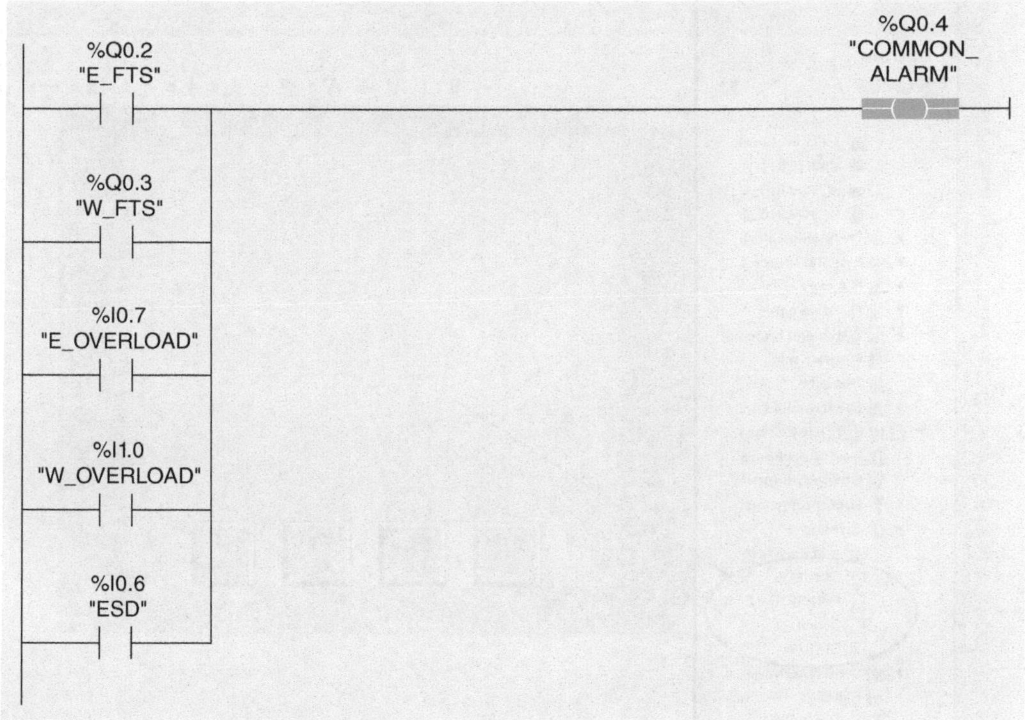

Figure 5.49 Common alarms.

- Right-click on the function key F2, and a red square appears on the status text. Under Properties and then Events, select the activate screen and choose Status from the popup list (Fig. 5.52).

NOTE Before downloading the program, configure the FUNCTION KEYS page as the start screen.

- Repeat the preceding steps to program the CONTROL screen.

The concepts just discussed will be applied to a simplified pump station control. The pump station HMI consists of three pages (FUNCTION KEYS, PUMP ALTERNATION, and ALARM). The following is a brief description of the implementation of these pages:

- *Function Keys page.* The Function Keys page allows users to go to the ALARM, STATUS, or PUMP ALTERNATION pages, as shown in Fig. 5.53.
- *Pump Alternation page.* The Pump Alternation page is shown in Fig. 5.54. It shows the status of the East and West Well Pumps and allows the user to enter the calender time to run both pumps.
- *Alarm page.* The Alarm page shown in Fig. 5.55 displays the East Pump Fail-to-Start, West Pump Fail-to-Start, and Common alarms.

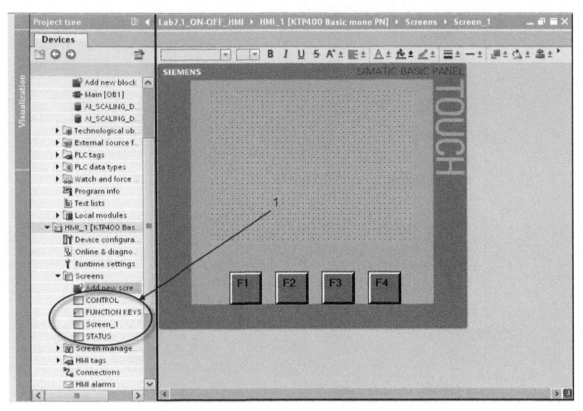

Figure 5.50 Create three pages (FUNCTION KEYS, CONTROL, and STATUS).

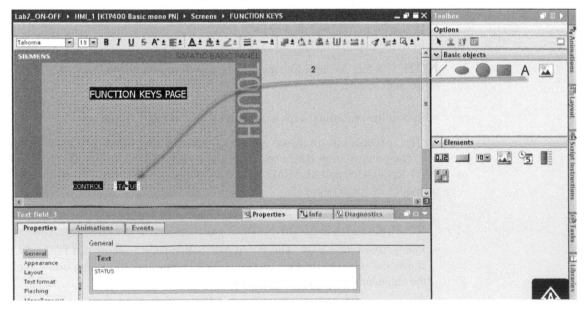

Figure 5.51

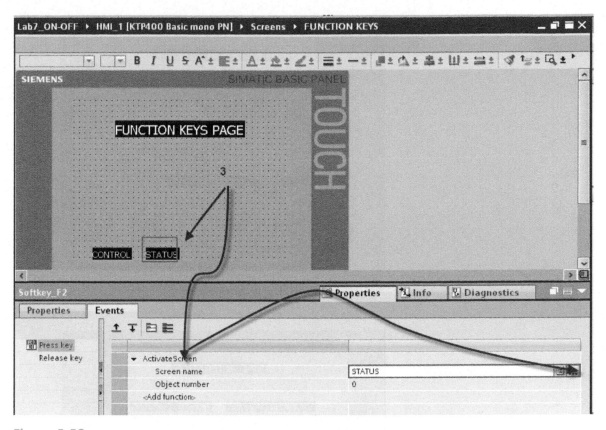

Figure 5.52

Homework Problems and Laboratory Projects

Problems

5.1 List four benefits of the use of HMIs in industrial automation and process control.

5.2 Explain how a PLC communicates with an HMI.

5.3 How many devices can be connected to an HMI?

5.4 What is the difference between direct communication and network communication?

5.5 What is a circuit object? Show a few examples.

5.6 What is a text field used for? Show a few examples.

5.7 What is an I/O field used for?

5.8 What are the properties of an I/O field?

5.9 What is a circuit object? Show the steps needed to establish such objects.

5.10 What is a Button field used for? Establish and show the steps used to configure a button.

5.11 What are the properties of the Button field? Explain the characteristics of a few of the HMI Button key properties.

5.12 List the steps required for the I/O preset event programming.

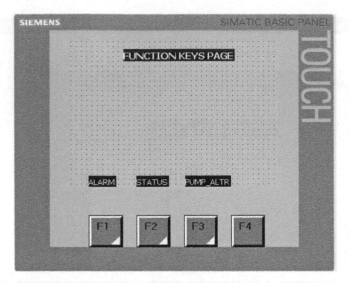

Figure 5.53 HMI Function Keys page.

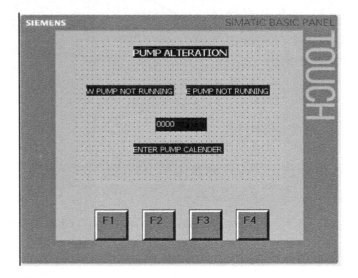

Figure 5.54 HMI Pump Alternation page.

5.13 Explain the use of an HMI in a small control application selected from one of previous chapters' laboratories.

5.14 Explain what a Switch field is used for, and give an example to demonstrate its use.

5.15 List the steps required to change the appearance of a Switch field.

5.16 Explain what a Bar field is, and give an example of where it can be used.

5.17 Explain the distributed-control-system approach, and give an example of where it can be used.

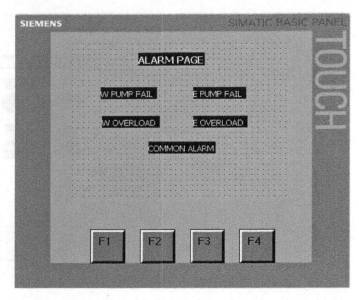

Figure 5.55 HMI Alarm page.

5.18 Distributed process control is becoming an attractive approach for implementing large automation projects. List and briefly discuss the technologies that support and recommend this approach.

5.19 HMIs are used for both status and system remote control. Other panels are typically available and provide the same functionality. How can conflict and race conditions be eliminated?

5.20 When using several HMI devices and PLCs, how is communication coordination task achieved?

5.21 PROFINET and Ethernet are multiple-accesse networks with collision-detect local-area-network protocols. How does this restrict the number of networked nodes in a real-time control application?

Projects

Laboratory 5.1: Merry-Go-Round HMI

This application was implemented in Laboratory 4.4, but the focus here is on the HMI task. It assumes four motors represented by four pilot lights on the trainer unit. A push-button START switch is used to start a sequence of motor operations, which can be terminated at any time by pressing a push-button STOP switch. The sequence starts Motor 1, then Motor 2, then Motor 3, and then Motor 4. The same sequence will repeat until the process is stopped. Each selected motor will run for a period of 5 seconds while the other motors are idle. This timer project was named by our students as the "Merry Go Round" project.

Laboratory Requirements
- Write a documented ladder-logic program.
- Configure the HMI page to display the four motor status represented by PL1, PL2, PL3, and PL4, as shown in Fig. 5.56. Configure a button to allow users to enter the timer preset value in seconds.
- Download the program, and perform the laboratory checkout.

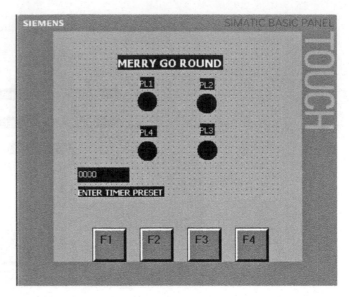

Figure 5.56

Laboratory 5.2: Controlling Feed Flow into a Digester

On completion of this laboratory, you will learn how to communicate with the HMI, and configure Control, Status, and Alarm pages to display/control the number of gallons fed to the digester. This laboratory will focus on the use of an HMI and its configuration to test and verify Laboratory 4.2 in Chap. 4.

Laboratory Requirements
- Load the ladder-logic program for the digester from Laboratory 4.2.
- Configure the HMI Function Keys page to allow the operator to move between Control, Alarm, and Status pages as shown in Fig. 5.57(*a*).
- Configure the HMI Control page to allow the operator to enter the total number of gallons and the high/low limits for the total number of gallons required to enter the digester as shown in Fig. 5.57(*b*).
- Configure the HMI Status page to monitor the PV1 ON/OFF status as shown in Fig. 5.57(*c*).
- Validate the number of gallons entered by the operator between the high/low limits.
- Configure the message, "Wrong number of gallons entered. Enter a valid number of gallons" on the Control page.

Laboratory 5.3: Downstream Lock/Vertical Gate Monitoring

A lock/vertical gate is equipped with one motor and two limit switches (LS1 and LS2). The gate moves inward and outward to regulate navigation in an irrigation canal. When the gate is fully open, LS1 is closed and LS2 is open. When the gate is fully closed, the limit-switch logic is reversed.

Laboratory Requirements
Develop a ladder-logic program using the training unit or the Siemens simulator to do the following:

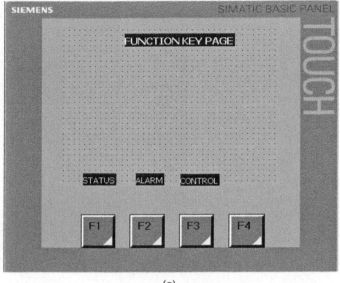

(a)

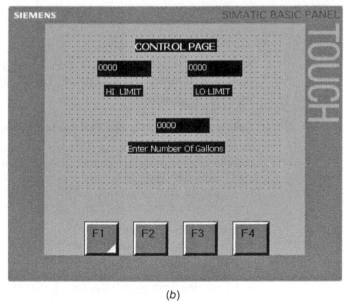

(b)

Figure 5.57

- Turn on pilot light 1, indicating that the gate is fully open.
- Turn on pilot light 2, indicating that the gate is fully closed.
- Flash pilot light 3, indicating that the gate did not reach the fully open or the fully closed position during the expected time.
- Configure the HMI Alarm page to display the message, "Vertical Gate 1 Stuck in Between" if the gate fails to reach final position in 15 seconds.

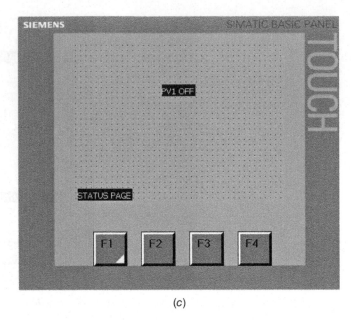

(c)

Figure 5.57 (*Continued*)

NOTE Figure 5.58(*a*) shows an easy way for linking ladder-program variables into an HMI newly created dynamic tag. Figure 5.58(*b*) also shows the creation of the HMI VG1 alarm.

Laboratory 5.4: Motor Fail-to-Start Alarm

To confirm that a motor is running after it receives the start command from the PLC, an input is received from the motor magnetic starter within 5 seconds from the time the motor start is activated. If the input is not received within the time limit, a fail-to-start alarm must be sent to the HMI to indicate that the motor failed to start. The ladder program for this laboratory was discussed in Chap. 3 and is listed in part in Fig. 5.59.

Laboratory Requirements
Program and test the ladder program to achieve the following:

- List the conditions for successful pump start.
- Change the timer preset value for fail-to-start operations.
- Use large values for the timer preset value, and monitor the program operation in the PLC.
- Simulate and monitor the pump fail-to-start alarm on the training unit pilot light or the Siemens simulator.
- Configure the HMI to flash a circle object every second, using red color indicating that a motor failed to start.

Laboratory 5.5: Storage-Area Monitoring

Figure 5.60 shows a system with two conveyor belts and a temporary storage area between them. Conveyor belt 1 delivers packages to the storage area. A photoelectric barrier at the end of conveyor belt 1 near the storage area detects how many packages are delivered to the

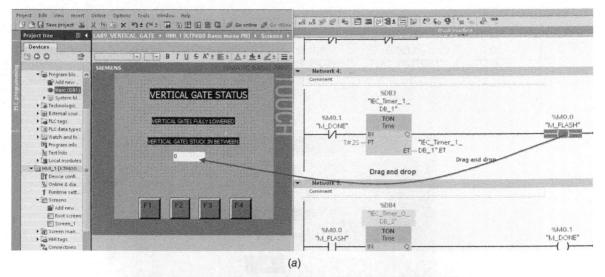

(a)

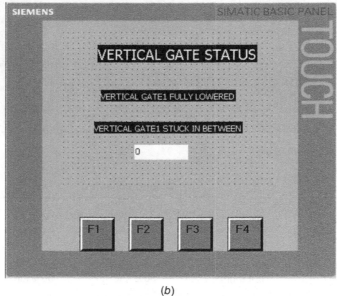

(b)

Figure 5.58

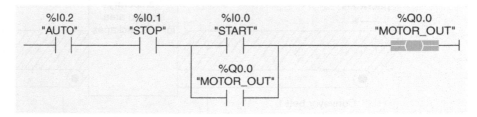

Figure 5.59

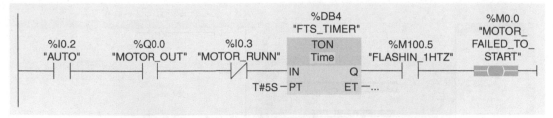

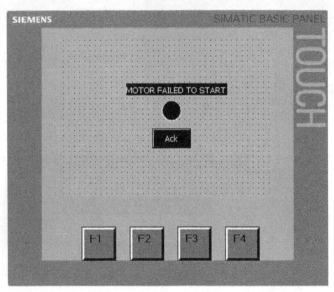

Figure 5.59 (*Continued*)

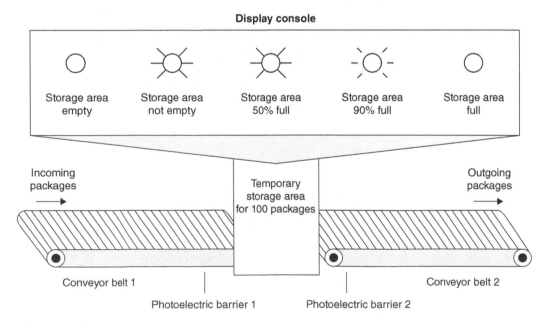

Figure 5.60

storage area. Conveyor belt 2 transports packages from the temporary storage area to a loading dock, where the packages are loaded on a truck for delivery to customers. A photoelectric barrier at the storage-area exit detects how many packages leave the storage area to go to the loading dock. A display panel with five lamps indicates the use of the temporary storage area. When a conveyor belt is restarted, the current count value is set to the number of packages available in the storage area.

Laboratory Requirements

Implement the application ladder program in Fig. 5.61, and do the following:

- Configure the HMI Status page as shown in the figure.
- Simulate the six situations displayed on the Status page
- Monitor the status, and report the results.
- Activate the conveyor restart push-button switch, and comment on the result.

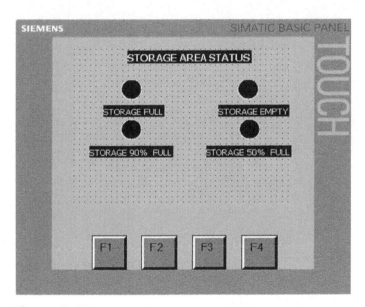

Figure 5.61

Laboratory 5.6: Local/Remote Data Entry

This application assumes one two-position switch (LEFT/RIGHT). In the left position, it displays the value entered from the PLC (VALUE_LOCAL_STATUS), and in the right position, it displays the value entered from the HMI (VALUE_REMOTE_STATUS). Implement the application described in Example 5.5 using the HMI Switch field.

6

Process-Control Design and Troubleshooting

This chapter covers PLC process-control design and troubleshooting. The system presented here is formed of three layers: process-control overview, process-control implementation, and process-control checkout and startup.

Chapter Objectives

▶ Document process description, components, and control.

▶ Implement process logic diagrams/function blocks.

▶ Be able to perform process-control checkout/startup.

▶ Understand safety standards.

This chapter covers the fundamentals of process-control/automation design techniques and industry-accepted standards. As with other established and successful computer systems, such as computer operating systems and computer network protocols, a layered hierarchal system is used. The simplified system presented here is made up of three layers: process-control overview (layer 1), process-control implementation (layer 2), and process-control checkout and startup (layer 3). The general guidelines for the three layers will be discussed in the first three sections of this chapter. Design reviews and approvals are required for each layer before proceeding to the next level and committing requested resources. Special attention in this chapter is paid to standard checkout techniques (layer 3) because design and implementation issues (layers 1 and 2) were covered in previous chapters' examples and in the comprehensive case study included in Chap. 8.

6.1   Process-Control Overview, Layer 1

Layer 1 provides a process-control/automation overview for a project, including a comprehensive definition of the level of automation and what will be controlled. It is used as the basis of preliminary costs, resources needs, and overall project schedule estimates. Specific issues to be covered in this layer include

- Control-system requirements
- Start and stop procedures
- Disturbance/alarm-handling strategies
- Process constraints
- Regulatory and safety requirements

The documentation in layer 1 will serve as the basis for later development of the detailed process-control strategy, layer 2. Layer 1 describes the intended scope and purpose of the project. It includes relevant information on the current process if the project is an enhancement, replacement, or revision of an existing system. Specialized or abbreviated terms should be avoided unless fully defined because this document is intended for use by professionals who are stakeholder in the overall system operation and well-being but not necessarily versed in process control/automation. Reference sources for the information supplied and used in this document should be provided whenever possible. Required details of layer 1 include process descriptions, level of control/automation, and control-system components, and all will be discussed in the next three subsections.

6.1.1   Process Descriptions

This subsection conveys in a very general way the functionality and details of the particular automation system or process control. It should clearly identify all components/processes of the system and list the important attributes of each process. Significant concerns must be documented, such as quality issues, safety concerns, energy consumption, and possible disturbances. Relevant information that may enhance the general understanding of the process must be included or referenced.

The information in this section is typically the result of collaborations between the process-control lead and other process/operation experts. It also can include references to already-existing documents.

Describe the specific process control/automation objectives for the project. The following are typical objectives:

- *Minimize manual interactions.* A standard objective of automation is to eliminate or minimize manual operations. This includes the automation of container loading or unloading, the use of computerized data acquisition, automating actuation control, adding safety functions, and other quality-enhancement functions.

- *Protection of equipment.* This includes operational restrictions for key equipment, safeguards during startup and shutdown, and other standard operation and maintenance procedures.

- *Quality control.* This includes ensuring correct raw materials handling, controlling recipe implementation, instrumentation adequacy for measurements, meeting product specifications, and providing adequate monitoring for continuous quality-control improvements.

- *Safety and the environment.* This includes handling of hazardous and toxic materials, operations at high pressure or temperature, leakage and spill avoidance, risk minimization, and safety procedures.

- *Energy use and recycling.* This includes energy-use reduction, improved reuse/recycling yield, elimination or reduction of waste, and wastewater/material treatment and recycling.

- *Asset utilization.* This includes increased capacity, reduced process cycle time, added flexible production capabilities, improved process performance, and reduced system down time.

- *New technology.* This includes implementation of new technologies safely and effectively to enhance process performance and product quality.

6.1.2 *Level of Control/Automation*

This subsection explicitly defines the recommended level of automation for the control system. It includes process start, run, stop, and potential upsets/disturbances. Recommendations must take into account the complexity, integration level, safety requirements, hazards, and quality needs of the required process flexibility/restrictions during the whole cycle of operation. Recommended level of control must add value to the existing system, and its associated cost benefit must be justified.

This section must define the startup philosophy, which can be either human-driven, totally automated, or a mixture of both. The following are examples of issues to be considered:

- Process-control utilities and associated functionality
- Subsystems coordination needs and sequencing
- Alarm and interlocking mechanisms and associated procedures
- Process control basic constraints and override procedures
- Emergency shutdown and power-failure restarting and recovery functions

This document can include additional details of both the functionality and primary objectives for the "run" and "shutdown" states of the process-control system. It also might include some specific constraints or limitations that need to be applied for successful and safe system operation. The level of control/automation also should identify any governmental requirements or professional standards that apply to the control and operation of the system. Standards include applicable internal organization practices and guidelines as well as specific contractual agreements. Social implications of the automation project and its impact on all stakeholders must be documented.

A wide variety of national and international standards apply to the design and implementation of automation and process-control projects. This include U.S. Food and Drug Administration (FDA) regulations, the Toxic Substances Control Act (TSCA), the International Standard for Developing an Automated Interface between Enterprise and Control Systems (ISA-95), Good Automated Manufacturing Practice (GAMP-4), the *Code of Federal Regulations* (CFR), and the American National Standards Institute (ANSI) and other specialized regulations.

6.1.3 Control-System Components

This subsection includes a high-level scope description of the process-control project and supporting systems. The type of control system and vendor should be identified with its specific components, if decided; otherwise, general requirements of the system must be given. Include any special requirements such as interfaces to vendor or customer systems, operator panels, remote input-output (I/O), field stations, control-room displays, and communication with other control systems. Define the basic system architecture, identifying and explaining any deviations from standard models.

The following is a sample of typical control-system components:

- Complete listing of control-system inputs and associated sensors
- Complete listing of control-system outputs and associated final control elements/actuators
- Basic system control structure and alarming parts
- Safety control-system requirements
- Human-machine interface (HMI) display and control interface
- Process information system and data acquisition
- Procedural control recipes, coordination, and sequencing
- Communication and networking interfaces

This subsection also identifies known software requirements with the versions selected for the control system, for the operator interface, for HMIs, for process modeling or advanced control, for data acquisition, and for simulation activities. The protection of the process from inadvertent operator actions needs to be addressed. Specific consideration needs to be employed in applying protection to the control system, including manual intervention procedures.

6.2 Process-Control Implementation, Layer 2

This document includes all the information associated with implementation of the process-control system, including software design, hardware configuration, and communication protocols used. It is intended to provide explicit documentation of all implemented tasks and to provide collaboration/communication avenues to all project stakeholders. This collaboration is essential and often produces changes and enhancements in the implementation. Stakeholders often include domain experts (e.g., electric, mechanical, chemical, civil, environmental, and industrial engineers/technicians), operations and maintenance personnel, and the owner.

Assumptions made during the implementation of this section (layer 2) that require clarification, confirmation, or additional information in the latter stages of the project in layer 3 must be fully documented. Outstanding issues that require follow-up or consultation must be clearly highlighted. These issues must be resolved and removed from the layer 2 section prior to start of the final checkout and startup, layer 3. The remainder of this section will detail the steps carried out during the process-control implementation, which include I/O detailed listings, data-acquisition tasks, closed-loop control, project logic diagrams/flowcharts, specification of ladder-program function blocks, and overall system documentation.

6.2.1 I/O Listing

All inputs and outputs associated with the process-control system must be listed in the I/O worksheet/action table along with the associated instrumentation. Required accuracy and resolution of relevant I/O are also documented in this table. The process-control designer identifies any special issues associated with the instrumentation in the I/O listing, including safe-fail position for outputs, calibration requirements for inputs, and redundancy considerations. The I/O action table captures most details necessary for the coding of all inputs and outputs, both digital and analog, as briefly described as follows:

- *Digital outputs.* Identify the position of each digital output in all process phases and against all possible interlock actions. Fail-position disconnect actions must be documented. Show the template used to create the code and program the output. The purpose and action for each digital output need to be understood and communicated.

- *Digital inputs.* Identify the position of each digital input in all process phases and against all possible interlock actions. Fail-position disconnect actions must be documented. Show the template used to create the code and program the input. The purpose and action for each digital input need to be understood and communicated.

- *Analog outputs.* Analog outputs are listed in the I/O action table, which captures most details necessary for coding. Identify the control action of each analog output in all process phases and against all possible interlock actions. The analog-output action table is also used to identify fail-position disconnect actions, the template used to create the code, and other additional logic required to program the output.

- *Analog inputs.* Analog inputs are listed in the I/O action table, which captures most details necessary for coding. Identify the control action of each analog input in all process phases and against all possible interlock actions. The analog-input action table is also used to identify fail-position disconnect actions, the template used to create the code, and other additional logic required to program the input.

Special types of I/O are common in some process-control systems. This includes a high-speed pulse counter for measuring flow rate and digital stepper motor output count. All special I/O must be clearly configured and documented. Also, associated interfaces and specification must be listed, and any deviations from default standards must be highlighted. Scaling must take into consideration instrument miscalibrations and potential signal deviations from the pre-specified range.

6.2.2   ### Data-Acquisition and Closed-Loop-Control Tasks

Data acquisition is the process of sampling analog input signals that measure real-world physical conditions and convert the resulting samples into digital numeric values that can be manipulated by the programmable logic controller (PLC) or, most commonly, by a computer system more situated for modeling, simulation, and database tasks. Most analog inputs are part of the data-acquisition task, with smaller number used in closed-loop control. Every closed loop has at least two analog variables assigned: the controlled and controlling values. The components of data-acquisition and closed-loop-control-system tasks include

- Sensors that constitute a simple data-acquisition task must be documented in both the functionality and the detailed programming implementation. This includes details of the method used, as in redundant sensory data validation or complementary sensory data fusion techniques.

- Data-acquisition tasks associated with HMIs must be detailed with the associated communication and graphics tasks. Linkage between the ladder program and the HMIs is very critical for system operation, future enhancements, and overall system maintenance.

- Some data-acquisition applications/tasks are often developed using various general-purpose programming languages or software tools, such as Visual Basic, Java, C++, and LabVIEW. These tasks need to be documented as part of the overall control-system design.

- Closed-loop-control tasks should clearly identify the controlled variable, the controlling variable, digital count format, engineering units, type of control, and the algorithms used to achieve control. ON/OFF and proportional-integral-derivative (PID) control are the most common process-control techniques in use today, with an increasing trend toward fuzzy-logic control in recent years.

Detailed coverage of process-control techniques and instrumentation is provided in Chap. 7. A case study in Chap. 8 will demonstrate the design and implementation of basic data-acquisition and process-control tasks for a real-time industrial project.

6.2.3   *Project Logic Diagrams and Ladder-Function Blocks*

Historically, flowcharts were often used to document logic flow prior to actual programming implementation. With the advancement of PLCs and distributed digital process control, pseudocodes and logic diagrams are becoming the tool of choice. Logic diagrams are used in this book and were introduced in Chap. 2. A structured approach must be enforced when a given task is divided into a number of smaller interconnected subtasks. Logic diagrams are effective visual documentation that directly maps into the ladder program and can be used with any PLC/control-system platform. Pseudocode tools are platform-dependent but can automatically generate the ladder or executable control code.

Although ladder programming may be the most widespread language for the implementation of PLC control, function-block diagrams/logic diagrams are probably the second most used language. This graphic language resembles a wiring diagram, with the blocks wired together into a sequence that is easy to follow. It uses the same notations/instructions of ladder diagrams but is visually more understandable to a viewer who is not versed in relay logic. Logic diagrams are not ideal for very large programming tasks, which include the use of special functions and I/Os. Implementing a process-control program using ladder diagrams requires more preparation up front to ensure full understanding of the program structure and the control flow before any code is committed.

Siemens' ladder-program structure is based on function blocks, as was described in Chap. 5. Typically, a process-control application uses a varying number of blocks, including an initialization block, a data-acquisition block, an HMI block, an alarms block, a communication block, a diagnostic block, or other major processing blocks. Chapter 8 will cover these concepts in a comprehensive industrial process-control case study. A small example will be used here to illustrate the concept.

6.2.4   *Control-System Preliminary Documentation*

The purpose of the system preliminary document is to communicate the process-control implementation details to all stakeholders before the final checkout and startup. The document essentially serves as the base for critical design reviews and potential revisions/enhancements. It also can be used by a wide range of audiences, including facilities, partners, customers, process leaders, or anyone who is involved in any of the processes outlined in the document. Always be aware of potential liabilities by documenting and seeking validation or clarification on all safety/hazard-related issues.

The following are some of the basic tasks involved in creating process-control preliminary documentation:

- Process scope and goal statement
- Inputs and outputs map
- Hardware configuration
- Communication protocol and configuration
- Ladder-program function blocks
- Process logic diagrams
- Safety and operational hazard issues

- Process- control management-system procedures
- Exception-management process
- Ladder and HMI listing

Each of these tasks will be a section within your document, which you can create in Microsoft Office Word or other word-processing program. We recommend using a program that automatically generates a table of contents based on the use of heading styles. Also, it's important to assign a clear numbering system within the document. For instance, the process scope and goal statement might be 1, whereas items included with that statement might be 1.1, 1.2, 1.3, and so on. The inputs and outputs map might be 2. A numbering system accomplishes two goals in that it makes information easy for readers to find and allows easy tracking of changes to the document by different individuals. Changes will be documented by date, person, and the revisions made. This is particularly critical for effective collaboration among a team of participants and stakeholders in the design and implementation process.

6.2.5 *Program Documentation Using Cross Reference*

There are several ways of displaying cross-references depending on whether you are in the Portal view or the Project view and which object you have selected in the project tree. In the Portal view, you can only display cross-references for the entire CPU; in the Project view, you can, for example, display cross-references for the following objects:

- PLC tags folder
- PLC data-types folder
- Program blocks folder
- Tags and connections folder
- Individual tags
- Individual PLC data types
- Individual blocks
- Technological objects

We will use an example (`La1_Comb_Logic`) to illustrate the subject of program documentation. The program has four logic functions (AND, OR, XOR, and XNOR). We will illustrate how to display the cross-reference information for an AND logic. Use the following steps to display the cross-reference:

- From Portal view, click Show cross-reference (Fig. 6.1).
- From Project view, you can monitor `AND_LOGIC` (address, date created, and date last modified) (Fig. 6.2).

6.3 Process-Control Checkout and Startup, Layer 3

The final step in the design is the checkout and startup of the process-control system, layer 3. Preliminary testing of hardware, software, communications, and user interfaces is a major part of the system implementation, Layer 2. This layer is

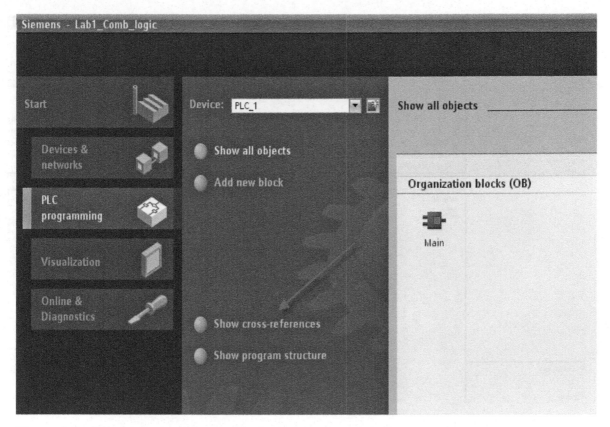

Figure 6.1 Portal view.

concerned with final system troubleshooting and commissioning. Technology transfer to the owner/customer is a key issue in completed automation projects. It must be planned and budgeted in the early stages of the project, layer 1. All checkout activities are performed in the field using the system physical resources in real time. Stakeholders, including operations and maintenance personal, are all critical parts of this step, which is led by the process-control/automation lead who designed and implemented the system. The process includes standard procedures and steps that are detailed in this section. Also, standard safety procedures will follow the checkout discussions.

The S7-1200 PLC provides three different methods for the testing of the user program: testing using program status, testing using the watch table, and testing using the force table. Each of the three techniques is briefly discussed as follows:

- *Testing with the program status and system diagnostics.* The program status allows users to monitor the running of the program. You can display the values of operands and the results of logic operations (RLOs), which allow for the detection and fixing of logical errors in the program. The PLC system diagnostic tools allow for the use of a wide variety of information during checkout.

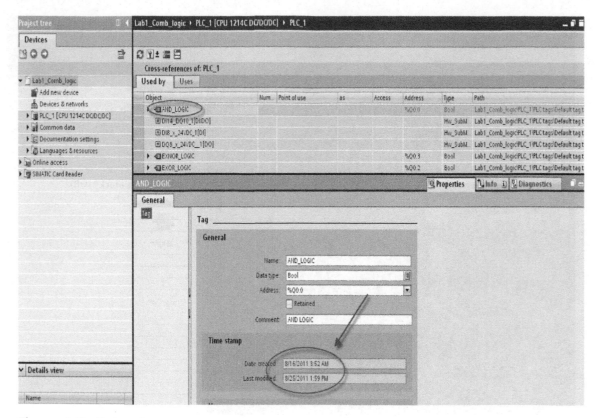

Figure 6.2 Project view.

- *Testing with the watch table.* With the watch table, current values of individual tags in the user program or on a CPU can be monitored and modified. Values can be assigned to individual tags for testing and program runs in a variety of different situations. Fixed values can be assigned to the I/O tags of a CPU in the STOP mode, which is typically used during the system static checkout of wiring.

- *Testing with the force table.* With the force table, current values of individual tags in the user program or on a CPU can be monitored and forced. Forcing an individual tag will overwrite that tag with the specified value. This allows program testing and run under various situations and scenarios.

6.3.1 *Checkout Using Forcing Functions*

Great care and precautions must be taken during the use of PLC forcing functions. Because a forcing function allows a user to intervene permanently in the process by assigning an arbitrary status/value to a control variable, observance of the following notices is essential:

- *Prevent potential personal injury and material damage.* An incorrect action while executing the Force function can harm people or pose a health hazard. It also can produce damage to machinery or the entire plant.

- Before starting the Force function, you should ensure that no one else is currently executing this function on the same CPU.

- Forcing can only be stopped by clicking the Stop Forcing icon or using the Online → Force → Stop forcing command. Closing the active force table does not stop the forcing.

- Forcing cannot be undone through the program logic execution.

Force tables can be used to monitor and force the current values of selected I/O in the program. When you force an input or an output, you overwrite the status of the selected I/O. This allows the logic designer to test the program while the system is running. The following safety precautions must be taken when forcing tags:

- Before starting the Force function, the user should ensure that no one else is currently executing this function on the same CPU.

- Forcing can only be stopped by clicking the Stop Forcing icon on the force table tool bar. Closing the active force table does *not* stop the forcing.

- Forcing is not affected by previous rung logic.

- Modifying tags are different from forcing tags. The modifications can be changed by the program logic or through the updated I/O memory image, whereas forcing is permanent. Users should check the Siemens technical manuals before using the Force function.

Forcing Outputs ON/OFF

The first step in the checkout process includes testing of all input and output device wiring to the PLC modules. This section covers output forcing in the Siemens S7-1200 system. As shown in Fig. 6.3, from the Project Tree menu under Watch and Force Table, click Force Table, and then enter the address you need to force and then enter Force value (TRUE or FALSE).

In Fig. 6.4, the physical switch SS2 is open, and output coil Q0.1 with tag name **PL2** is forced ON. Physical pilot light (PL2) in the Network 1 illuminates as indicated. The normally open contact Q0.1 in Network 3 is not affected by the force ON applied to the output Q0.1 in Network 1, and thus pilot light PL4 is not turned ON. The figure also shows PL3 in Network 3 OFF because SS2 is OFF.

Figure 6.5 illustrates the preceding output forcing example in relation to the actual physical inputs and outputs used. The figure shows the status of the pilot lights (PL2, PL3, and PL4) in relation to input switch SS2 and the ladder program, as described earlier. The forcing function is intentionally designed to affect only the selected output, but none of the remaining logic in the program will be affected. This function is used primarily during static checkout of all outputs wiring.

Notice the status of the normally open contact SS2 with address I0.1, which is OPEN. Output coil Q0.1 is physically connected to PL2, and this output coil is being forced ON/TRUE. PL2 turns ON as a result of the forcing of coil Q0.1 to TRUE status. The normally open contact associated with PL2 stays OFF. Only the forced output Q0.1 is ON; thus PL4 stays OFF. PL3 is OFF because SS2 is also OFF, which does not allow power to propagate to the coil associated with PL3 with address Q0.2. Of course, switching SS2 ON will cause all lights to go ON regardless of the forcing action.

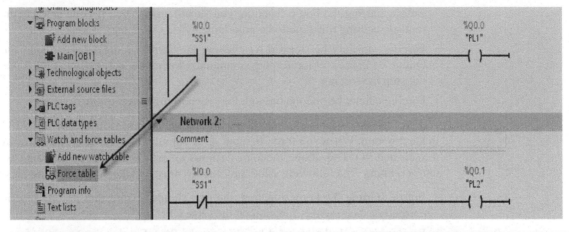

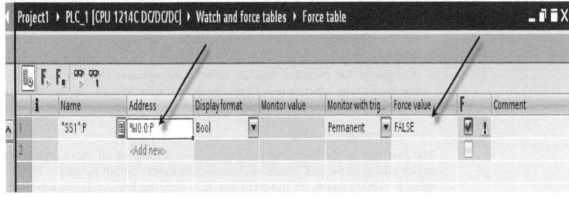

Figure 6.3   Force table.

Forcing Inputs ON/OFF

Figure 6.6 shows the physical switch SS1 in the open position, and the normally open input I0.0 instruction with tag name **SS1** is forced ON. Output Q0.0 with tag name **PL1** is physically ON when 0 to 1 is detected by RLO (rung logic output). Output coil Q0.1 with tag name **PL2** is OFF because normally closed instruction I0.0 with tag name **SS1** is forced ON. In the last rung, the normally open instruction Q0.0 with tag name **PL1** is not affected by the forced ON, and a 0 to 1 is not detected by RLO, causing output coil Q0.2 with tag name **PL3** to turn OFF, which physically turns **PL3** OFF.

As stated earlier in our coverage of output forcing functions, the same rules apply to the input forcing functions. A forced input turns that input to the selected force value. Associated networks with the forced input will behave according to the forced value. This function is used to verify all input wiring and associated sensor operations. This static checkout process is performed entirely from the PLC programming terminal with assistance from a technician or an operator in the field where the actual sensors and actuators are located. Communication between the person in the field and the PLC operator is essential to the checkout process

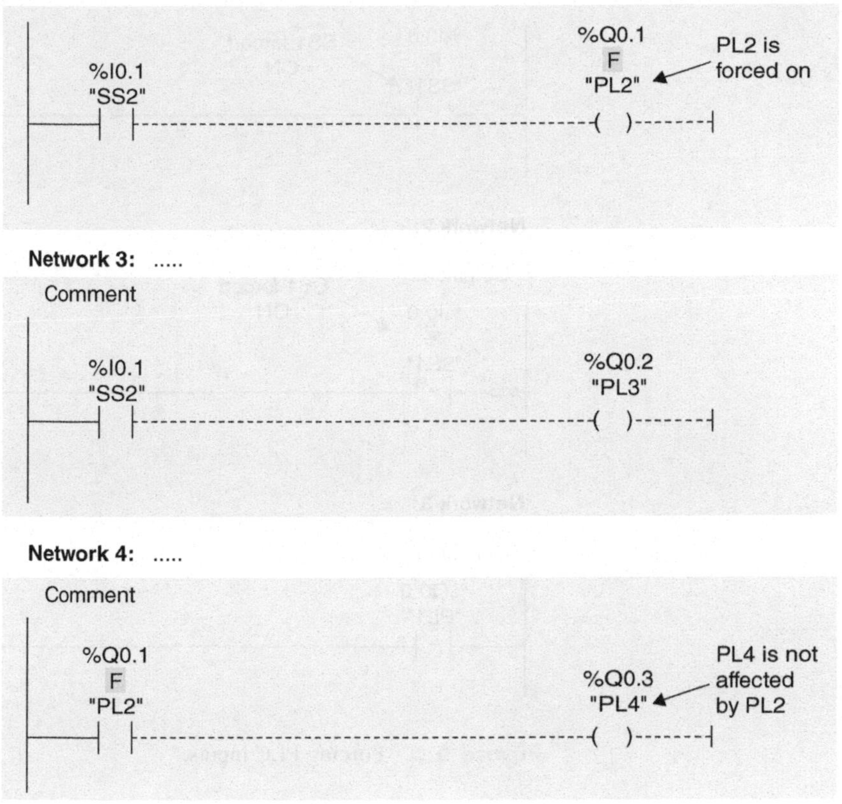

%I0.1
"SS2"
%Q0.1
F
"PL2"
PL2 is forced on
Network 3: .....
Comment
%I0.1
"SS2"
%Q0.2
"PL3"
Network 4: .....
Comment
%Q0.1
F
"PL2"
%Q0.3
"PL4"
PL4 is not affected by PL2

Figure 6.4 Forcing a PLC output.

L1
SS2
SS2 IS OPEN
I0.1
SS2 is OFF
F
PL2 is forced ON
SS2
I0.1
PL2
Q0.1
SS2
I0.1
PL3
Q0.2
F
PL2
Q0.1
PL4
Q0.3
Q0.1
Q0.2
Q0.3
PL2
PL3
PL4
L2

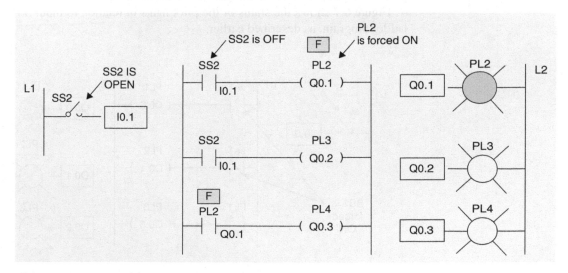

Figure 6.5 Forcing output in relation to physical inputs and outputs.

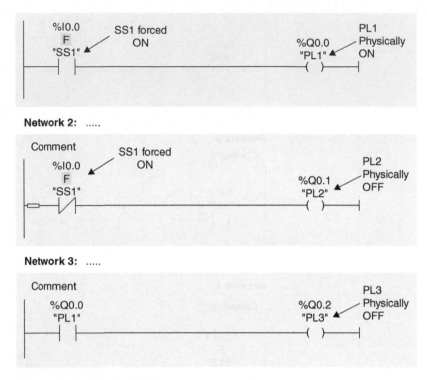

Figure 6.6 Forcing PLC inputs.

using the I/O forcing functions. It is important to execute the forcing in sequential order and remove the forcing of the completed item before moving on to the next I/O in the checkout list. There is no sense in checking the implemented control-program logic before completing a successful static checkout.

Figure 6.7 shows the status of the pilot lights in relation to input SS1 and the ladder program, as described earlier.

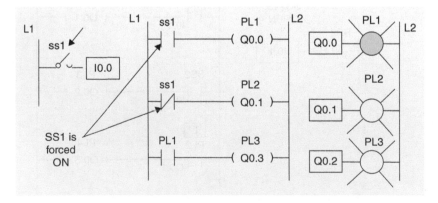

Figure 6.7 Forcing inputs in relation to physical inputs and outputs.

6.3.2   *Checkout Using Watch Tables*

Using output coils in more than one location in a program can be a source of trouble during checkout. In this section we will demonstrate checkout using watch tables in conjunction with this problem. Output coil addresses can't be repeated due to the way the processor scans the program (left, up, down, right), as shown in the ladder diagram and watch table in Fig. 6.8; PL1 will be OFF when SW1 is ON, SW2 is ON, and SW3 is OFF.

Unpredicted results can be reached because of the repeated coil address in the same program. As shown in Fig. 6.9, another combination can produce the same

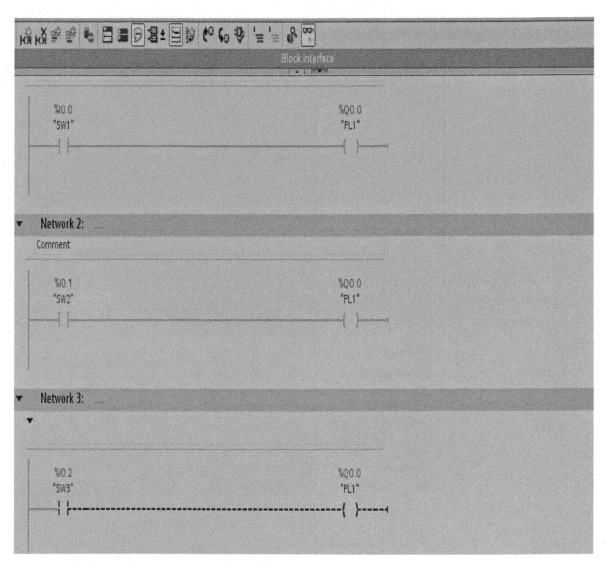

Figure 6.8   Unexpected results due to repeated use of an output.

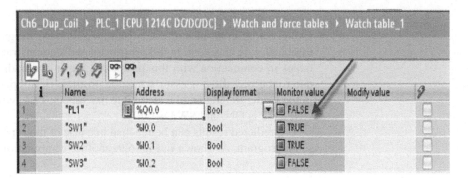

Figure 6.8 (*Continued*)

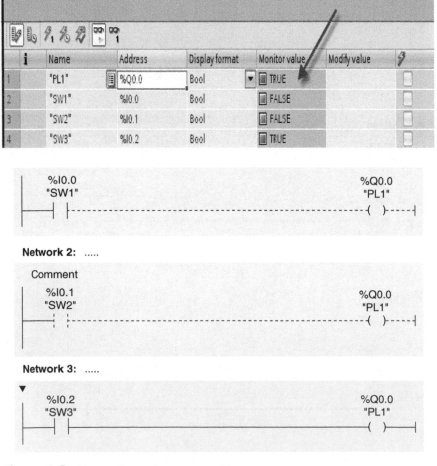

Figure 6.9 Repeated use of an output and its consequences.

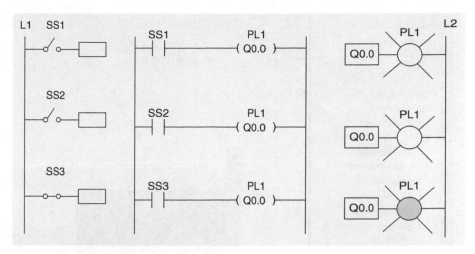

Figure 6.9 (*Continued*)

confusing results. The ladder logic in the figure demonstrates this situation. As shown, PL1 will be OFF when SW1 is OFF, SW2 is OFF, and SW3 is ON in the ladder diagram and the Watch table.

6.3.3   ### Checkout Using Cross-Reference, Program Status, and System Diagnostics

The Siemens software performs diagnostic tasks every ladder program scan and updates the hardware status. It also provides a log for all important and relevant events. This information is available, and users need to be aware of the best way to make use of such information during program checkout. Program status and cross-reference information is also available and must be accessed and used during program, hardware, communication, and graphical user interface (GUI) debugging. This section introduces these tools, but more details are available in the technical manuals or, more conveniently, online using the Help menus. The Appendix covers some of the key diagnostic tasks and screens.

As shown in Fig. 6.10, use the following tools to maneuver around the project:

1. Device configuration shows the hardware configured in your system.
2. Arrow up shows a device overview.
3. Arrow down gets you back to device configuration view.

Debugging Using Cross-Reference

There are several ways of displaying cross-references depending on whether you are in the Portal view or the Project view and which object you have selected in the Project Tree. In the Portal view, only cross-references for the entire CPU can be displayed. In the Project view, cross-references for the following objects can be displayed:

- PLC tags folder
- PLC data types folder

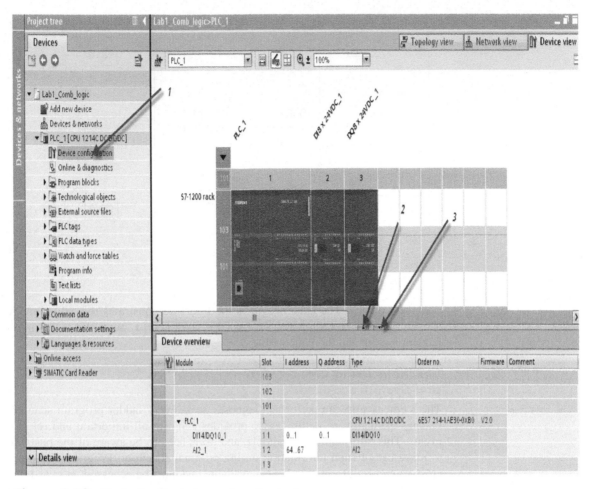

Figure 6.10 Navigating the device configuration tools.

- Program blocks folder
- Tags and connections folder
- Individual tags
- Individual PLC data types
- Individual blocks
- Technological objects

A program titled **Lab1_Comb_Logic**, as shown in Fig. 6.11, is used to illustrate the subject of program documentation using cross references. The program implements four logic operations (AND, OR, XOR, and XNOR). AND logic is used below to illustrate the use of cross-reference information. The following steps can be used to display the program cross-reference:

1. From the Portal view, click Show cross-references.
2. From the Project view, you can monitor the **AND_LOGIC** (address, date created, and date last modified), as shown in Fig. 6.12.

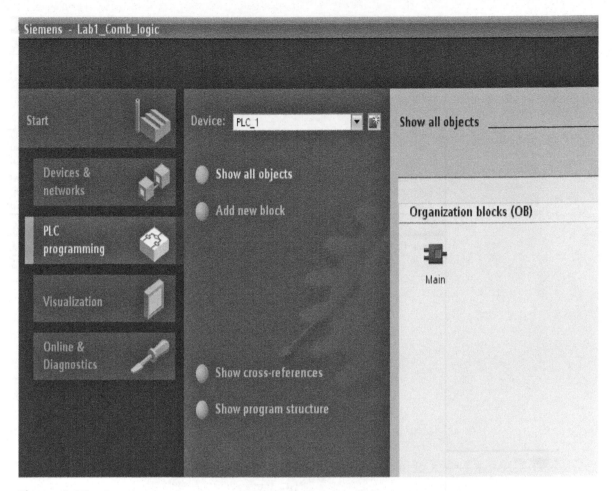

Figure 6.11   Cross references.

Debugging Using Program Status and Diagnostic Information

A small program is shown in Fig. 6.13 that which has two networks implementing combinational logic AND and OR operations. This program will be used to demonstrate debugging using program status and diagnostics tools. Two switches, SW1 and SW2, are used to provide the input to the logic operation. Two coils with tag names `AND_LOGIC` and `OR_LOGIC` are used to output the results of the two logic operations. This program is extremely small and simple but is used here to illustrate the checkout and troubleshooting techniques. Close SW1, and monitor the normally open contact energizes.

The CPU supports a diagnostic buffer that contains an entry for each diagnostic event. Each entry includes the date and time the event occurred, an event category, and an event description. The entries are displayed in chronological order, with the most recent event at the top. While the CPU maintains power, up to 50 most recent events are available in this log. When the log is full, a new

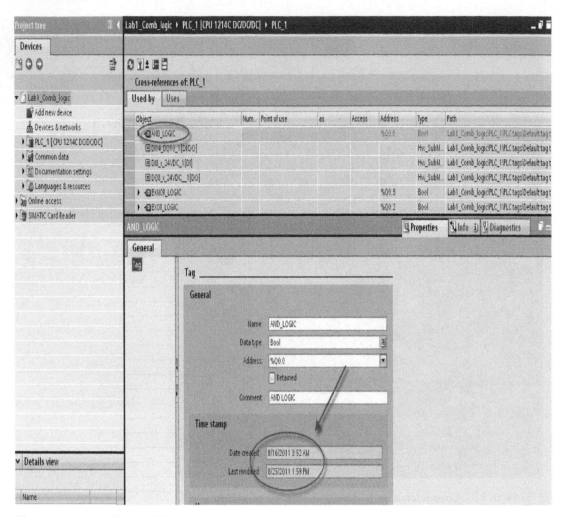

Figure 6.12 AND_LOGIC cross references.

event replaces the oldest event in the log. When power is lost, the 10 most recent events are saved.

The following types of events are recorded in the diagnostics buffer:

- Each system diagnostic event, for example, CPU errors and module errors
- Each state change of the CPU, for example, power up, transition to STOP, and transition to RUN
- Each change to a configured object except changes issued by the CPU and the user program

To access the diagnostic buffer, you must be online. Locate the log under Online & Diagnostics → Diagnostics → Diagnostics buffer (Fig. 6.14).

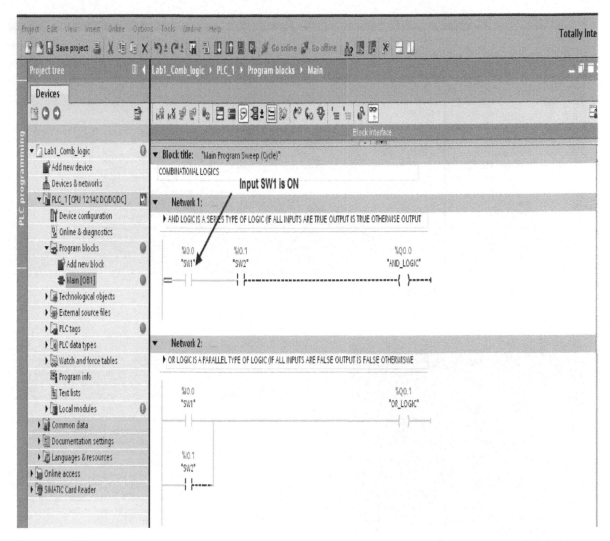

Figure 6.13 Combination logic AND and OR program.

The following are the diagnostic functions for the Sematic-1200:

1. From the Portal view, open the project **LAB6_AIRPORT_CONVEYOR** (→ Open existing project → **LAB6_AIRPORT_CONVEYOR** → Open) (Fig. 6.15).
2. Go online (→ **LAB6_AIRPORT_CONVEYOR** → Go online) (Fig. 6.16).
3. Next, select PG/PC interface for online access (→ PG/PC interface for online access → Go online) (Fig. 6.17).
4. As you go online, you can START or STOP the PLC, as shown in Fig. 6.18.
5. On the bottom to the right, the Diagnosis symbols can be combined with smaller supplementary symbols that indicate the result of the online/offline comparison (Fig. 6.19).

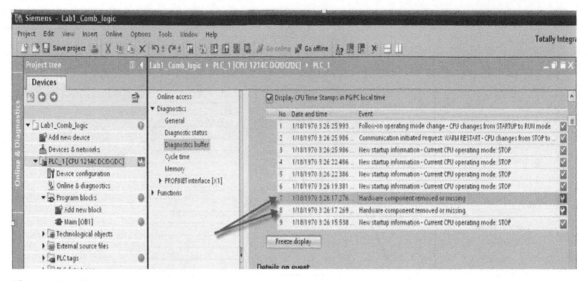

Figure 6.14 Diagnostic buffer.

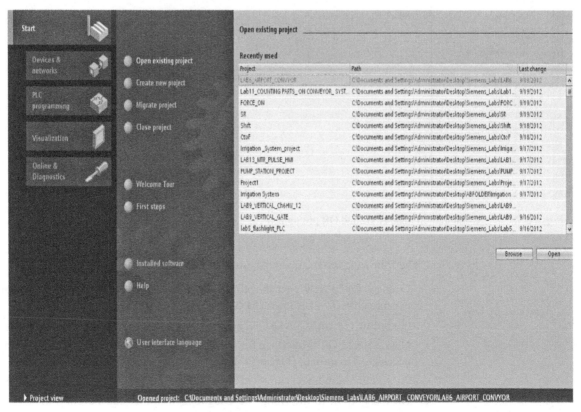

Figure 6.15 Open existing project.

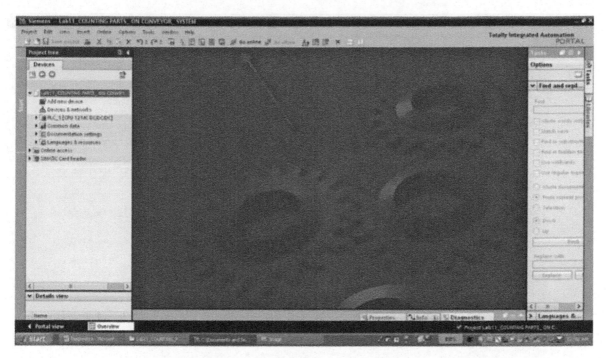

Figure 6.16 Go online.

Combined Diagnosis and Comparison Symbols
See Fig. 6.20.

Diagnosis Symbols for Modules and Devices
See Fig. 6.21.

Figure 6.22 shows a system configuration with two I/O modules added but that do not exist.

Error is detected because of the wrong configuration, online and offline diffrences, or hardware problems. Error is shown as a flashing red square on the PLC error tag, in the Project Tree, and also under Diagnostic in the software diagnostic buffer. Refer to the symbol tables for error details (Figs. 6.23 through 6.25).

Open Device configuration under Diagnosis, and check the status of the individual components, such as memory usage, cycle time, and communication ($\rightarrow$ Device configuration $\rightarrow$ Diagnosis) (Fig. 6.26).

6.4 System Checkout and Troubleshooting

System checkout and troubleshooting are critical tasks before the final deployment and commissioning of any process-control/automation system. These tasks must be performed according to well-established rules and documented procedures. These rules/procedures are designed carefully to prevent any potential problems that might cause personnel injuries or damage to resources or

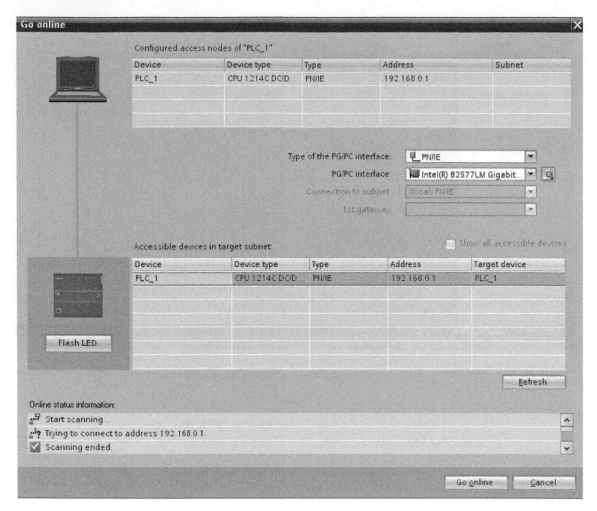

Figure 6.17 Selecting and configuring online access device.

compromise any of the predefined process/end-product qualities. Simulation and emulation techniques are used throughout the implementation phases of the project prior to the final checkout. Simulation tools allow designers to examine system-performance situations under varying scenarios before or aside from the actual implementation. These steps also can be used to efficiently verify design concepts prior to the commitment of resources. Emulation phases are accommodated in PLC software-development tools that allow the user to emulate I/O hardware, HMIs, control logic, and communication facilities. The initial phase of emulation checks only the control logic, whereas the second phase includes actual and simulated hardware/interfaces. The final emulation phase uses actual system hardware, interfaces, and communication resources.

Figure 6.18 Start/stop the PLC.

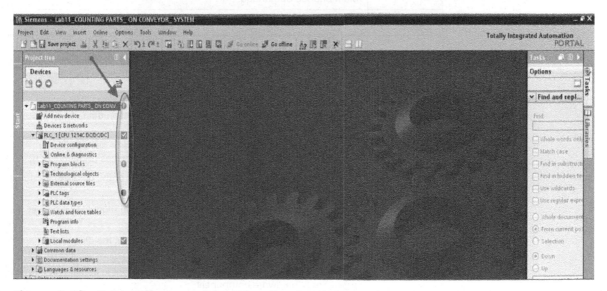

Figure 6.19 Online/offline comparison and associated symbols.

Symbol	Description
!	Folder contains objects whose online and offline versions differ
?	Comparison results are not known
■	Online and offline versions of the object are identical
◐	Online and offline versions of the object are different
◖	Object only exists offline
◗	Object only exists online

Figure 6.20

Icon	Meaning
	The connection with a CPU is currently being established.
	The CPU is not reachable at the set address.
	The configured CPU and the CPU actually present are of incompatible types.
	On establishment of the online connection to a protected CPU, the password dialog was terminated without specification of the correct password.
✔	No fault.
	Maintenance required.
	Maintenance demanded.
	Error.
	The module or device is deactivated.
	The module or the device cannot be reached from the CPU (valid for modules and devices below a CPU).
!	Diagnostics data are not available because the current online configuration data differ from the offline configuration data.
	The configured module or device and the module or device actually present are incompatible (valid for modules or devices under a CPU).
?	The configured module does not support display of the diagnostics status (valid for modules under a CPU).
	The connection is established, but the state of the module has not yet been determined.
⊘	The configured module does not support display of the diagnostics status.
!	Error in subordinate component: An error is present in at least one subordinate hardware component.

Figure 6.21   Diagnosis symbols for modules and devices.

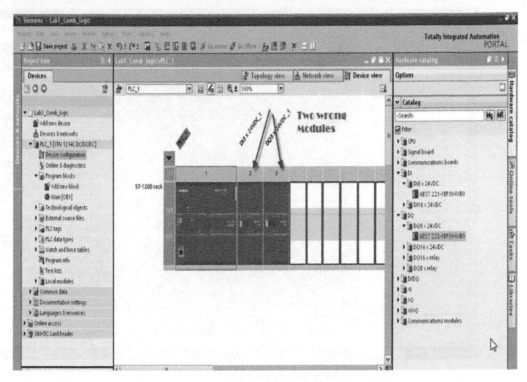

Figure 6.22 Configuration with two nonexisting I/O modules.

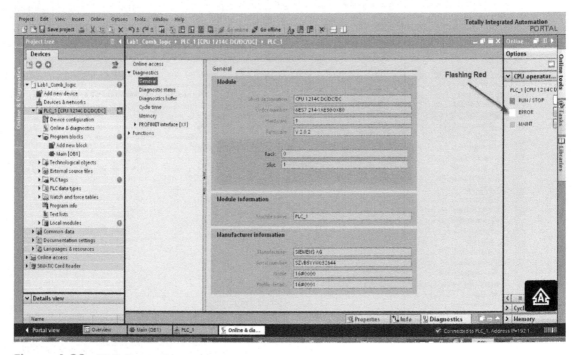

Figure 6.23 PLC diagnostics screen.

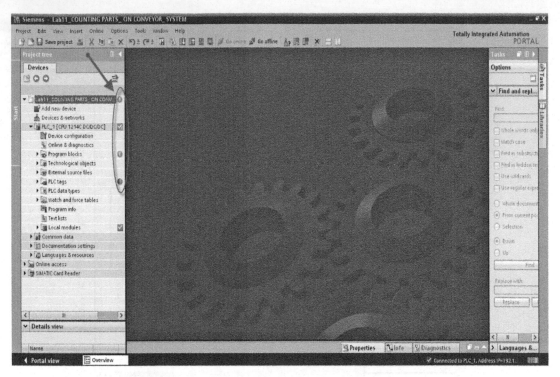

Figure 6.24 PLC project diagnostics symbols.

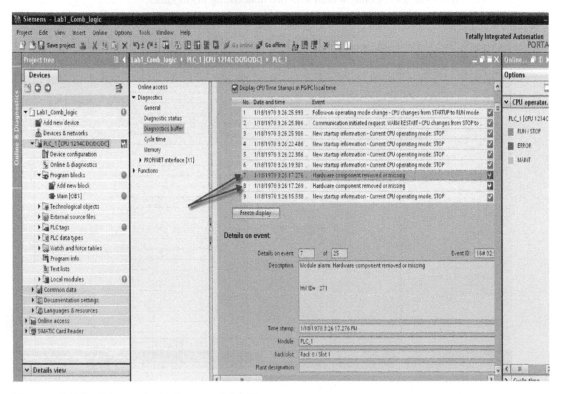

Figure 6.25 Diagnostics buffer recorded online events.

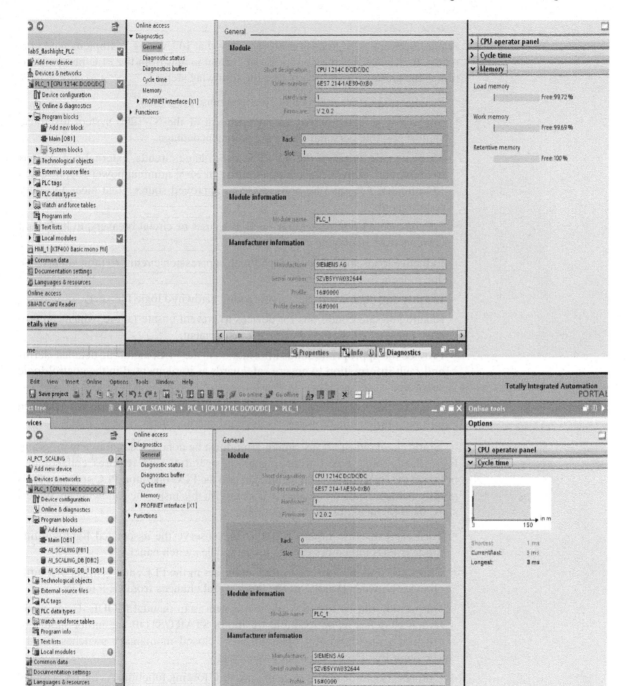

Figure 6.26 PLC memory and CPU cycle time diagnostics.

6.4.1   *Static Checkout*

Static checkout is a process used to verify that I/O wiring is implemented correctly. It also verifies the continuity of current to and from the PLC/instrumentation. This process must incorporate the following steps:

- Provide proper grounding and electrical noise protection.
- Implement safeguards that are independent of the S7-1200 system to protect against possible personal injury or equipment damage.
- Maintain safe status of the S7-1200 low-voltage circuits, external connections to communication ports, analog circuits, all 24-V nominal power supplies, and I/O circuits. All must be powered from approved sources that meet all safety requirements.
- Provide overcurrent protection, such as a fuses or circuit breakers, to limit fault currents from damaging supply wiring.
- Inductive loads should be equipped with suppression circuits to limit voltage rise due to inductive reactions.
- Identify any equipment that might require hardwired logic for safety.
- Ensure fail-safe conditions for devices to prevent unsafe failure, which can produce unexpected startup or damage to equipment.
- Provide the appropriate status information to the PLC, including detailed alarms and fault information to ensure safeguards as implemented in the control-logic software.
- Provide the appropriate system-status information to the assigned HMIs to enforce implemented operator-safeguard interactions.

Static checkout of all instrumentation must be conducted before any program-logic debugging. It is a time-consuming operation and requires collaboration between the instrumentation technician and the PLC operator. PLC software tools greatly reduce the effort needed to accomplish this task. The following are typical steps used to checkout I/O connections:

- Turn each discrete input ON/OFF, and observe the associated light-emitting diode (LED) indicators on the input-module switch panel.
- Check the assigned discrete input addresses in the PLC, and confirm that input changes between 0 and 1 as the input signal changes from OFF to ON.
- Outputs should be tested while the system is in manual MODE. Using push-button switches as an example for motor START/STOP, the motor should be checked using normally open/normally closed momentary switches and by verifying that the motor responds correctly.
- Outputs should be tested by using the PLC forcing function. As the bit is forced from the PLC from 0 to1, output should turn OFF to ON. Safety precautions should be taken before forcing I/O.
- Check all analog input signals, and make the needed adjustments in the sensor offset and span values. This analog signal calibration requires coordination and

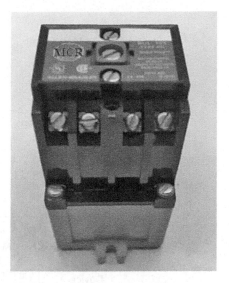

Figure 6.27 Typical industrial-type MCR.

is time-consuming but essential to achieving accurate data acquisition and overall process control.

6.4.2 *Safety Standards and Precautions*

Master control relays are used as a standard hardware safety technique to protect the system under emergency conditions. A master control relay is operated by START and STOP push-button switches. Figure 6.27 shows a typical industrial-type MCR. Pressing the START push button energizes the master control relay and provides power to the I/O modules. Pressing the ESD or E-stops push button, as shown in the figure, will disconnect the power to the I/O modules, but the processor will continue to operate, allowing operator access to the PLC programs and software tools with complete isolation from all I/Os. MCR is a hardwired relay that is used to provide a controlled safe stop. The master control relay deenergizes the supply power to the outputs and the machine power to the controlled devices. An industrial-type master control relay is also shown in Fig. 6.28 along with the wiring circuit diagram.

The following are a few examples of safety-standard routine practices used in implementing process-control systems. As stated, these are universal standards, not optional actions or choices, and they must be followed:

- A step-down transformer provides isolation from high main power to the 120 Vac going to the PLC power supply (L1, L2).

- A normally closed contact Emergency Stop push button is wired in series with the MCR coil to deenergize the coil in case of an emergency and disconnects the power to all I/O modules (CPU still receives power, and its LEDs provide status information).

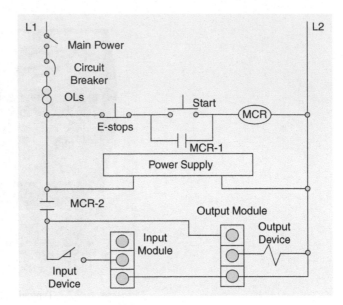

Figure 6.28 Master control relay circuit.

- Normally open/normally closed momentary push-button contacts are used for motor START/STOP operation in order to protect against accidental restart of motors after a power outage is restored.

- Two START push buttons are used simultaneously to start a machine to protect the operator's hand during startup.

- An interlocking switching mechanism is used for motors running in forward/reverse directions. This system requires motor stopping before the reversal of running direction and must have the highest possible reliability.

- The use of an emergency shutdown (ESD) switch with all power connection to PLC modules conducted through its contacts is necessary. Activation of the ESD stops all outputs in case of any emergency situation.

- Fault-detection hardware is deployed in critical areas, such as in a hazardous chemical possibly overflowing a reactor tank, to provide backup in case of sensor or control-logic failure. Redundant PLCs are often used in critical control systems with automatic switchover for uninterrupted operation.

- PLCs have hardware and online impeded diagnostic tools, including a combination of both software and hardware. LEDs are used to indicate the status of the PLC, CPU, and I/O modules.

- Diagnostic PLC software is an essential part of every program scan cycle. The CPU provides the following diagnostic status indicators:
  - STOP/RUN
    - Solid orange indicates the STOP mode.
    - Solid green indicates the RUN mode.
    - Flashing (alternating green and orange) indicates that the CPU is starting up.

- ERROR
  - Flashing red indicates an error, such as an internal error in the CPU, an error with the memory card, or a configuration error (mismatched modules).
  - Solid red indicates defective hardware.

6.5 Safeguard Implementation Examples

This section presents a small sample of examples showing safety implementations in Siemens S7-1200 PLC systems. All principles included in these examples are practical situations independent of the PLC platform used for the process-control implementation. A few minor syntax modifications may be needed for other vendors' PLC environments.

Example 6.1

A given machine-tool process requires an operator to start the conveyor system by pushing two START push-button switches at the same time in order to restrict the location/position of the operator's hands and thus guarantee maximum human protection during the startup process. The network in Fig. 6.29 assumes two normally open START1 and START2 push buttons, one normally closed STOP push button and one output coil for Motor 1 start.

Once the START1 and START2 push buttons are pressed, I3.0 and I3.1 become TRUE, which, in turn, makes Q3.4 TRUE, and Motor 1 runs. In the next scan, Q3.4 will latch the START1 and START2 push buttons and maintain the network TRUE status, which keeps Motor 1 running. Once the STOP push button is pressed, Motor 1 loses power and goes OFF.

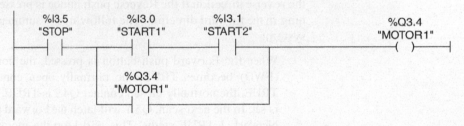

Figure 6.29 Two simultaneous motor's start push buttons.

Example 6.2

Figure 6.30 illustrates a line diagram of a magnetic reversing starter that uses an interlock to control the motor in forward and reverse directions. Please refer to the book's website, www.mhprofessional.com/ProgrammableLogicControllers, for an *interactive simulator* illustrating the operation of this starter and the motor control.

Circuit operation: Pressing the Forward push button completes the forward coil circuit from L1 to L2, causing coil F to be energized. Energizing coil F,

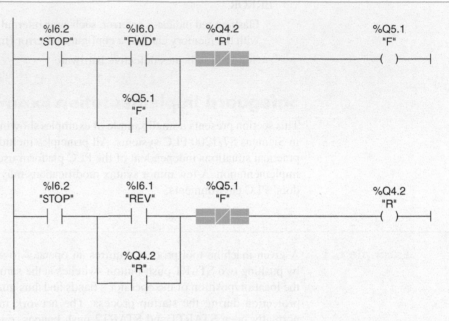

Figure 6.30 Forward/reverse interlocking.

in turn, energizes two auxiliary contacts, F-1 and F-2. The normally open contact F-1 provides a latch around the Forward push button, maintaining coil F energized. The normally closed contact F-2 will prevent the motor from running in the reverse direction if the Reverse push button is pressed while the motor is running in the forward direction. The following is a summary of the operation of this system:

1. When the Forward push button is pressed, the normally open contact I6.0 (FWD) becomes TRUE, the normally open contact I6.2 (STOP) is still TRUE, the normally closed contact Q4.2 is TRUE, and output coil Q5.1 (F) is set. In the next scan, Q5.1 will latch the Forward push button and maintain Network 1 TRUE status. This will keep the motor running in the forward direction.

2. To stop the motor, the operator presses the STOP push button. This causes the normally open contact I6.2 (STOP) to go FALSE momentarily, which drops the sealing around the normally open contact I6.0 (FWD), causing the output Q5.1 coil (F) to deenergize.

3. When the Reverse push button is pressed, the normally open contact I6.1 (REV) becomes TRUE, the normally open contact I6.2 (STOP) turns TRUE, the normally closed contact Q5.1 becomes TRUE, and the output coil Q4.2 (R) is set. In the next scan, Q4.2 will latch the Reverse push button and

maintain Network 2 TRUE status. This will keep the motor running in the reverse direction.

4. To run the motor in the reverse direction while it is running forward, the operator must press the STOP push button before pressing the Reverse push button.

5. The F and R coils are automatically interlocked with the normally closed contact Q5.1 (F) and the normally closed contact Q4.2 (R).

Example 6.3   The network shown in Fig. 6.31 was disscused in Chap. 5. Readers can refer to that chapter for further description. It shows the use of an emergency-shutdown selector switch (ESD) for protection against unusal conditions. Note the distinction from the MCR ESD contacts that are wired in series with the output and input modules. Once ESD is activated, it interrupts the power to the motor regardless of

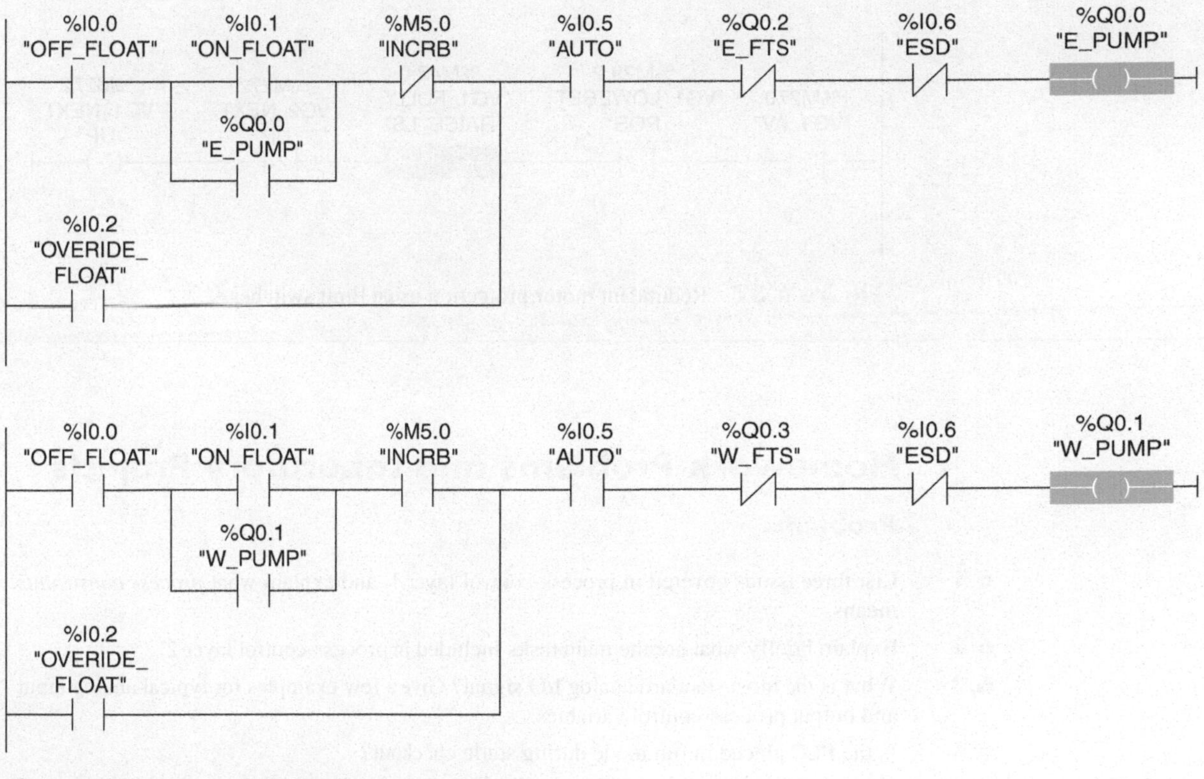

Figure 6.31   Emergency-shutdown selector switch.

the action of the PLC scanned logic. Two networks are designed, one for starting the East Well pump E_PUMP and the other for starting the West Well pump W_PUMP. The two pumps require simillar conditions for each to be selected to run. Common conditions for both require the system to be in AUTO mode, ESD not to be activitaed, and the pump motor not to fail to start. Notice that ESD has to stay inactive for either pump to run, even with all other conditions satisfied.

Example 6.4   The network shown in Fig. 6.32 was disscused in Chap. 5. Readers can refer to that chapter for further description. It shows the use of a limit switch that is placed in the upper position of vertical Gate 1. This limit switch, VG1_FULLY_RAISE_LS, ensures that power will disconnect (VG1_NEXT_UP is the input to the VG1_RAISE) when vertical Gate 1 does reach the upper position to protect both the motor and the structure from overload. Similar protection exists inside the motor, such as overload and temperature protection. Similar networks exist for the vertical Gate 1 next to go down.

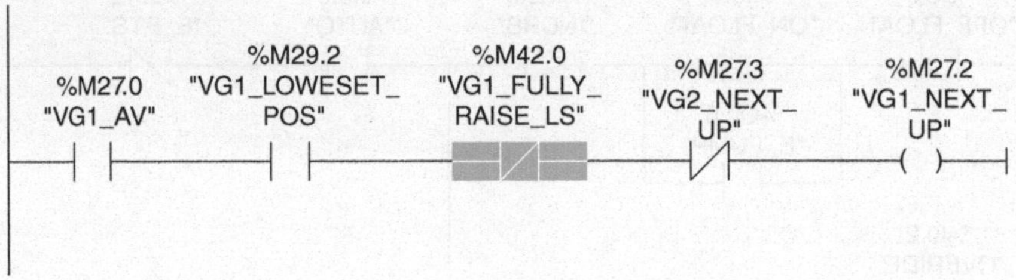

Figure 6.32   Redundant motor protection using limit switches.

Homework Problems and Laboratory Projects

Problems

6.1   List three issues covered in process-control layer 1, and explain what *process constraints* means.

6.2   Explain briefly what are the main tasks included in process-control layer 2.

6.3   What is the most standard analog I/O signal? Give a few examples for typical analog input and output process-control variables.

6.4   Is the PLC placed in run mode during static checkout?

6.5   List at least four of the basic tasks involved in creating the process-control preliminary documentation.

6.6   Explain briefly what are the main tasks included in process-control layer 3.

6.7 What are the basic three modes used in testing the S7-1200 user program?

6.8 A program malfunctioned, and the force output method was used to check out and debug its operation. The problem was fixed, but later during another run, the program malfunctioned again. What do you think might have caused this failure?

6.9 Is it possible to account for all possible future process scenarios during system checkout? How do HMIs contribute to overall system continuous enhancement?

6.10 Why do some machines require two START push buttons to start manually, whereas only one is necessary to start the machine?

6.11 How is the ESD (emergency shutdown) push button wired and included in the PLC system functionality?

6.12 What is the advantage of 24-Vdc I/O modules over 120-Vac modules?

6.13 Why does the *National Electrical Code* require normally open push buttons to start a motor and normally closed push buttons to stop a motor?

6.14 What are the components of a data-acquisition system? Explain briefly the function of each component.

6.15 A vertical gate is traveling between two limits, fully raised and fully lowered. One motor is used, which runs in the forward direction for raising the gate and in the reverse direction for lowering the gate. Write a ladder-logic program to protect the motor from moving in either direction if either of the two limits (forward/reverse) is reached. Add sensors as needed.

6.16 Explain the function of an HMI. How does it enhance the overall automation system operation and maintainability?

6.17 As shown in Fig. 6.33, the physical switch SS1 is open and in the ladder logic is forced ON. What is the status of PL1, PL2, and PL3?

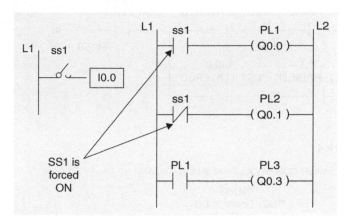

Figure 6.33 Homework problem 6.17.

6.18 Study the ladder program shown in Fig. 6.34, and answer the following:
   a. Explain the function of each network as programmed and documented.
   b. Check each network, and verify that it implements the desired specification as shown in the documentation.
   c. Troubleshoot the program, and reprogram the network (S) to work as it should based on shown and correct documentation. Use the three methods of troubleshooting described in the chapter.

▼ **Network 1:**

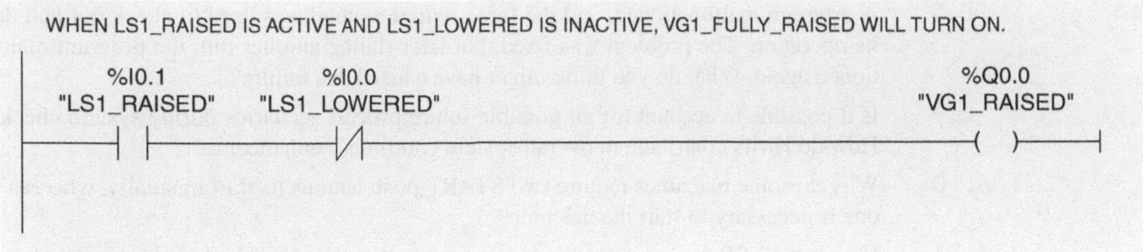

WHEN LS1_RAISED IS ACTIVE AND LS1_LOWERED IS INACTIVE, VG1_FULLY_RAISED WILL TURN ON.

```
    %I0.1                    %I0.0                                        %Q0.0
 "LS1_RAISED"            "LS1_LOWERED"                                 "VG1_RAISED"
──────┤ ├──────────────────┤/├──────────────────────────────────────────( )──────
```

▼ **Network 2:**

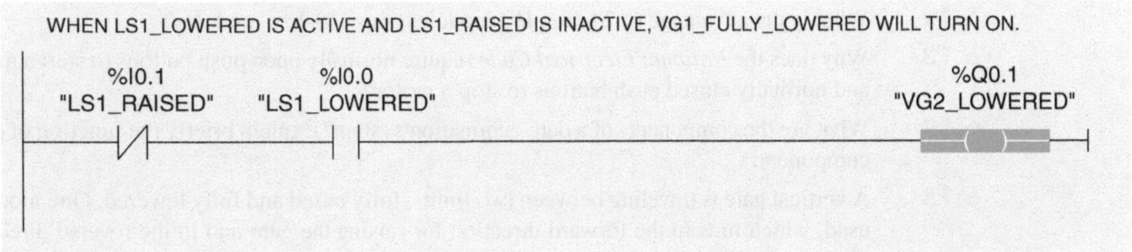

WHEN LS1_LOWERED IS ACTIVE AND LS1_RAISED IS INACTIVE, VG1_FULLY_LOWERED WILL TURN ON.

```
    %I0.1                    %I0.0                                        %Q0.1
 "LS1_RAISED"            "LS1_LOWERED"                               "VG2_LOWERED"
──────┤/├──────────────────┤ ├──────────────────────────────────────────( )──────
```

▼ **Network 3:**

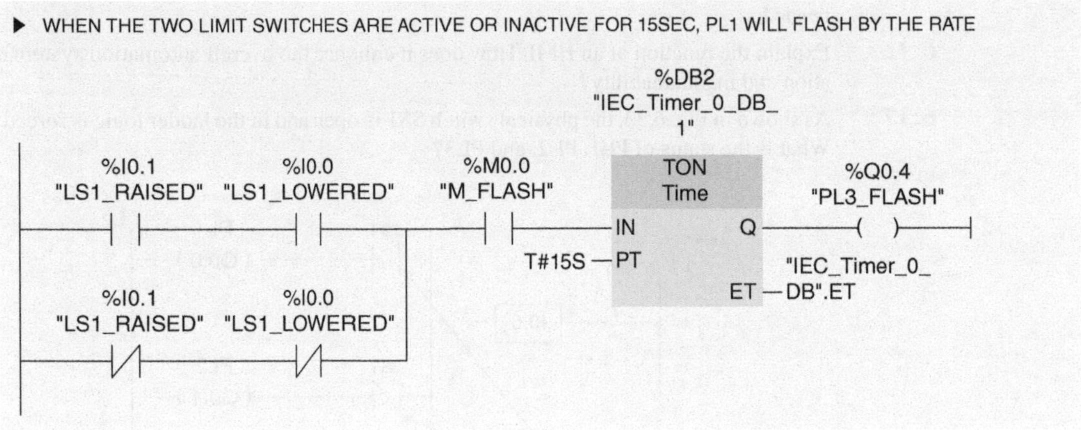

▶ WHEN THE TWO LIMIT SWITCHES ARE ACTIVE OR INACTIVE FOR 15SEC, PL1 WILL FLASH BY THE RATE

```
                                                        %DB2
                                                 "IEC_Timer_0_DB_
                                                       1"
    %I0.1          %I0.0          %M0.0          ┌──────────────┐         %Q0.4
 "LS1_RAISED"  "LS1_LOWERED"    "M_FLASH"        │     TON      │      "PL3_FLASH"
──────┤ ├──────────┤ ├────────────┤ ├───────────│     Time     │──────────( )──────
  │                                           ───│IN          Q│
  │                                      T#15S ───│PT           │    "IEC_Timer_0_
    %I0.1          %I0.0                          │          ET │───  DB".ET
 "LS1_RAISED"  "LS1_LOWERED"                      └──────────────┘
──────┤/├──────────┤/├───────────
```

▼ **Network 4:**

M1_FLASH COIL SHOULD FALSH EVERY 2SEC

```
                           %DB3
                    "IEC_Timer_1_DB_
                          1"
    %M0.1          ┌──────────────┐                              %M0.0
 "M1_DONE"         │     TON      │                            "M_FLASH"
──────┤ ├──────────│     Time     │──────────────────────────────( )──────
                ───│IN          Q│
           T#2S ───│PT           │       "IEC_Timer_1_DB_
                   │          ET │───      1".ET
                   └──────────────┘
```

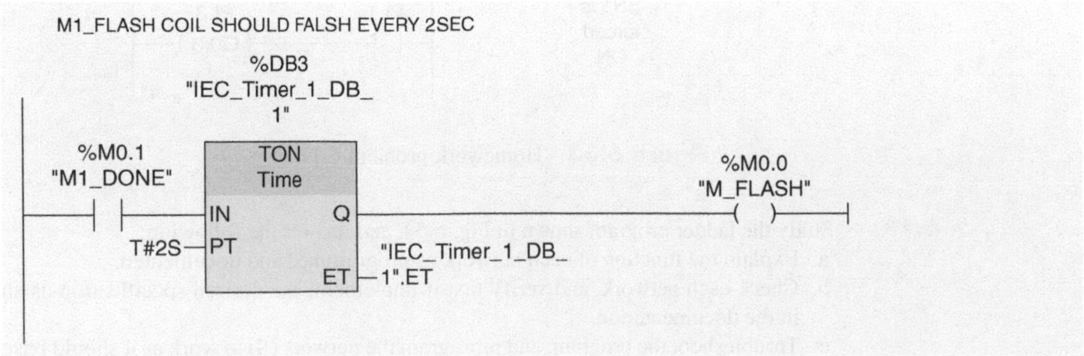

Figure 6.34 Homework problem 6.18.

6.19   In Fig. 6.35, the physical switch SS2 is open, and output coil Q0.1 with tag name `PL2` is forced ON. What is the status of the physical outputs PL2, PL3, and PL4?

```
        %I0.1                                                %Q0.1
        "SS2"                                                  F
                                                             "PL2"
      ──┤ ├─────────────────────────────────────────────────( )──────┤


        %I0.1                                                %Q0.2
        "SS2"                                                "PL3"
      ──┤ ├─────────────────────────────────────────────────( )──────┤


        %Q0.1                                                %Q0.3
          F                                                  "PL4"
        "PL2"
      ──┤ ├─────────────────────────────────────────────────( )──────┤
```

Figure 6.35   Homework problem 6.19.

6.20   What is the advantage of an Alarm acknowledge page on the HMI. If the alarm sounds, how would you proceed to solve the problem?

6.21   What kind of problems arises if the analog input signal is coming to the PLC in a non-standard format? Explain how can you solve this problem using scaling?

6.22   What is the most important tool(s) in a control system to identify a fault? List three such tools.

Projects

Laboratory 6.1: Basic Logic Functions Program Debugging

This laboratory will focus on the use of debugging and troubleshooting techniques covered in the chapter to test and verify Laboratory 2.2 of Chap. 2.

Laboratory Requirements

- Program the required networks, as shown in Fig. 2.70.
- Download the program to the PLC, and go online.
- Start program debugging using the training unit or the Siemens simulator.
- Configure the Watch and Force tables to record the value of `TAG_VALUE1`, `TAG_VALUE2`, and TAG_RESULT shown in the following table.
- Use the Watch and Force tables to enter the values of `TAG_VALUE1` and `TAG_VALUE2`.
- On the same page, configure `TAG_RESULT` to display the result of the word logic AND, OR, XOR, and XNOR.

Logic AND

Tag_value1	Tag_value2	Tag_result
15h	30h	
25h	1Ah	
5Bh	11h	

Logic OR

Tag_value1	Tag_value2	Tag_result
15h	30h	
25h	1Ah	
5Bh	11h	

Logic XOR

Tag_value1	Tag_value2	Tag_result
15h	30h	
25h	1Ah	
5Bh	11h	

Logic XNOR

Tag_value1	Tag_value2	Tag_result
15h	30h	
25h	1Ah	
5Bh	11h	

Laboratory 6.2: Conveyor-System Control

The main objective of this laboratory is to learn about the methods of troubleshooting used in industrial applications. A part is moving on a conveyor line crossing a photoelectric cell. The photoelectric cell functions primarily to count parts. The conveyor line stops after 100 parts are counted.

Detailed Descriptions

* The START push button starts the conveyor motor after a 5-second delay. The motor must start only if the AUTO/MANUAL switch is placed in the AUTO position. The photoelectric cell should count only when the motor is running.
* When the count reaches 100, stop the conveyor after a 3-second delay.
* ON pilot light indicates the end of the sequence.
* STOP switch, when activated, takes the system to the initial condition.
* The Operator can restart the same sequence by activating the START switch.

NOTE Use SS1 for the photoelectric cell input signal, SS2 for the motor running input indication signal, PL1 for the motor starting output signal, and PL2 for the end-of-sequence pilot light output signal.

Testing Inputs and Outputs

The following are typical steps used to check out I/O connections:

- Turn each discrete input ON/OFF, and observe the associated LED indicators on the input-module switch panel.

- Check the assigned discrete input addresses in the PLC, and confirm that input changes between 0 and 1 as the input signal changes from OFF to ON.

- Outputs should be tested while system is in manual MODE. Using push-button switches as an input for motor START/STOP, the motor should be checked using normally open/normally closed momentary switches and by verifying that the motor responds correctly.

- Outputs also should be tested using the PLC forcing function. As the bit is forced from the PLC from 0 to1, output should turn ON from the OFF state. Safety precautions should be taken before forcing I/O.

Laboratory Requirements

- Assign the system inputs.
- Assign the system outputs.
- Enter the networks to implement the above requirements.
- Add comments to each network.
- Download the program, and go online.
- Simulate the program operation using the training unit's I/O or the Siemens simulator, and verify that the program is running according to the process description.

Laboratory 6.3: Cascaded-Tanks Reactor

Three tanks are cascaded through a series of outlet solenoid valves. Material is discharged from tank 1 to tank 2 during the first 7 hours after the start of the process. At the end of the 7 hours, tank 1's valve closes, and Tank 2's valve opens to allow material to flow into tank 3 for 8 additional hours. In the final stage, tank 2's valve closes, and tank 3 is cleared by keeping its outlet valve open for additional 5 hours. The entire reaction takes 20 hours, after which all valves are closed. Activating the STOP push button will terminate the process and return all valves to the default closed position. The START push button initiates the entire reactor process. Please refer to Example 4.17 for the listing of the four networks used to implement this application.

Laboratory Requirements

- Assign the system inputs.
- Assign the system outputs.
- Enter the needed networks (refer to Chap. 4)
- Download the program, and go online.
- Configure a watch table to monitor the three valves and the timing operation. Use scaled-down times in seconds instead of hours during the simulation.
- Simulate the program operation using the training unit's I/O or the Siemens simulator, and verify that the program is running according to the process description.

Instrumentation and Process Control

This chapter covers PLC process-control and associated instrumentation funda-mentals. Process-control strategies, control modes, and common types of control are covered. Instrumentation and system performance are examined.

Chapter Objectives

▶ Understand instrumentation basics, digital and analog.

▶ Identify process-control elements and associated uses.

▶ Understand signal-conversion and quantification errors.

▶ Understand process-control types and common techniques.

A programmable logic controller (PLC) is designed to regulate real-time processes having both discrete and analog variables. So far we have covered the discrete/digital programming side; we will do the same for analog programming in this chapter. Analog signals are brought to a PLC from sensors through an analog-to-digital (A/D) converter interface, whereas analog output signals are produced by a PLC through a digital-to-analog (D/A) converter interface. All analog signals are connected to the PLC through standard input-output (I/O) modules, which incorporate the A/D and D/A conversions, signal conditioning/scaling, and electronic isolation. Sensors measure and provide small-current or small-voltage signals to the analog input module, whereas actuators receive the output analog signal from the PLC output module. Sensors, actuators, and analog I/O modules are available in different standard formats and can meet the requirement of any process.

7.1 Instrumentation Basics

An important part of a control system is the incorporation of sensors. Sensors translate between the physical real-time world and the standardized world of PLCs. This section will explain the common types of sensors used in process control and the basic operation fundamentals of sensors. It will briefly cover common analog and digital sensors.

7.1.1 Sensors Basics

Sensors translate physical process attributes into values that the I/O modules can accommodate. The translation produces sensor standard output that interfaces with the PLC. In general, most sensors fall into one of the two categories: analog sensors and digital sensors. An analog sensor, such as a thermostat, might be wired into a circuit and calibrated to produce an output that ranges from 0 to 10 V. The analog signal can assume any value within the available range, 0 to 10 V in this case, as defined by the sensor resolution. In order for the PLC to deal with analog signals, those signals must be converted to a digital format in a simple and transparent way. The transformation from analog to digital and digital to analog also must follow predefined universal standards.

Digital sensors generate discrete signals that typically have a stair-step shape in which every signal has a predefined relation with the values preceding and following it. A push-button switch is one of the simplest sensors with two discrete signal values: ON or OFF. Other discrete sensors might provide a binary value in a given range. A stepper-motor position encoder, for example, may provide the motor's current position by sending a 10-bit value with a range from 0 to 1023. In this case, the discrete signal has 1024 possibilities. Much of our discussion assumes a digital/discrete signal to be binary.

7.1.2 Analog Sensors

Analog sensor signals must be converted to a digital format. Sensor output circuits are designed to be connected to an A/D converter port. Most standard microcontrollers and PLCs, such as the Siemens S7-1200 system, have built-in A/D ports in the I/O module interface.

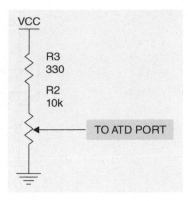

Figure 7.1 Potentiometer sensor wiring diagram.

An often overlooked but extremely popular sensor is the potentiometer. A potentiometer is a resistive sensor. Almost all resistive sensors are wired in a similar fashion as part of a voltage divider. Figure 7.1 shows an example of a potentiometer that is connected to the VCC (assumed to be +10 V in this section) and the GND (assumed to be 0 V). The potentiometer must be carefully selected to ensure adequate current limitation. Notice in this circuit that the current-limiting resistor R3 is connected to limit the output current when the sweep on the potentiometer is turned all the way to the top position. There are two types of potentiometers on the market: audio and linear. A linear potentiometer changes its value at a linear rate. An audio potentiometer changes its value on a logarithmic scale.

Using the resistor values shown in the figure, the voltage ranges the potentiometer will provide can be calculated. With the sweep all the way to the top position, the value for R2 at the top sweep is 10 kΩ. The voltage drop across R2 = VCC × [R2/(R2 + R3)] = 10.0 × [10 kΩ/(10 kΩ + 330)] = 9.68 V. Assuming an A/D converter resolution of 0.01 V, the highest digital value will be 9.68/0.01 = 968. The lowest value should be zero because with the sweep all the way to the bottom, the A/D port will be connected to GND. Thus the limiting resistor has reduced the useful range of the potentiometer to 9.68 V instead of 10 V. The range can be increased by selecting a larger R2. For example, using a 100-kΩ potentiometer means a top sweep voltage across R2 of 10.0 × [100 kΩ/(100 kΩ + 330)] = 9.94 V. Thus the highest digital count value is 9.94/0.01 = 994.

7.1.3 *Digital Sensors*

There are many different types of digital input sensors. Many of them are wired in the same form, which uses a pull-up resistor to force the line voltage high and to limit the amount of current that can flow to the A/D converter circuit. One of the most basic of all sensors is a simple switch. Switches are used to detect limits of motion, proximity to an object, user input, and a whole host of other things.

Switches come in two types: normally open (NO) and normally closed (NC). Many microswitch designs actually have one common terminal, and both an NO

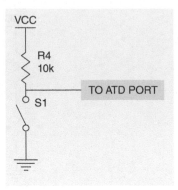

Figure 7.2 NO switch sensor connection.

and an NC terminal. The PLC wiring diagram for a switch is simple, as discussed in Chap. 2. NO switches are recommended to limit the amount of power consumed (Fig. 7.2). With a 10-kΩ pull-up resistor, the amount of current is small, but many switches can add up to some noticeable power.

7.2   Process-Control Elements

A simple process-control loop consists of three elements: the measurement, the controller, and the final control element. Measurement is one of the most important elements in any process-control plant. Decisions made by the controller are based on the real-time measurements information received. Regardless of system type, all controller decisions are similarly based on measurements, control strategy, and the desired process response/performance. Final control elements can refer to actuators such as control valves, heaters, variable-speed drives, solenoids, and dampers. In most chemical process plants, a final control element is often a control valve. In manufacturing assembly lines, final control elements mostly will include variable-speed drives, solenoids, and dampers. Final control elements receive command signals from the controller/PLC in real time to bring about the desired changes in the controlled process. As with sensors/measurement elements, final control devices interface with the PLC output modules in a similar way. The PLC digital-signal outputs are transformed to the actuator-required digital- or analog-signal format, which might require a D/A conversion or coupling isolation.

7.2.1   Basic Measurement System

A basic instrument/measurement system consists of three elements:

- *Transducer/sensor.* The transducer is the part of the measurement system that initially converts the controlled variable into another form suitable for the next stage. In most cases, conversion will be from the actual variable into some form of electrical signal, although there is often an intermediate form, such as pneumatic.

- *Signal conditioning.* In computer process control, signal conditioning is used to adjust the measurement signal to interface properly with the A/D conversion system.

- *Transmitter.* The transmitter has the function of propagating measurement information from the site of measurement to the control room where the control function is to occur. Usually pneumatic or electronic signals are used.

A simplified block diagram of a basic measurement system is shown in Fig. 7.3.

Figure 7.3   Instrument block diagram.

Most modern analog instruments use the following standard signal ranges:

- Electric current of 4 to 20 mA
- Electric voltage of 0 to 10 V
- Pneumatic pressure of 0.2 to 1.0 bars (The bar is a unit of pressure equal to 100 kPa and roughly equal to the atmospheric pressure on earth at sea level.)
- Digital with a built-in binary digital encoder so as to provide a binary digital output

Having a standard instrument range or using digital signals greatly contributed to the advancement of digital process control and the evolution of modern PLCs. The following are a few of the primary advantages of such instruments:

- All instruments can be easily calibrated.
- The signal produced is independent of the physical measurement. For example, the minimum signal (e.g., temperature, speed, force, pressure, acidity, and many other measurements) is represented by 4 mA or 0.2 bars, and the maximum signal is represented by 20 mA or 1.0 bar.
- The same PLC hardware-interface modules are used for all measurements.
- Users can select instruments from a large number of competing vendors; all must comply with universal standards.

7.2.2   *Process-Control Variables*

Process-control variables that are commonly either measured by sensors or regulated through actuators (*final control elements*) include temperature, pressure, speed, flow rate, force, movement, velocity, acceleration, stress, strain, level, depth, mass, weight, density, size, volume, and acidity. Sensors may operate simple ON/OFF switches to indicate certain events or detect objects (proximity switch), empty or full (level switch), hot or cold (thermostat), high or low pressure (pressure switch), and other overload conditions.

German physicist Thomas Johann Seebeck discovered the conversion of temperature differences directly into electricity in 1821. The block diagram of a

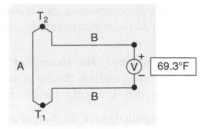

Figure 7.4 Temperature sensor.

temperature sensor is shown in Fig. 7.4. This phenomenon takes place when two wires with dissimilar electrical properties (*A* and *B* wires) are joined at both ends, and one junction is made hot (T_1) and the other cold (T_2). A small electrical voltage is produced proportional to the difference in temperature between the two junctions. This voltage can be calibrated to indicate the measured temperature value, as shown in the figure. A typical industrial temperature probe with a flexible extension and standard plug is shown in Fig. 7.5.

The final or correcting control element is the part of the control system that acts to physically change the process behavior. In most processes, the final control element is a valve used to restrict or cut off fluid flow, pump motors, louvers used to regulate airflow, solenoids, or other devices. Final control elements are typically used to increase or decrease fluid flow. For example, a final control element may regulate the flow of fuel to a burner to control temperature, the flow of a catalyst into a reactor to control a chemical reaction, or the flow of air into a boiler to control boiler combustion. In any control loop, the speed with which a final control element reacts to correct a variable that is out of set point is very important. Many of the technological improvements in final control elements are related to improving their response time.

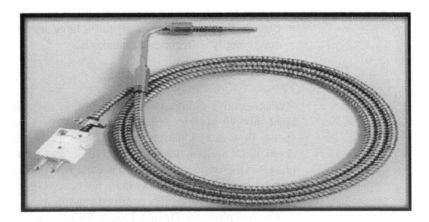

Figure 7.5 Industrial temperature probe.

7.2.3 *Signal Conditioning*

The signal-conditioning element changes the characteristic of the sensor/transducer-measured signal. One example is the square-root extractor. For example, differential-pressure flowmeters/sensors produce an output that is directly proportional to the square of the flow. A signal conditioner/processor is used to extract the square root so that the resulting signal delivered to the transmitter element is directly proportional to the actual flow rate. Other signal-conditioning elements include integrators, differentiators, and a wide variety of signal filters. Filters are used with electrical signals to remove unwanted parts, noise, or interferences. For example, a signal may contain unwanted frequencies or undesired direct-current (dc) components as a result of external sources or the inherent characteristic of the sensor technology. Some signal conditioning is performed to prepare for successful transmission.

7.2.4 *Signal Transmitters*

Industrial sensory information needs to be sent from one place to another using either a wireless or wired transmission medium. Transmitters are used to send the measured and conditioned signals to the controller using a single frequency of transmission or a single pair of wires such as the twisted pair used in wired telephone lines. Modems are simple devices that can receive and transmit information through direct connection or the telephone network. They require a marker signal to let each other know when to receive or transmit. They use "handshaking" to regulate and enforce the predefined communication protocol. A time-division multiplexing (TDM) for digital signals or frequency-division multiplexing (FDM) for analog signals can be used to accommodate four sensory communication channels, as shown in Fig. 7.6.

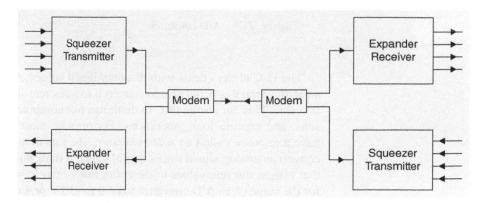

Figure 7.6 Communication frequency-division multiplexor.

7.3 Signal Conversion

The real word is a mix of analog and digital variables combining to describe a process or a system behavior in time. The PLC side uses digital format in both variables measurements and control. This section will cover simplified A/D and

D/A converters. It will also briefly discuss issues associated with these converters, including resolution, sampling rate, and quantification errors.

7.3.1 Analog-to-Digital Conversion

Analog-to-digital (A/D) conversion can be achieved in different ways. One type of converter uses a synchronous counter, as shown in Fig. 7.7. The output of the counter, which is driven by a clock of a fixed frequency, is a digital pattern. This count is converted back into an analog signal by the D/A converter and compared with the input analog signal. Once the two signals match, the counter is disabled by the end-of-conversion signal, and the digital count is latched. The counter stops, and the digital value of the counter output, which is equivalent to the analog signal, is acquired by the PLC/computer-input interface. The higher the number of counter bits, the higher is the resolution/accuracy of the A/D converter and the longer is the average conversion time.

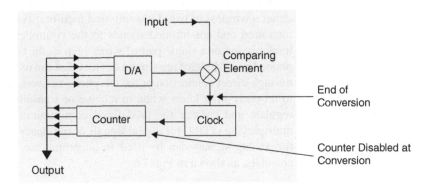

Figure 7.7 A/D converter.

The PLC always deals with discrete/digital values. An important part of using an analog signal is being able to convert it to a discrete signal, such as a 10-bit digital value. This allows the PLC to do things like compute values, perform comparisons, and execute logic operations. Fortunately, most modern PLCs/controllers have a resource called an *A/D converter*. The function of the A/D converter is to convert an analog signal into a digital value. It does this with a mapping function that assigns discrete values to the entire range of voltages or currents. It is typical for the range of an A/D converter to be 0 to +10 V or 4 to 20 mA.

The A/D converter will divide the range of values by the number of discrete combinations. For example, Table 7.1 shows eight samples of an analog signal that have been converted into digital values. The range of the analog signal is 0 to +10.24 V. It is a 10-bit A/D converter, which has 1024 discrete values. Therefore, the A/D converter divides 10.24 V by 1024 to yield approximately 0.01 V per step. The table shows how voltages map to specific conversion values. The values shown only included the first six samples, but the table would continue up to the conversion value of 1023.

Table 7.1   10-Bit A/D Conversion Pattern

From (V)	To (V)	Conversion (decimal)	Conversion (binary)
0.00	0.01	0	0000000000
0.01	0.02	1	0000000001
0.02	0.03	2	0000000010
0.03	0.04	3	0000000011
0.04	0.05	4	0000000100
0.05	0.06	5	0000000101
0.06	0.07	6	0000000110
0.07	0.08	7	0000001000

There are many types of A/D converters on the market. An important feature of A/D converters is their resolution, which is proportionate to the number of bits used by the A/D converter to quantify and store the analog signal samples. An 8-bit converter is used widely on microcontrollers, whereas 12- and 16-bit A/D converters are common in PLCs. A 16-bit A/D converter will use 65,356 discrete values. The resolution required for an application depends on the accuracy the sensor requires and the transient nature of the process. The higher the resolution, the greater is the accuracy of the signal representation in digital format and the lower the quantization error. In this example, the worst-case quantization error is 0.01 V, whereas the average value is 0.005 V.

Example 7.1   A 10-bit A/D converter has a 10-V reference V_r and a digital output count of 0010100111.

a. What is the A/D resolution R in volts per bit?

b. What is the digital output count N in hex for an analog input of 6 V?

c. What is the average quantization error?

Solution

a. $R = V_r/2^n = 10/1024 = 0.0098$ V/bit

b. $N = 2^n \times V_{in}/V_r = 1024 \times 6/10 = 614 = 266$ H

c. $R/2 = 0.0098/2 = 0.0049$ V/bit

7.3.2   *Digital-to-Analog Conversion*

Wikipedia, the free encyclopedia, states that a digital-to-analog (D/A) converter is a device that converts a binary digital code to an analog signal (current or voltage). An A/D converter performs the reverse operation. Signals are easily stored and transmitted in digital form, but a D/A converter is needed for the signal to be recognized by human senses or other nondigital systems. A common use of D/A converters is generation of audio signals from digital information in music players. Digital video signals are converted to analog in televisions and cell phones to display colors and shades. D/A conversion can degrade a signal, so conversion details are normally chosen so that the errors are negligible.

Given the cost and need for matched components, D/A converters are almost exclusively manufactured on integrated circuits (ICs). There are many D/A converter architectures that have different advantages and disadvantages. The suitability of a particular D/A converter for an application is determined by a variety of measurements, including speed and resolution. As documented in the Nyquist and Shannon sampling theorems, a D/A converter can accurately reconstruct the original signal from the sampled data provided that its bandwidth meets certain requirements. Digital sampling introduces quantization error that manifests as low-level noise added to the reconstructed signal. Figure 7.8 shows a simplified functional diagram of an 8-bit D/A converter.

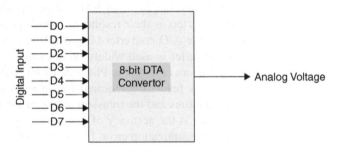

Figure 7.8 Simplified D/A conversion.

PLCs use a wide variety of analog input and output modules that accommodate different current, voltage, and high-speed pulse signals. These modules come in different sizes (the number of I/O points per module) and also A/D versus D/A resolutions. A typical analog I/O module uses 12-bit resolution with signed or unsigned integer representation. The internal operation of the PLC analog modules is independent of the type of physical sensors/actuators interfaced. This simplifies the analog module configuration and its associated application programming/deployment.

Example 7.2

For a 12-bit D/A converter with a 10-V reference voltage that is used to convert digital counts to analog output voltage, answer the following questions:

a. What is the analog output voltage for a digital input = 0A3h H?

b. What is the input digital count N for an analog output voltage of 8 V?

Solution

a. $V_{out} = (N/2^n) \times V_r = (163/4096) \times 10 = 0.398$ V

b. $N = (2^n \times V_{in})/V_r = (4096 \times 8)/10 = 3276$

7.3.3 *Quantification Errors and Resolution*

The resolution of an A/D or a D/A converter indicates the number of discrete values it can produce over the range of analog values. The values are usually stored electronically in fixed-length binary form, so the resolution is usually expressed in bits. The number of discrete values or levels available is a power of 2. For example, an A/D converter with a resolution of 8 bits can encode an analog input to one in 256 different levels because $2^8 = 256$. The values can represent the ranges from 0 to 255 (unsigned integers) or from -128 to 127 (signed integer) depending on the application. Figure 7.9 illustrates the D/A conversion for a 3-bit resolution and a normalized 1-V range. The binary count ranges from 000 to 111, which represents 2^3, or 8, different levels equivalent to the analog range from 0 to 1 V. The least significant bit (LSB) in this example is equivalent to a D/A converter resolution of 0.125 V, which is equal to the worst-case quantization error. The average quantization error in this case is 0.0625 V.

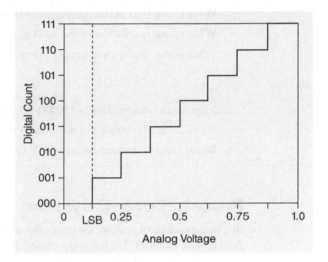

Figure 7.9 Three-bit D/A conversion.

Resolution also can be defined electrically and expressed in volts or current. The minimum change in voltage required to guarantee a change in the output code level is called the *least-significant-bit* (LSB) *voltage*. The resolution R of the A/D converter is equal to the LSB voltage. The voltage resolution of an A/D converter is equal to its overall voltage measurement range divided by the number of discrete voltage intervals, as shown below:

$$R = \text{full-scale range}/N$$

where N is the number of voltage intervals, and full-scale range is the difference between the upper and lower extremes, respectively, of the voltages that can be coded.

The number of voltage intervals is given by

$$N = 2^M$$

where M is the A/D converter's resolution in bits.

Quantization error, also known as *quantization noise*, is the difference between the original analog signal and the digitized binary count. The magnitude of the average quantization error at the sampling instant is equal to half of one LSB voltage. Quantization error is due to the finite resolution of the digital representation of the signal and is an unavoidable imperfection in all types of A/D converters.

Example 7.3   An 8-bit A/D converter with a 10-V reference converts a temperature of 0°C into 00000000 digital outputs. If the temperature transducer outputs 20 mV/degree, answer the following questions:

a. What is the maximum temperature that the converter can measure?

b. What is the resolution of the A/D converter in millivolts per bit?

c. What is the worst-case quantization error in degrees Celsius?

Solution

a. Maximum temperature = 10,000/20 = 500°C

b. Resolution = 10,000/256 = 39.06 mV/bit

c. Worst-case quantization error = 500/256 = 1.952°C/bit

7.4   Process-Control System

In a process-control system, the controller is the element linking the measurement and final control element. Traditionally, closed-loop proportional-integral-derivative (PID) controllers are used. These controllers are designed to execute PID control functions. Other types of control are also common, including ON/OFF and fuzzy logic. Advancement in computer hardware and software has a led to a wide deployment of PLCs and distributed digital control systems as a replacement of the hardwired analog relay controllers. This section focuses on controller types typically implemented using PLCs.

Analog controllers use mechanical, electrical, pneumatic, or other type devices that cause changes in the process through the final control element. The controller moving mechanical parts are subject to wear and tear over time that affect the response/performance of the process. Also, analog controllers regulate the process by continuously providing signals to the final control element. Digital controllers do not have mechanical moving parts. Instead, they use processors to calculate the output based on the measured values. Because they do not have moving parts, they are not susceptible to deterioration with time. Digital controllers do not regulate continuously, but they execute at very high rate, usually several times every second. Pneumatic controllers use instrument air to pass measurement and controller signals instead of electronic signals. These controllers have the disadvantage of longer dead time and lag owing to the compressibility of the instrument air.

7.4.1 Control Process

In the industrial world, the word *process* refers to an interacting set of operations that lead to the manufacture or development of some products. In the chemical industry, process means the operations necessary to take an assemblage of raw materials and cause them to react in some prescribed fashion to produce a desired end product, such as gasoline. In the food industry, process means to take raw materials and operate on them in such a manner that an edible product results. In each use, and in all other cases in the process industries, the end product must have certain specified properties that depend on the conditions of the reactions and operations that produce them. The word *control* is used to describe the steps necessary to ensure that the conditions produce the correct properties in the product.

A process, as shown in Fig. 7.10, can be described by an equation. The process has m variables v_1 to v_m. Suppose that we let a product be defined by a set of properties $P_1, P_2, \ldots, P_n$. Each of these properties must have a certain value for the product to be correct. Examples of properties are things such as color, density, chemical composition, and size.

$$P_i = f(v_1, v_2, \ldots, v_m, t)$$

where P_i is the ith property and t is time.

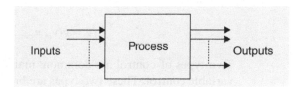

Figure 7.10   Process block diagram.

7.4.2 Controlled Variables

To produce a product with the specified properties, some or all the variables must be maintained at specific values. Figure 7.11 shows an unregulated tank with open flow coming in and out. Fluid level in the tank is expected to vary as long as a difference in flow in and out exists. A controlled/controlling variable can be used to exert control over this simple tank process. Choices of such a variable can include the flow rate in or flow rate out of the tank. The controlled variable must be accessible and easy to change. It also must have adequate influence on the regulation of the selected control variable, the tank level in this case.

Some of the variables in a process may exhibit the property of self-regulation, whereby they will naturally maintain a certain value under normal conditions and

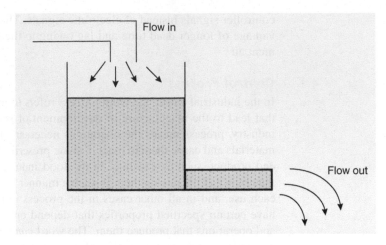

Figure 7.11 Unregulated tank process.

very small disturbances. Control of variables is necessary to maintain the properties of the product within specification.

7.4.3 *Control Strategy and Types*

The value of a variable v_i actually depends on other variables in the process, as well as on time. Typically, one or a few variables dominate and define the dependency relationship. This relation can be expressed as shown in the following equation:

$$v_j = g(v_1, v_2, \ldots, v_c, \ldots, v_m, t)$$

Two types of control are common: mainly the single-variable control and multivariable control. These two types are briefly defined below.

Single-Variable Control

We will demonstrate single-variable control using the tank process with minor modification. Two valves are added, one at the inlet flow to the tank and one at the output. The level in the tank varies as a function of the flow rate through the input valve and the flow rate through the output valve. The level in the tank is the control variable, which can be measured and regulated through either inlet- or output-valve control and adjustment. Figure 7.12 illustrates the modified tank process, which can accommodate the implementation of single-variable control. Only one of the two valves available in this mode can be selected as a controlling variable to regulate the tank level, the control variable. The next mode of control, multivariable control, is more complex and can use more than one controlling variable to regulate the control variable, the tank level in our process. For example, regulation of the two valves can be used to control the tank level at a desired set of values at different times.

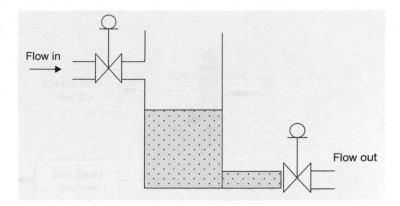

Figure 7.12 Regulated tank process.

Multivariable Control

Figure 7.13 shows a schematic diagram of an oven used to bake crackers in a system under multivariable control. The control variables that may be used include the feed rate, conveyor speed, oven temperature, cracker color, and cracker size. Other variables such as the temperature outside the oven are difficult to measure, control, and use in the control-system strategy. Multivariable control is more complex because of the strong and nonlinear interactions between variables.

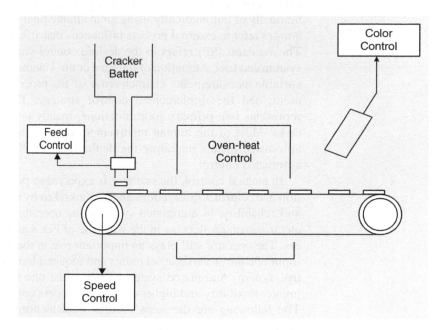

Figure 7.13 Multivariable oven control.

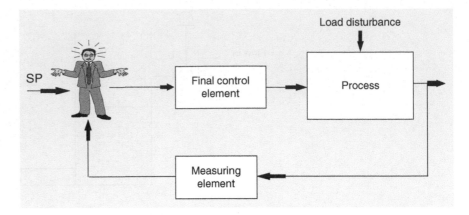

Figure 7.14 Process-control loop.

7.4.4 *Process-Control Loop*

Process-control loops are the core of all control systems and process-automation tasks. A schematic of a typical control loop is shown in Fig. 7.14. The loop consists of three blocks: process, measuring elements, and final control element. What comes from the process is what we referred to as the *control variable*, the tank level in the preceding example, which can be easily measured and quantified in time. What goes as input to the process is the *controlling variable*, the inlet-valve opening or flow rate in the tank process. The valve opening can be changed manually or automatically using an available final control element. *Load disturbances* refer to external process influences that affect the behavior of the process. The *set point* (SP) refers to the desired control-variable value as defined by the system end user. Actuation of the final control element is based on the SP, control-variable measurements characteristic of the process, type of final control element, and the implemented control strategy. The man shown in the loop represents two primary loop functions, mainly error detection and the control tasks. Most of the human involvement in the control loop can be replaced by automated means including the deployment of PLCs, PCs, and other types of automated control.

In manual control, the operator is expected to perform the task of error detection and control. Observations and actions taken by operators can lack consistency and reliability. In automated systems, the operator is removed and replaced by electronic controllers, as in the wide use of PLCs and specialized digital computers. The operator still plays an important role in the automated system, including input of control-variable set points and required human interactions with the control system. Automated systems such as the one shown in Fig. 7.15 allow for greater flexibility and higher degrees of process control, quality, and consistency. The following are the steps used in implementing single-variable closed-loop control:

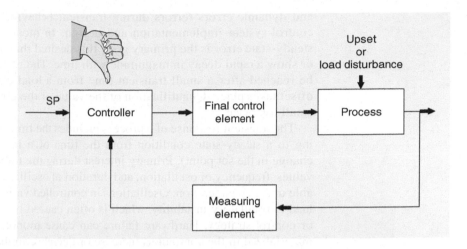

Figure 7.15 Automated process-control loop.

1. Select one variable to be a controlled variable (the controlled variable in this case is labeled v_c).
2. Make a measurement of the controlled variable v_c to determine its present value. The desired value is called the *set point* of the controlled variable.
3. Compare the measured value of the controlled variable with the desired value (set point).
4. Determine a change in the controlling variable that will correct any deviation or error in the controlled variable.
5. Feed back this changed value of the controlling variable to the process through the final control element to create the desired correction in the controlled variable.
6. Go to step 2 and repeat.

7.4.5 *Control-System Error Quantification*

Perfect regulation of a process variable by any control system is not possible. Errors can be measured in three ways:

- *Variable value.* Set point = 230°C; measured value = 220°C; range = 200 to 250°C; and Error = 10°C.
- *Percent of set point.* The error is expressed as a fraction or percent of the controlled variable SP. Error = $(10/230) \times 100 = 4.4\%$.
- *Percent of range.* The error is expressed as a fraction or percent of the controlled variable range. Error = $[10/(250 - 200)] \times 100 = 20\%$.

Two types of errors are of great importance in system performance, as observed in all selected control variables: mainly the steady-state residual errors and the transient dynamic of such errors. Both residual (the steady-state error) and dynamic errors (errors during transient behavior) are used to evaluate a control-system implementation and design. In most control applications, the steady-state error is the primary goal. It is desired that this error would be small or show a rapid decay in magnitude with time. The residual error is expected to be reached after a small transient time from a load change or when a system offset takes place. Quantification of the value of the error is always subject to a small tolerance.

The transient response of a process includes the time interval prior to and leading to a steady-state condition from the time of a load change or an offset (a change in the set point). Primary interest during the transient time includes error values, frequency of oscillation, and duration of oscillation for the controlled variable under consideration. Oscillations in controlled variables are expected but can lead to oscillatory instability, which is often caused by an ill-designed controller or control strategy. Hardware failure can cause monotonic instability, which is materialized in the continuous increase or decrease in the value of the controlled variable resulting in an overall system failure. For example, if the transmitter for the tank-level signal fails, it can cause the level to increase and eventually overflow the tank. The tank can run dry if the last correct measurement before the transmitter failure required a decrease in tank level. This type of failure can be detected and prevented by using backup hardware such as limit switches and redundant sensors.

7.4.6 *Control-System Transient and Performance Evaluation*

The quality of a control-system performance is based on many factors, including transient response, steady-state errors, stability, scalability, user interface, continuous quality improvements, and ease of maintenance. Controller response to errors depends on the control-system strategy. The objective of a control system is to minimize, not to eliminate, the error without affecting the overall system stability and performance.

As stated in Sec. 7.4.5, perfect regulation of a process variable by any control system is not possible, which means that we have to live with some errors, hopefully small. Control strategy and adequate tuning of existing control loops are very critical to the performance of the whole system. This task is one of the most challenging, and it can be easier to perform if the error behavior is well understood.

Figure 7.16 shows an ill-designed control-system behavior or one completely out of tuning. As shown in the figure, the controlled variable has gone wild, with increasing amplitude oscillations ultimately leading to instability and system shutdown. This type of behavior is called *oscillatory instability* and can be eliminated during implementation of the controller and overall strategy.

As stated in Sec. 7.5.1, a tolerance around the set point where no controller action is needed is essential to protect the final control element (actuator) from failure owing to excessive control actions. This is true in all kinds of process regulations: manual, PID, or ON/OFF control. Figure 7.17 shows the transient behavior of an overdamped process where T_D is the total delay/transient time. There are no overshoots or oscillation in this case, but the controller action is slow. Figure 7.18 shows a faster controller for an underdamped

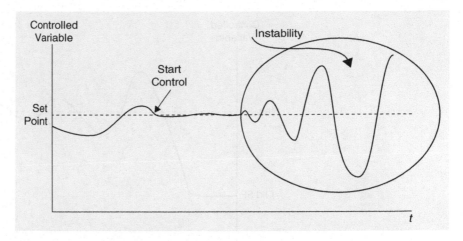

Figure 7.16 Oscillatory instability.

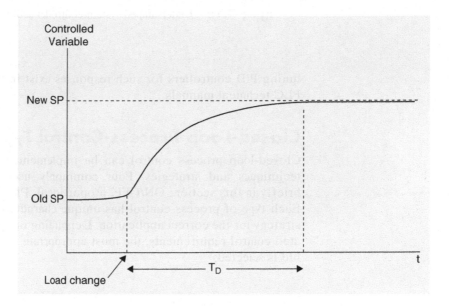

Figure 7.17 Overdamped system-controller action.

process resulting in controlled variable oscillations and shorter transient/delay time T_D. Configuring the controller to be very aggressive in reacting to real-time controlled-variable errors can lead to instability and the situation shown in Fig. 7.16. The most desired controlled-variable response is known as *quarter decay*, which means that the ratio of consecutive overshoots is approximately 4, which represents a 0.25 decay ratio. Guidelines and techniques for

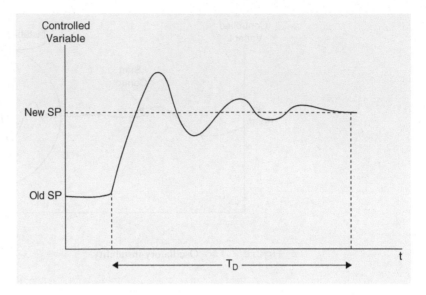

Figure 7.18 Undersamped system-controller action.

tuning PID controllers for such responses exist in the literature and in most PLC technical manuals.

Closed-Loop Process-Control Types

Closed-loop process control can be implemented using a wide variety of techniques and strategies. Four commonly used techniques are covered briefly in this section: ON/OFF, proportional, PID, and supervisory control. Each type of process control has unique characteristics and can be the best strategy for the correct application. Depending on the process and the associated control requirements, the most appropriate and simple technique possible is selected.

ON/OFF Control Mode

ON/OFF control is the simplest mode of closed-loop control, but it is a good match for many applications. The final control element is either ON or OFF depending on the controlled-variable measured value. A *dead band* (a tolerance around the set point where no control action is needed) is used to prevent the final control element from being switched ON/OFF excessively. The controller output is ON if the error value is greater than the set point $+ \varepsilon$ and is OFF if the error value drops below the set point $- \varepsilon$, where ε is one-half the dead band. No control action is needed while the controlled-variable measurement lies within the implemented dead band (DB). Figure 7.19 illustrates the general behavior of an ON/OFF control action.

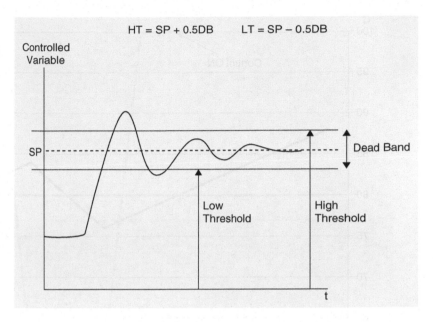

HT = SP + 0.5DB   LT = SP − 0.5DB

Figure 7.19   ON/OFF controller action.

Example 7.4   Assume a temperature cooling control process with a set point of 80°C and a dead band of 6°C. The system cools at −2°C/min once the controller output is ON. The system heats at +4°C/min when the output is OFF. The ON/OFF controller response is shown in Fig. 7.20. Control action is ON while the temperature is higher than the upper threshold of the dead band (83°C) and turns OFF once the temperature dips below the low threshold of the dead band (77°C).

The figure shows two functions of time: the measured temperature variation and the controller output. The first function is continuous in time, whereas the second is discrete—the controller is either ON or OFF. Notice that the control action only takes place while measured temperature is outside the specified dead band. More precise control can be achieved, if needed, using other, more sophisticated techniques.

7.5.2   *Proportional Control Mode*

Proportional control mode assumes a correction strategy based on the calculated error, the difference between the measured controlled variable and the set point. The controller output (controlling variable) is proportional to the amount of error in addition to the fixed amount needed to support the process during the time the control variable stays within the defined dead band. This amount is known as the *controller output with zero error*. The following is the mathematical formulation for the proportional control mode:

$$Cp = Kp \times Ep + Co$$

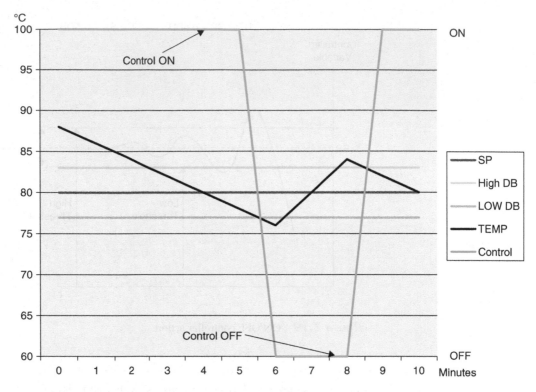

Figure 7.20 ON/OFF temperature control.

where Cp = controller output in percent
Kp = proportional gain in percent of output/percent of error
Ep = error in percent of range
Co = controller output with zero error

Example 7.5   A control system is to control pressure in a range from 120 to 240 lb/in² with a 180 lb/in² set point. If the proportional gain is 2.5 percent and the zero-error output is 65 percent, the error as a percentage of range is given by the following formulation:

$$Ep = (P - 180)/(240 - 120) \times 100$$

$$= 0.833 \times (P - 180)$$

where P is the measured pressure. The controller output is given by the following formulation:

$$Cp = 2.5 \times Ep + 65$$

$$= 2.5 \times 0.833 \times (P - 180) + 65$$

$$= 2.0825 \times (P - 180) + 65$$

The controller output varies from 0 to 100 percent, which corresponds to an error range defined as in the following formulation:

$$Ep \text{ (at } Cp = 0) = -26\% \qquad \text{and} \qquad Ep \text{ (at } Cp = 100\%) = 14\%$$

The range of errors covering the entire available controller output is known as the *controller proportional band*, which is calculated in this example as follows:

$$\text{Proportional band} = 14\% - (-26\%) = 40\%$$

Notice that the higher the proportional band, the lower is the controller proportional gain because the following relation is true at all times:

$$\text{Proportional band} \times \text{proportional gain} = 1$$

This type of control is more complex than ON/OFF control because it requires experience in implementing and tuning both the proportional gain and the controller zero-error output.

7.5.3 Composite Control Mode

Closed-loop process controllers can be designed to respond to the history of error during a prespecified time period (integral mode), the forecast of the error behavior in the near future (derivative mode), and the current instantaneous value of errors (proportional mode). These controllers are commonly labeled in the following three types:

- Proportional-integral (PI) mode
- Proportional-derivative (PD) mode
- Proportional-integral-derivative (PID) mode

Figure 7.21 shows a simplified schematic of PID control while the AUTO/ MANUAL mode switch is placed on AUTO, which is the universal format for the composite controller. All composite controllers must include the proportional mode.

7.5.4 PLC/Distributed Computer Supervisory Control

The initial use of computers in process control was in support of the traditional analog-system process control. This type of application of computers still exists because many industries use analog control systems and will no doubt continue to do so. Generally, large- or medium-scale computers provide the support activities, including data acquisition, human interface, simulation/modeling, communication, and digital control. A simplified block diagram for a supervisory control system is shown in Fig. 7.22.

Distributed computer control/supervisory control has evolved as the choice for automation and process-control implementation in the past 20 years. This revolution

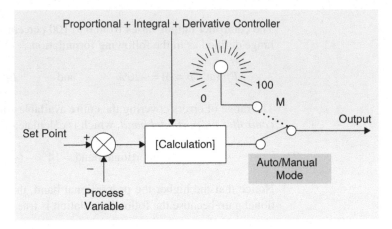

Figure 7.21   PID composite controllers.

greatly benefited from and made use of huge advancements in technology, including universal standards, digital hardware, real-time operating systems, communication and networking, human-machine interfaces (HMIs), remote sensing, sensory fusion, redundancy and safety tools, and the widespread use of open-system architectures. Some of the world's largest international chemical and petroleum corporations own and operate the largest global computer networks. Every process controller, data-acquisition system, HMI, actuator, sensor, and communication device has to be accessible from any location in the world with proper authorization.

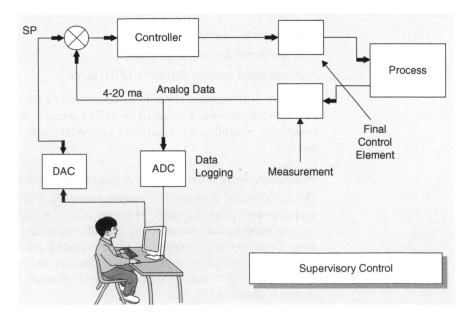

Figure 7.22   Supervisory control.

Supervisory distributed control allows for a more efficient modular design and far easier overall system cost in all phases of system development, implementation, deployment, enhancement, expansion/scalability, and maintainability. It also allows for greater operator, designer, and user interactions, leading to the realization of an overall system of continuous quality improvement. Most large systems are composed of several highly interactive and connected subsystems. The set point for a given single-variable closed-loop control might be a function of other variables belonging to different subsystems. Distributed control systems demand larger and more complex development with higher cost, but they are more effective and less expensive in the long term.

Homework Problems and Laboratory Project

Problems

7.1 Define the following terms:
   a. Signal conditioner
   b. Transmitter
   c. Multiplexors
   d. Modems
   e. Quantization error

7.2 What is the difference between the following?
   a. Digital sensors and analog sensors
   b. Sensors and actuators

7.3 Which statement is *true* about instruments?
   a. All instruments can be easily calibrated.
   b. All instruments produce a signal that is independent of the physical measure.
   c. Users can select instruments from a large number of competing vendors; all must comply with universal standards.
   d. All of the above.

7.4 What are the elements of a basic measurement system?

7.5 What does the term *Seebeck effect* refer to?

7.6 Draw the basic elements of a process-control loop, and describe the function of each element.

7.7 List two standard analog signals.

7.8 For an 8-bit analog input module (A/D), what is the range of values it can represent (signed/unsigned)?

7.9 What does multivariable control mean?

7.10 What methods are used in the process-control loop to provide optimal control to a process-control system?

7.11 Define the following terms:
   a. Data logging
   b. Digital process control
   c. Supervisory control
   d. A/D converter
   e. D/A converter

7.12 For which mode of control is the rate of change of the controller output determined by the amount of error, ON/OFF, proportional, derivative, or integral control?

7.13   For which mode of control is the number of outputs determined by the rate of change of the error, ON/OFF, proportional, derivative, or integral?

7.14   Explain what the word *process load* means, and give an example.

7.15   Which of the following statements is *true* in proportional control?
   a. Proportional control usually produces zero error when stability is reached after a change in the load.
   b. Proportional control usually produces an offset when stability is reached after a change in the load.

7.16   Define the following terms:
   a. Process transient time
   b. Process load time
   c. Process regulation time
   d. Process lag time

7.17   A temperature sensor is used to measure an oven temperature between 50 and 300°F. The output of the sensor is converted to a signal in the range of 0 to 5 V. The signal is connected to a 12-bit A/D converter of a PLC system. Answer the following questions:
   a. What is the sensor resolution in °F/volt?
   b. What is the A/D converter resolution in °F/bit?
   c. What A/D converter digital count corresponds to 100°F?
   d. What is the resolution of the A/D converter?
   e. Calculate the average quantization error of the A/D converter.

7.18   A sensor provides real-time temperature measurements of an oven in the range of 0 to 10 V, which represents engineering unit values of 50 to 400°F. The oven temperature is desired to maintain a 200°F set point by sending an ON/OFF signal to the oven heater. The dead band allowed for the ON/OFF oven control is 2°F. Calculate the high and low thresholds around the dead band.

7.19   In a home heating system, what is the effect of increasing or decreasing the dead band?

7.20   What is the advantage of set-point automatic generation in supervisory control systems?

7.21   What is the proportional band of a temperature controller having a 0.75 proportional gain and a set point of 300°F?

7.22   A 12-bit A/D converter with 10 V reference has an input signal of 2.69 V. What is the digital count for this input signal? What is the equivalent analog input signal to a 3A5 hex digital count?

7.23   A sensor provides real-time temperature measurements of an oven in the range of 4 to 20 mA, which represents engineering unit values of 40 to 350°C. An analog input module is interfaced with the sensor output. The oven temperature is desired to maintain a 250°C set point using ON/OFF control and an oven heater. The dead band allowed for the ON/OFF oven control is 4°C. Answer the following questions:
   a. What is the sensor resolution in °C/mA?
   b. What is the A/D converter resolution in °C/bit?
   c. What is the digital count of the A/D converter for a 159°C measurement?
   d. What is the resolution of the A/D converter?
   e. Calculate the maximum quantization error of the A/D converter.

7.24   A sensor provides real-time measurement of a tank level in the range of 4 to 20 mA, which represents engineering unit values of 20 to 500 m. An analog input module with 12-bit resolution is used to acquire this signal. What range of level does a $(256)_{10}$ count represent?

Project

Laboratory 7.1: ON/OFF Temperature Control

The objective of this laboratory is to get hands-on knowledge of the ON-OFF process control in commercial and industrial application.

Process Description
A sensor provides real-time temperature measurement of an oven in the range of 0 to 10 V, which represents engineering unit values of 50 to 400°F. The oven temperature is desired to maintain a 300°F set point by sending an ON/OFF signal to the oven heater. The dead band allowed for the ON/OFF oven control is 1°F.

Implementation Specifications

* Configure the PLC analog module attached to the CPU for a 0- to 10-V input signal range.
* Use a 10-V potentiometer to supply the analog input signal (0 to 10 V). Apply the signal to the analog input module connected to the main CPU of the training unit.
* Change the potentiometer setting in the range from 0 to 10 V to represent an oven temperature in engineering units in the range of 50 to 400°F.
* Configure a new HMI Function Keys page. This page will allow the operator to access two other pages: Status and Control pages. Define this page as the HMI start page.
* Configure a Status page in the HMI to display the oven temperature (in °F) as you change the potentiometer from the minimum to the maximum voltage (0 to 10 V), representing the engineering units (50 to 400°F). Also, define two text objects: Heater ON and Heater OFF. The function key F1 should allow users to go back to the Function Keys page.
* Configure a Control page in the HMI to allow an operator to enter the oven set point in the range of 50 to 400°F.

Laboratory Requirements

* Assign the system inputs.
* Assign the system outputs.
* Program the required networks.
* Download the program, and go online.
* Simulate the program using the training units or the Siemens simulator. Configure the Watch table to display the raw input digital count (0 to 27,648), temperature in engineering units (50 to 400°F), and the 12-bit analog input/digital count (0 to 4095) as the potentiometer setting is adjusted from 0 to 10 V.
* Simulate the program using the HMI. Verify that the program is running according to the process description.

Laboratory Modifications
Document all the networks shown in Fig. 7.23. Add the network needed to validate that the operator-entered set point is within limits (50 to 400°F). If the set point is outside the limit, send a message to the HMI: "Invalid Set Point! Re-Enter." Implement the HMI requirements, and make the needed modifications to the S7-1200 PLC ladder program.

Network 1:

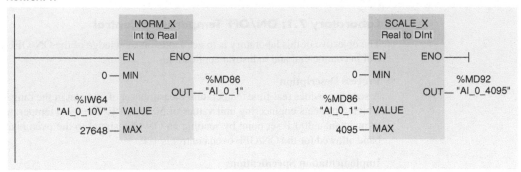

Network 2:

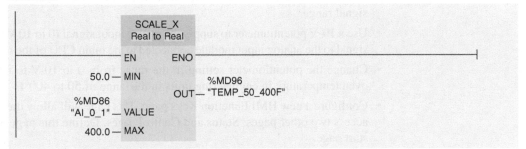

Network 3:

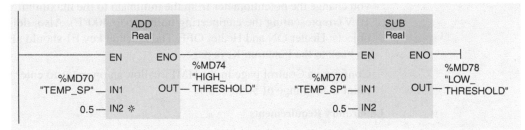

Network 4:

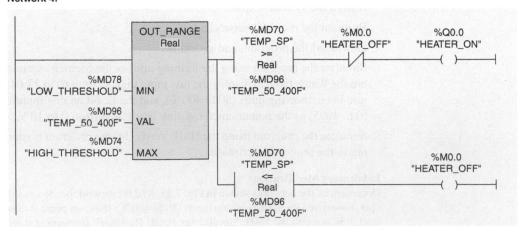

Figure 7.23 ON/OFF temperature control.

8

Analog Programming and Advanced Control

This chapter will examine the process of interfacing analog input and output variables to the Siemens S7-1200 processor. It will cover the fundamentals of analog I/O programming and its use in process control. An advanced industrial process-control implementation will be briefly discussed.

Chapter Objectives

▶ Perform analog module configuration and diagnostic.

▶ Perform and debug Siemens S7 analog I/O programming.

▶ Perform Siemens S7 PID configuration/programming.

▶ Understand PID control structure and performance.

Programmable logic controllers (PLCs) are the result of evolution of hardwired analog control systems. *Analog* is a keyword in the world of process control and automation. Most physical entities we would like to control in the real world are analog in nature. These include variables such as temperature, pressure, speed, acidity, position, level, flow rate, viscosity, displacement, weight, frequency, and many others. All can be measured and quantified in time, and some exhibit self-regulation ability. Self-regulation is the ability of a control variable to make adjustments and arrive at a stable state under small disturbances, which is a desired characteristic in the selected control variables. This chapter examines the process of interfacing analog input and output variables with the Siemens S7-1200 processor. It will cover the fundamentals of analog input-output (I/O) programming and its use in process control. Advanced process-control techniques will be discussed with a small industrial implementation.

8.1   Analog I/O Configuration and Programming

Standard modules are available off the shelf and can be interfaced with the Siemens S7-1200 PLC. The CPU used in this chapter has an analog input module with two inputs and an analog output module with one output. One of the available analog inputs and the analog output are configured and used in our coverage of analog programming and the implementation of closed-loop process control. Some of the Siemens S7-1200 typical analog I/O modules will be listed and followed by the steps to configure the module, scale the I/O, and program it.

8.1.1   Analog I/O Modules

The following is a brief list of typical analog I/O modules:

SB 1232 Analog Output (AQ). The S7-1200 Analog Output Module 6ES7 232-4HA30-OXB0 has one output with 12-bit resolution and can plug into the front of the CPU. It can be configured for ±10 V or 0 to 20 mA. This module is the one used and referenced in this chapter's programming examples.

SM 1231 Analog Input (AI). The S7-1200 Analog Input Modules are of two types: SM1231AI 4 (13-bit), which has four inputs with 12-bit resolution + sign bit, and SM1231AI 8 (13-bit), which has eight inputs with 12-bit resolution + sign bit and full-scale range −27,648 to 27,648. One AI4 module with two analog inputs is embedded in the CPU module.

SM 1232 Analog Output Module (AQ). The Analog Output Modules are also of two types, SM1232AQ 2 (13-bit), which has two outputs, and SM 1232AQ 4 (13-bit), which has four outputs with 12-bit resolution + sign bit and full-scale range −27,648 to 27,648.

SM 1234 Analog Input Output Module (AI/AQ). The Analog Input Output Module is SM 1234 AI 4 (13-bit)/AQ 2 (14-bit), which has four inputs with 12-bit resolution plus one sign bit and two outputs with 13-bit resolution plus one sign bit.

Range Selection for Analog I/O Modules

Given the many analog devices available and the wide variety of applications, the analog I/O modules come with a variety of signals to accommodate voltage or current (±10 V, ±5 V, ±2.5 V, or 0 to 20 mA).

8.1.2   *Configuring Analog I/O Modules*

Signal modules including analog input and output can be added to the PLC processor. Figure 8.1 shows the device-configuration screen. To start the configuration process, open Device Configuration for the PLC. Add your module, as shown in Fig. 8.2, and then configure each channel's properties, as illustrated in Fig. 8.3. Note that for current-based signals, the only range available is 0 to 20 mA. If your signal is 4 to 20 mA, still configure it as 0 to 20 mA, and scale the input value as demonstrated in the figures.

Configuring the channel property for an analog input module to ±10 V will include a few menu clicks. Under the Project View screen, click on Device Configuration, and then add your module. Configure each channel's properties assuming the ±10-V format.

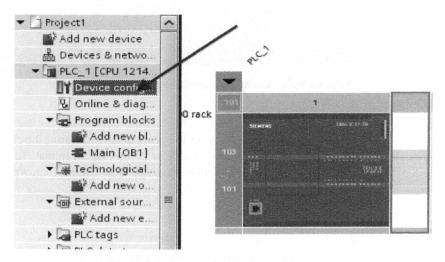

Figure 8.1   Device configuration.

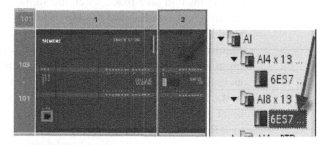

Figure 8.2   Adding an analog input module.

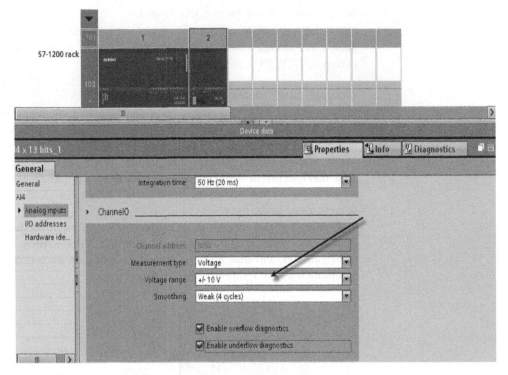

Figure 8.3 Configuring analog input properties (±10 V).

Figure 8.4 shows the process of configuring the channel property for an analog input module using a 0- to 20-mA signal format. To configure the current analog input module, follow the same steps used earlier, which are marked on the screen using arrows.

8.1.3   *Analog I/O Diagnostic Configuration*

Analog module diagnostic configuration initiates the protocol for handling signal exceptions during execution of the process-control program. Figure 8.5 shows the screen for enabling the analog output module overflow diagnostics. Overflow and underflow can take place when the signal value exceeds the defined range, below the minimum and above the maximum values. As shown in the figure, click the Enable overflow and the Enable underflow in the Diagnostics checkbox. Sensors require frequent calibrations to maintain accurate input signals to the PLC analog input module. Occasionally, an out-of-calibration sensor can cause an overflow or an underflow. Enabling overflow/underflow will automatically trigger such events to the ladder program. The program also can manually check for out-of-range and reset values to the predefined limits.

The power circuit of an analog output must provide continuity in order to maintain current flow. Wire breaks cause the interruption of current flow to the analog input or from the analog output modules. Short circuits also will cause

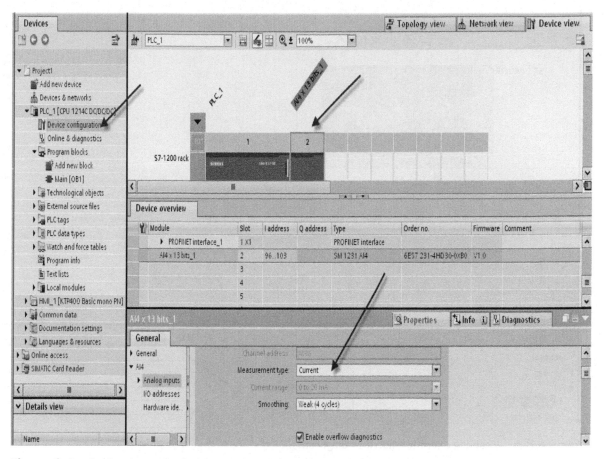

Figure 8.4 Configuring analog input properties (0 to 20 mA).

analog signal interruption. Wire breaks, exceeding the limit, and short circuits all will cause the following S7-1200 PLC light-emitting diodes (LEDs) to blink (color of blinking LEDs: red):

- The ERROR LED on the CPU
- The DIAG LED if it is a signal module
- The LED of the associated channel

Enabling the broken-wire diagnostic will require the following steps shown in Fig. 8.6:

- Browse the Project Tree for the Device configuration of your S7-1200 PLC.
- Click on your signal module in the Device configuration window.
- Select the Properties tab, and click AI4/AO2 in the navigation area.
- Scroll down to the channel you want to monitor.
- Choose the Current analog output type.
- Click the Enable wire-break diagnostics checkbox.

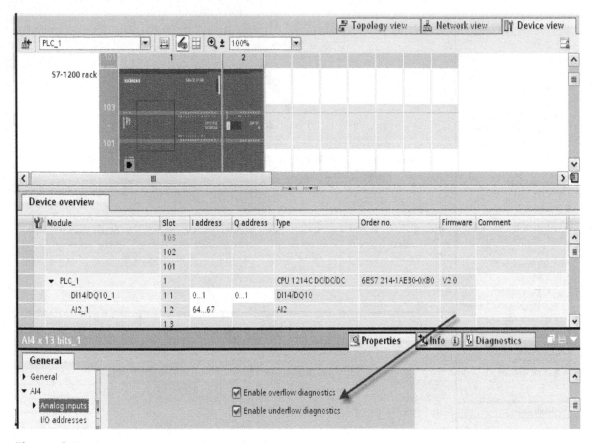

Figure 8.5 Enabling overflow/underflow diagnostics.

Enabling short-circuit diagnostics will require the following steps shown in Fig. 8.7:

- Browse the Project Tree for the Device configuration of your S7-1200 PLC.
- Click on your signal module in the Device configuration window.
- Select the Properties tab, and click AI4/AO2 in the navigation area.
- Scroll down to the channel you want to monitor.
- Choose the Voltage analog output type.
- Click the Enable short-circuit diagnostics checkbox.

8.1.4 Analog Input Range and Scaling

The low limit of an assumed 16-bit (1 implicit sign bit) analog I/O module is exceeded once the count value is less than 0 for current and less than or equal to −4865 for voltage. The high limit is exceeded once the count value is more than or equal to 32,512. The analog input and output signals can be scaled to 0- to

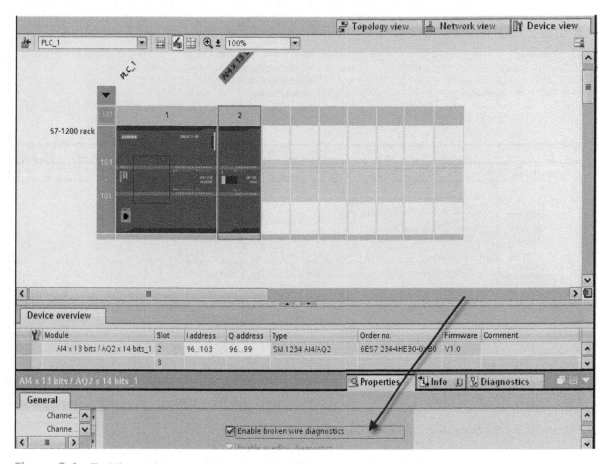

Figure 8.6 Enabling analog output broken-wire diagnostics.

20-mA input and output signals using the predefined global library functions `Scale_current_input` and `Scale_current_output` or by using the SCALE instruction in the Siemens ladder software. Figure 8.8 graphs the scaling operation. The actual count values used by the Siemens analog modules are listed below:

Decimal range = 0 to 27,648

Overflow = 32,512 to 32,767

Undershoot = 0 to −4864

Underflow = −4865 to −32,768

The preceding figure maps the 4 mA to a 0 count and the 20 mA to 27,648 count using a linear scaling function. The mapping eliminates an area labeled "Wire break" from 0 to 4 mA. Figure 8.9 shows the initialization of the watch table tags for monitoring analog variables. The display format can be adjusted by selecting the desired format. The monitor icon at the top of the screen,

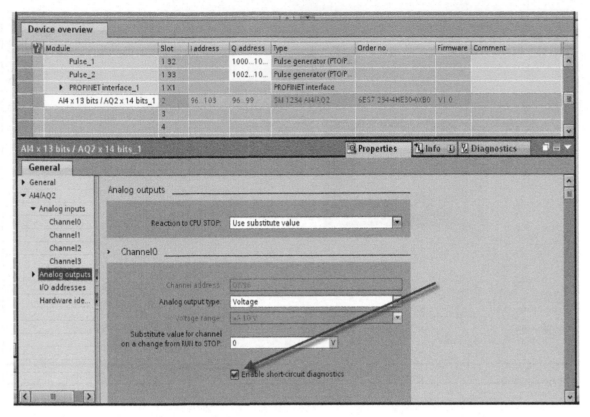

Figure 8.7 Enabling short-circuit diagnostics.

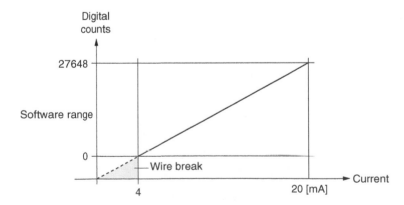

Figure 8.8 Scaling 4- to 20-mA signal count.

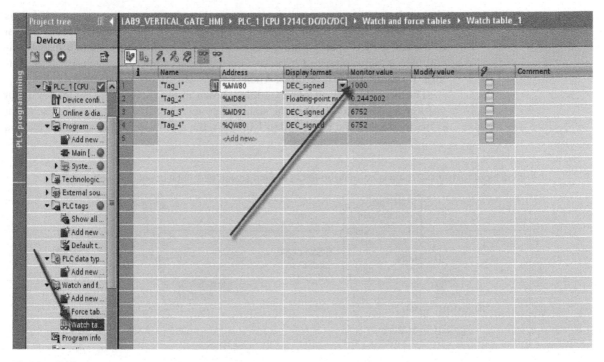

Figure 8.9 Monitoring analog signal count.

shown shaded, must be activated to start the real-time monitoring. All monitored or forced values must have a prior definition in the PLC tags with the correct addresses and the associated matching formats. The software will validate inputs and prompt the user to make the needed corrections.

8.1.5 *Analog I/O Programming*

This section demonstrates fundamental issues associated with analog I/O programming. Two analog signals are configured and used in all discussions; both are connected to the main CPU module. Configuration is not required for the two analog inputs and the analog output connected to the processor module. Other analog modules must be added and configured before use. AI2_1 is the analog input module (available on the processor module) used with two channels, each using two memory words (starting at IW64 and IW66). Each channel is coded in 16 bits using the format covered in the preceding section. Our programming will make use of the first analog input channel, IW64-IW65. AQ2_1 is the analog output module used with one channel employing two memory words (starting at QW80). The output channel is coded in 16 bits using the same format used by the analog input signal. The SCALE and NORM instructions covered in Chap. 4 will be used repeatedly in the next analog programming example. The reader is advised to review and practice these instructions before proceeding.

The next example uses a 10-V potentiometer to supply the analog input signal. The analog output is interfaced with a small voltmeter. The potentiometer setting is changed, and the corresponding changes in the voltmeter reading are recorded. This setting is very simple to implement in the laboratory or even at home, but it is typical of the actual PLC analog setting and interfacing in a more elaborate industrial application. No high-voltage or excessive-current devices/interfaces are used in this setting. Precautions must be taken when dealing with high-voltage/current circuits.

The first network of our example, shown in Fig. 8.10, reads an analog input voltage signal (0 to 10 V) in the standard count range from 0 to 27,648. The count is normalized in the range 0.0 to 1.0. Next, the normalized value is scaled back in

Analog input (0-10V/0-27648) is normalized (0.0-1.0) then scaled to a 12-bit count (0-4095).

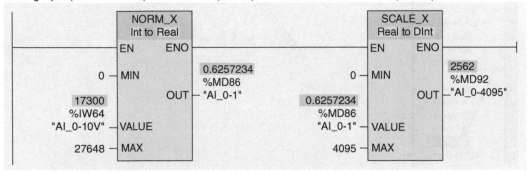

The standard (A 12-bit AI module is used) analog input count is halfed to produce the output count.

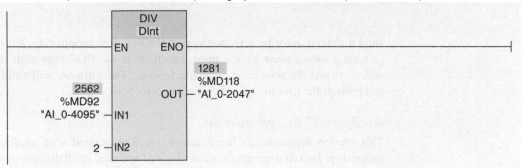

The output count is normalized and scaled to produce the standard analog output signal (0-10V/0-27648).

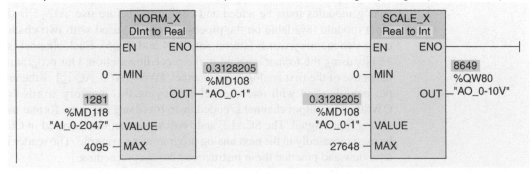

Figure 8.10   Analog I/O programming.

the range from 0 to 4095. Three tags are used in this network: `AI_0-10V`, the raw analog input (IW64); `AI_0-1`, the normalized intermediate tag (MD86), and `AI_0-4095`, the 12-bit scaled count (MD92). Only IW64 is the physical I/O address among the three used.

The second network produces half the value of Network 1 output count and stores it in the `AI_0-2047` tag. The snapshot shown for Networks 1 and 2 is taken at an input analog count of 17,300, which is approximately 6.26 V. Network 3 normalizes the count produced by Network 2 and then scales it in the range from 0 to 27,648. The newly scaled value is stored in the `AQ_0-10V` physical output tag (QW80).

Figure 8.11 shows the PLC watch table snapshots for the six tags used in our analog I/O example. The first screen shows a full watch table with all used fields highlighted and pointed to by an arrow. The PLC Watch and Force table is an

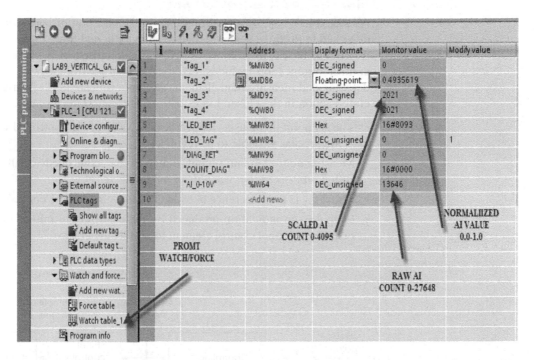

Figure 8.11 PLC Watch and Force table snapshots.

important tool frequently used during the process-control software development, implementation, and checkout.

8.2 PID Control Configuration and Programming

Proportional-integral-derivative (PID) closed-loop control is one of the most widely used process-control techniques. It reacts to process errors or deviations from desired behavior using a strategy based on knowledge about error values, the history of error, and the future forecast of errors. This section will cover PID closed-loop control with practical aspects of its implementation in industrial automation.

8.2.1 Closed-Loop Control System

As stated in Chap. 1, a *control system* is a collection of hardware and software designed to produce desired process behavior in real time. Figure 8.12 shows the functional block diagram for single-variable closed-loop process control. The control-system loop includes the process, final control element, controller, error detector, and measurement element. The following are the key variables and associated action in the closed loop:

- The set point is the user-defined desired value of the controlled variable, which is the process variable to be regulated.
- Measurement-element output indicates the controlled-variable status, also called *process output variable* or *feedback signal.*
- The error is the difference between the set point and the controlled variable.
- Output is an indication produced by the controller to initiate the action to be taken by the final control element to correct the error.

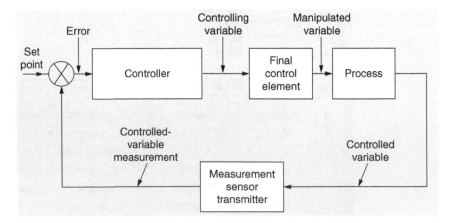

Figure 8.12 Single-variable closed-loop control.

- The final control element is the actuator that brings about the changes in the process and selected controlled variable. The output of the final control element is called the *process controlling* or *manipulating variable*.

8.2.2 *Control-System Time Response*

Time response of the control system can be determined based on the time characteristic of the process-controlled-variable value x following a step change in the controlling-variable output value y. Most control systems are self-regulating, which means that a new controlled-variable equilibrium is reached immediately or shortly after a process change or disturbance. For example, the speed of an induction motor settles at the value that produces the required load torque. The motor will run at this equilibrium point under the same load conditions (for a rotating object, equal force and load lead to constant speed). A small increase in load due to disturbances will force the motor to decelerate (load greater than motor torque) and reach a lower speed point of equilibrium after a short period of oscillation. Thus the speed of the induction motor is an excellent candidate for selection as controlled variable. Motor speed control is a very common application in process control.

Figure 8.13 plots a process-controlling variable output response to a step input change in the controlling variable for a possible controller implementation. Other responses are possible through a different controller design.

The time response can be determined using the controlled-variable delay time T_u, recovery time T_g, and final value x_{max}. The variables are determined by applying a tangent line to the maximum value and the inflection point of the step response. In many systems it is not possible to drive the process and record the response characteristic up to the maximum value because the process-controlled

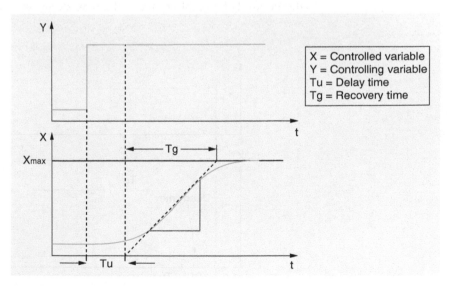

Figure 8.13   Step-change responses.

Table 8.1 System Controllability

Process Type	T_u/T_g	System Controllability
Slow reacting	<0.1	Very good
Reacting	0.1 to 0.3	Can be controlled
Fast Reacting	>0.3	Difficult to control

variable cannot exceed a specific limit. In this case, the rate of controlled-variable rise is used to identify the system. The controllability of the system can be defined based on the ratio T_u/T_g, as shown in Table 8.1.

A controlled system with dead and recovery times reacts to a step change in the controlling-variable process input, as shown in Fig. 8.14. T_d is the dead time, which is the time interval from initiating a change in the process input signal (controlling variable) to the point where we observe the start of the process output (controlled-variable) response. The controllability of a self-regulating controlled system with dead time is determined by the T_d/T_g ratio. The dead time T_d must be a small fraction of the recovery time T_g for good controllability. Processes with large dead times are extremely difficult to control and include large boiler drums' temperature, cascaded-tanks' liquid volume, and irrigation canal's downstream water-level controls. In a typical temperature-control application, as shown in Fig. 8.15, the temperature inside a tank is selected as the controlled variable, whereas the heater current represents the controlling variable of the closed-loop feedback controller.

Figure 8.15 shows an oscillatory control-system response with no dead time. The controlled variable reacts immediately after a set-point change, reaching and then exceeding the new set point. Finally, it goes through attenuating oscillations, reaching a steady-state value within the allowable dead band. The total time for the controlled variable to reach the new steady-state value is called the *settling*

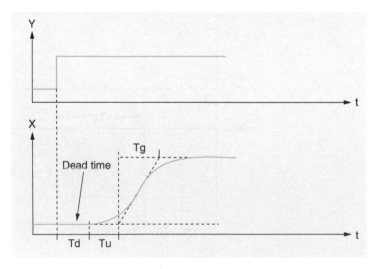

Figure 8.14 Control process with dead-time response.

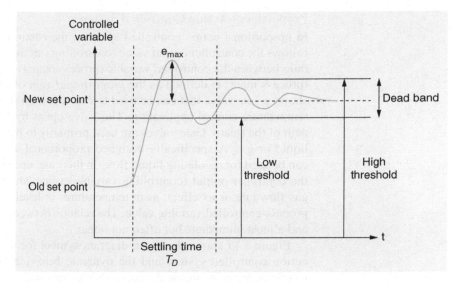

Figure 8.15 Typical oscillatory closed-loop control.

time T_D. In Fig. 8.14, response, $T_D = (T_d + T_g + T_u)$. Settling time is an important measure of control-system performance. Typical system behavior will show several and repeated oscillations before reaching a steady-state condition. The maximum error e_{max} in the controlled variable corresponds to the first overshoot response and is also a good measure of controller performance. In systems where oscillations are not permitted, different control strategies and controller designs are used. This includes the use of ON/OFF control, set-point ramping to the final target using small increments, and a slow response–tuned controller.

8.2.3   *Control-System Types*

Controlled systems are classified based on their time response to step changes in the controller output value. We distinguish between the following controlled systems:

- Self-regulating controlled systems
  - Proportional-action controlled systems
  - First-order controlled systems
  - Second-order delay-element controlled systems
- Non-self-regulating controlled systems
- Controlled systems with dead time

Processes with no self-regulation or with excessive dead time are difficult to control and require special and more complex control techniques, which include feed forward, cascaded control, and dead-time compensation. These techniques are not covered in this book. The following is a brief coverage of the three most common self-regulating controlled systems:

Proportional-Action Controlled Systems

In proportional-action controlled systems, the controlled-variable process value follows the controller output value (controlling variable) almost immediately. The ratio between the controlled variable (process output) and the controlling variable (process input) is defined as the *proportional gain* of the controller. Figure 8.16 shows a gate valve that can be used to control steam flow in a piping system in a temperature-control application. The valve opens by lifting a wedge out of the path of the steam. Gate valves are used primarily to permit or prevent the flow of liquid or gas. A specifically designed proportional-action controlled gate valve can be used for regulating liquid flow. In this case, opening of the valve represents the controller output (controlling variable) value, and the corresponding liquid/gas flow rate or its effect, as in temperature- or level-control applications, is the process-controlled variable value. The relation between the two variables is direct and almost immediate but often not linear.

Figure 8.17 shows the block-diagram symbol for the gate valve proportional-action controlled system and the dynamic behavior of the controlled variable (liquid temperature inside the tank) after a step change in the manipulated variable (valve position, controlling variable). Ideal control action without any lag is not possible in practical systems. The characteristic curves clearly show that a proportional-action controlled system exhibits self-regulation because a new equilibrium is reached immediately after a step change. The following is the ideal equation relating the temperature T to the valve position Y:

$$T = K_p \times Y$$

where K_p is the process proportional gain.

Figure 8.16   Industrial gate valve.

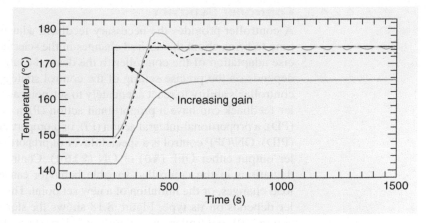

Figure 8.17 Proportional-action controlled system.

Small values for the proportional gain will produce slow response and large residual steady-state errors. Large proportional-gain values can speed up the response but also may cause large temperature oscillations. Excessive proportional gains can cause large and unsustainable oscillations, which can force a manual or an automatic system shutdown. Selecting and tuning the controller parameters, including the proportional gain, will be discussed later in this chapter.

First-Order Controlled Systems

In a first-order system-controlled application, the controlled-variable value initially changes in proportion to the change in the controller output, the controlling-variable value. The rate of change of the controlled-variable value is reduced as a function of the time elapsed until process steady state is reached. This type of control is also known as a *PT1 system*. A water container that is heated with steam is an example of a first-order controlled system. In simple controllers, the time behaviors for the heating and cooling processes are assumed identical. Control is clearly more complex with different process cooling and heating time characteristics.

Second-Order Delay-Element-Controlled Systems

In a second-order delay-element-controlled system, the process-controlled value does not immediately follow a step change in the controller output value. The process-controlled value initially increases in proportion to the positive rate of rise in the controller output value and then approaches the set point at a decreasing rate of rise. The controlled system shows what is known as a *proportional-response characteristic* with second-order delay element. Pressure, flow-rate, and temperature controls are all examples of this type, which is also known as *PT2 system*.

8.2.4 *Controller Behavior*

A controller provides the necessary feedback adjustments to the process in real time to bring about the desired changes in the selected controlled variable. A precise adaptation of the controller to the desired controlled-variable time response depends on the precise setting of the control strategy, parameter tuning, and the controller's ability to react adequately to set-point and load disturbances. Controller feedback can have a proportional action (P), a proportional-derivative action (PD), a proportional-integral action (PI), or a proportional-integral-derivative action (PID). ON/OFF control is a special flavor of proportional control but with controller output either OFF (%0) or ON (%100). Controller actions are triggered by deviations in the controlled-variable behavior caused by process disturbances, load changes, or the initiation of a new set point. The step response of the controller depends on its type. Figure 8.18 shows the step response of a proportional-action (P) controller, a proportional-derivative (PD) controller, and a PID controller. Proportional-integral (PI) control is most common, which can help to improve the control-system response and reduce the final steady-state errors. A controller with derivative action is not appropriate if the control system has pulsing measured quantities, as in the case of pressure or flow process control.

8.2.5 *Selection of the Suitable Controller Structures*

To achieve optimal control results, select a controller structure that is suitable for the controlled system and that you can adapt to the controlled system within specific limits. Table 8.2 provides an overview of suitable combinations of a controller structure and controlled system. Table 8.3 provides an overview of suitable combinations of a controller structure and physical quantity.

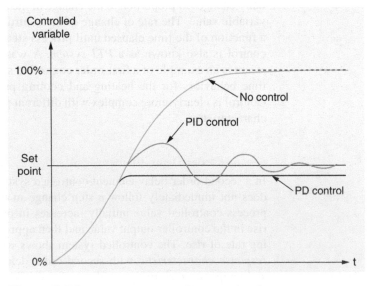

Figure 8.18 Closed-loop control types and performance.

Table 8.2 Controlled-System Structure Selection

Controlled System	Controlled Structure			
	P	**PD**	**PI**	**PID**
With dead time only	Unsuitable	Unsuitable	Suitable	Unsuitable
PT1 with dead time	Unsuitable	Unsuitable	Well suited	Well suited
PT2 with dead time	Unsuitable	Suited conditionally	Well suited	Well suited
Higher order	Unsuitable	Unsuitable	Suited conditionally	Well suited
Not self-regulating	Well suited	Well suited	Well suited	Well suited

Table 8.3 Physical-Quantity Control-Structure Choices

Physical Quantity	Controller Structure			
	P	**PD**	**PI**	**PID**
	Sustained Control Deviation		**No Sustained Control Deviation**	
Temperature	For low performance requirements and proportional action controlled systems with $T_u/T_g < 0.1$	Well suited	The most suitable controller structures for high-performance requirements (except for specially adapted special controllers)	
Pressure	Suitable, if the delay time is inconsiderable	Unsuitable	The most suitable controller structures for high-performance requirements (except for specially adapted special controllers)	
Flow rate	Unsuitable, because required GAIN range is usually too large	Unsuitable	Suitable, but integral action controller alone often better	Hardly required

In the case of PI and PID controllers, which represent the most common types of controllers, controlled-variable oscillations occur. Experience shows that extensive tuning of the initial parameters is always necessary. Level control and temperature control inside very large containers or a set of cascaded containers are examples of systems with dead time. Dead time represents the time that has to expire before a change can be measured at the system output. In systems with dead time, changes in the controlled variable owing to the controller action are delayed by the amount of the dead time. A system with $T_u/T_g > 0.3$ is typically difficult to control. Several technical sources provide a best starting guess for the PID parameters based on known process models. Most PID parameter-value recommendations are based on a one-quarter-decay controlled-variable behavior, which means that consecutive overshoot oscillations maintain this decay ratio. Additional tuning of selected controller parameters must be performed during system checkout. Other tuning parameters are available for different and commonly desired process-control behavior.

8.3 PID Instruction

PID closed-loop control is a common method of process control in most industrial and commercial applications. This section will cover the use of PID instruction blocks in the Siemens S7-1200 PLC system. Similar structures exist in most other PLC brands with slight changes in format but the same basic operation. Figure 8.19 shows the details of the PID instruction block. The PID block includes two parameter sets, one for input and one for output values.

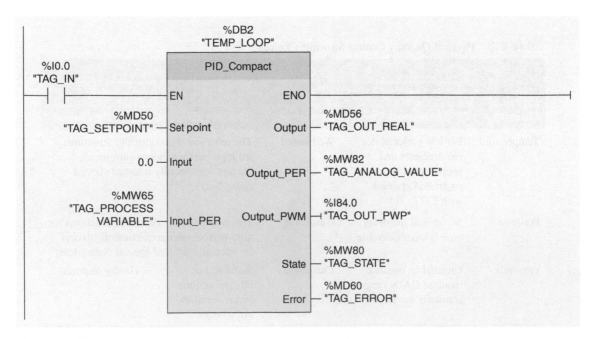

Figure 8.19 PID instruction block.

Table 8.4 PID Instruction Input Parameters

Parameter	Data Type	Default	Description
Set point	REAL	0.0	Set point of the PID controller in automatic mode
Input	REAL	0.0	A variable of the user program is used as source for the process value
			If you are using parameter input, then sPid_Cmptb_input_PER_On = FALSE must be set
Input_PER	WORD	W#16#0	Analog input as the source of the process value
			If you are using parameter input_PER, then sPid_Cmptb_input_PER_On = TRUE must be set
ManualEnable	BOOL	FALSE	• A FALSE -> TRUE edge selects "Manual mode," while State = 4, sReti_Mode remains unchanged
			• A TRUE -> FALSE edge selects the most recently active operating mode, State= sReti_Mode
			A changed of sReti_Mode will not take effect during ManualEnable=TRUE. The change of sReti_Mode will only be considered upon a TRUE -> FALSE edge at ManualEnable
Manual/Value	REAL	0.0	Manual value
			This value is used as the output value in manual mode
Reset	BOOL	FALSE	The Reset parameter restarts the controller

Table 8.4 lists the PID input parameters. Three of the parameters are internal to the PID. Reset is a Boolean parameter that, when set, will cause a restart of the PID controller. It is initially defaulted to FALSE. Manual enable is defaulted to FALSE, and when set, it disables the PID controller and forces a manual output value on the PID output. This output value is defined in the Manual value input parameter. The remaining three input parameters—Set point, Input, Input PER—are used during normal automatic operation of the PID controller. Input is the process variable to be controlled, as defined in the user-application program. Input PER refers to the actual analog input variable. The controller calculates the process-variable error relative to the desired set point. PID tuning parameters discussed earlier are not listed in this table but must be defined during PID configuration and programming. Desired PID control selection is accomplished through the definition of these tuning parameters. Proportional (P), proportional-integral (PI), proportional-derivative (PD), and proportional-integral-derivative (PID) control mode selections are all possible. Table 8.5 lists the PID output parameters. Three of these parameters are the scaled value of the input process variable Scaled input, the PID output in real format Output, and the output analog value Output PER. Other output parameters are well defined in the table.

8.3.1 *SIMATIC S7-1200 Tank-Level PID Control*

This section details the implementation of a PID control of a tank-level process using the Siemens Step 7 Basic software. Description of all the steps needed to program and implement the desired tank-level regulation is included, along with the associated commands and the SIMATIC software screens.

Table 8.5 PID Instruction Output Parameters

Parameter	Data Type	Default	Description
ScaledInput	REAL	0.0	Output of the scaled process value
Outputs "Output," "Output_PER," and "Output_PWM" can be used concurrently.			
Output	REAL	0.0	Output value in REAL format
Output_PER	WORD	W#16#0	Analog output value
Output_PWM	BOOL	FALSE	Pulse-width-modulated output value
			The output value is formed by minimum On and Off times.
SetpointLimit_H	BOOL	FALSE	If SetpointLimit_H=TRUE, the set point absolute high limit is reached. The set point in the CPU is limited to the configured set point absolute high limit. The configured process value absolute high limit is the default for the set point high limit.
			If you set sPid_Cmptr_Sp_Hlm to a value within the process value limits, this value is used as the set point high limit.
SetpointLimit_H	BOOL	FALSE	If SetpointLimit_L=TRUE, the set point absolute low limit has been reached. In the CPU, the set point is limited to the configured set point absolute low limit. The configured process value absolute low limit is the default setting for the set point low limit.
			If you set sPid_Cmptr_Sp_Lim to a value within the process value limits, this value is used as the set point low limit.

A sensor measures the tank level and converts it into a 0- to 10-V signal; 0 V corresponds to the level when the tank is empty (0 L), whereas 10 V is the level indication for a completely filled tank (1000 L). The sensor is connected to the first analog input of the SIMATIC S7-1200. An S1 switch (I0.0) is used to initiate a sudden set-point change in the tank level. This level is to be controlled alternatively for the 0 L level (S1 = 0) or 700 L level (S1 = 1).

A **PID_Compact** controller integrated in the Siemens Step 7 Basic V10.5 software is used. This PID controller, in turn, controls a pump as the manipulated variable by means of a 0- to 10-V analog output signal. The following is the simplified I/O assignment list for this control process:

Address	Symbol	Data Type	Comment
%IW 64	X_Level_Tank1	Int	Analog input actual level filling
%QW 80	Y_Level_Tank1	Int	Analog output manipulated value
%I 0.0	S1	Bool	Sudden set-point change enabler

For project management and programming, the software Totally Integrated Automation Portal is used. Here, under a uniform interface, components such as the controller, visualization, and networking of the automation solution are set up, parameterized, and programmed. Online tools are available for error diagnosis.

In the following 28 steps, a project can be set up for the SIMATIC S7-1200, and the solution for the task can be programmed:

1. The central tool is the Totally Integrated Automation Portal. It is called here with a click (→ Totally Integrated Automation Portal V11) (Fig. 8.20).

Figure 8.20 Totally integrated automation portal.

2. Programs for the SIMATIC S7-1200 are managed in projects. We are now setting up such a project (→ Create new project → `Tank_PID` → Create) (Fig. 8.21).

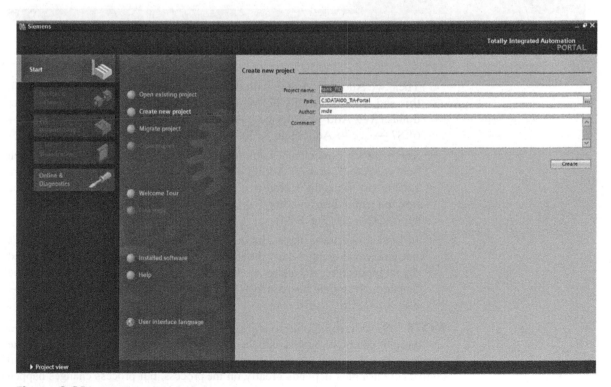

Figure 8.21 Creating a new project.

3. Next, First steps are suggested for configuration. First, we want to Configure a device (→ First steps → Configure a device) (Fig. 8.22).

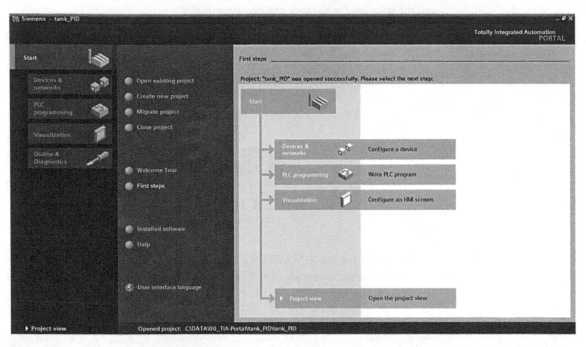

Figure 8.22 Configuring a new device.

4. Then, we Add new device with the Device name `controller_tank`. To this end, we select CPU1214C from the catalog, with the matching order number (→ Add new device → `controller_tank` → CPU1214C → 6ES7 → ... → Add) (Fig. 8.23).

5. The software now changes automatically to the Project view with the open hardware configuration. Additional modules can be added from the hardware catalog (to the right!). Here the signal board for an analog output is to be inserted using drag and drop (→ Catalog → Signal board → AO1 × 12 bits → 6ES7 232- ...) (Fig. 8.24).

6. In Device overview, input and output addresses can be set. Here the CPU's integrated analog inputs have the addresses `%IW64` to `%IW66`, and the integrated digital inputs have the addresses `%I0.0` to `%I1.3`. The address of the analog output on the signal board is `%QW80` (→ Device overview → AQ1 × 12 bits → 80 ... 81) (Fig. 8.25).

NOTE PID control applies to any system with a measurable process-controlled variable that can be regulated using a controlling variable to derive an appropriate actuator. The dynamic nature of the interactions between the controlling and controlled variables depends greatly on the process, but process changes happen so fast. Analog interface with the process is needed in order to allow the PLC to

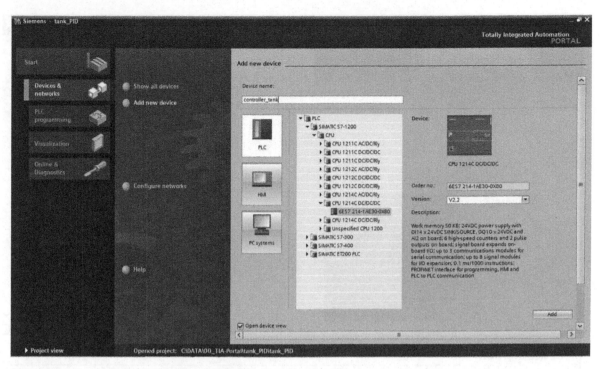

Figure 8.23 Selecting and adding the CPU.

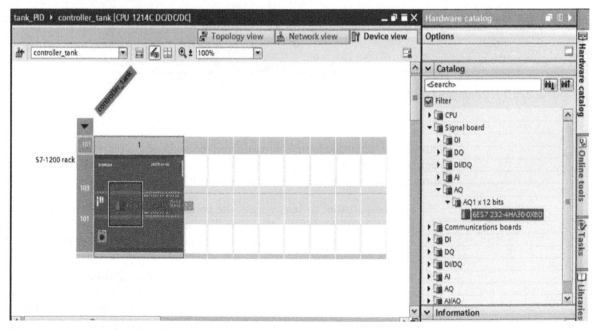

Figure 8.24 Adding the signal board for an analog output.

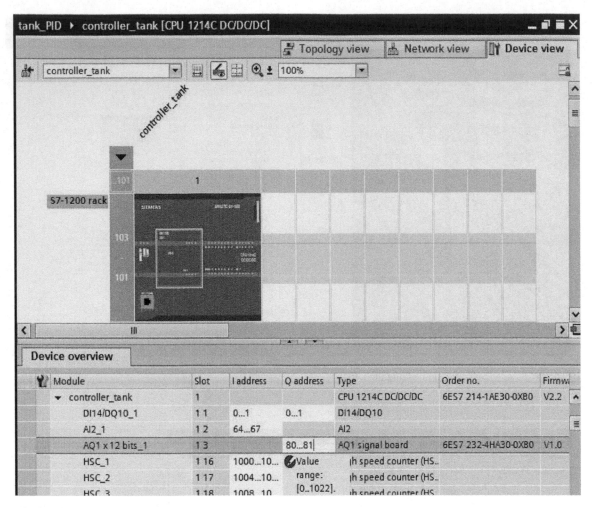

Figure 8.25 Configured devices overview.

capture the controlled-variable real-time measurements and provide the desired regulation through its output to the actuator (controlling variable). This will be shown in the final steps of this section during PID tuning and monitoring activities.

7. For the software to later access the correct CPU, its IP address and subnet mask have to be set (→ Properties → General → PROFINET interface → Ethernet addresses → IP address: 192.168.0.1 → subnet mask: 255.255.255.0) (Fig. 8.26).

8. Because modern programming uses tags instead of absolute addresses, the global PLC tags have to be specified here. These global PLC tags are descriptive names with a comment for the inputs and outputs that are used in the program. Later, global PLC tags can be accessed by means of this name. These global tags can be used in the entire program in all blocks. In program

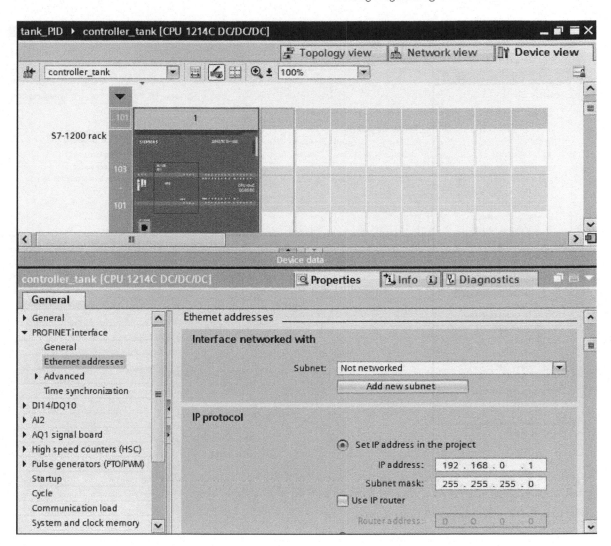

Figure 8.26 Setting the CPU PROFINET IP address.

navigation, select `controller_tank [CPU1214C DC/DC/DC]`, and then select PLC tags. With a click, open the PLC tags table, and enter, as shown below, the names for the inputs and outputs (→ `controller_tank [CPU1214C DC/DC/DC]` → PLC tags → Default tag table) (Fig. 8.27).

9. To create the function block FC1, in project navigation, select `controller_ tank [CPU1214C DC/DC/DC]` and then Program blocks. Next, click on Add new block.

(→ `controller_tank [CPU1214C DC/DC/DC]`→ Program blocks → Add new block) (Fig. 8.28).

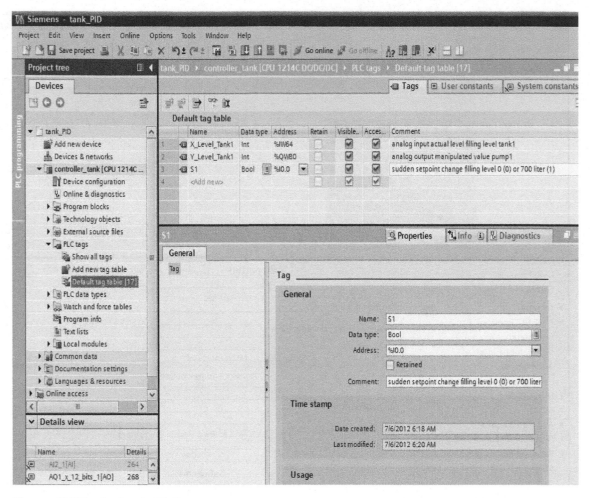

Figure 8.27 Setting the PLC tag names.

10. Select Organization block (OB), and type the name `Cyclic interrupt`. As programming language, specify FBD. Numbers are assigned automatically (OB100). We leave the permanent scan time at 100 ms. Accept the inputs with OK (→ Organization block (OB) → `Cyclic interrupt` → FBD → Scan time 100 → OK) (Fig. 8.29).

NOTE The PID controller has to be called with a permanent scan time (here 100 ms) because its processing is critical with respect to time. The controller could not be optimized if it would not be called accordingly.

11. The organization block `Cyclic interrupt [OB100]` will be opened automatically. Before the program can be written, its local tags have to be specified. In the case of this block, only one tag type is used:

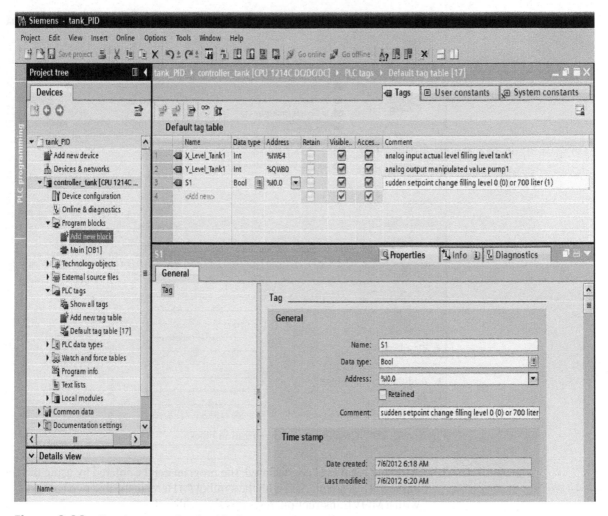

Figure 8.28 Creating a new function block.

Type	Name	Function	Available In
Temporary/ Local data	Temp	Tags that are used to store temporary intermediate results. Temporary data are retained only for one cycle.	Functions, function blocks, and organization blocks

12. In our example, only one local tag is needed: `Temp: w_level_tank1`, with data type Real. This tag stores the set point for tank1 as an intermediate value. It is important in this example to use the correct data type Real; otherwise, it is not compatible in the following program with the PID controller block used. For the sake of clarity, all local variables also should be provided with sufficient commentary (Fig. 8.30).

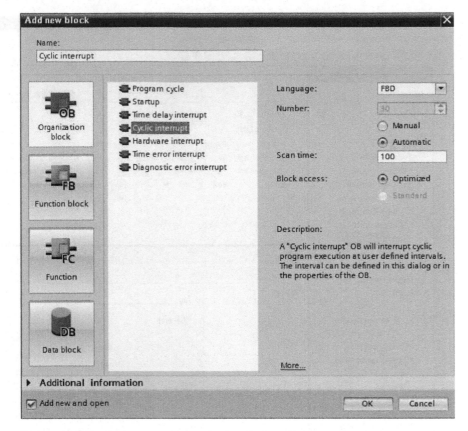

Figure 8.29 Creating a new organization block.

13. After the local tags are declared, the program can be entered by using the tag names. (Tags are marked with the symbol #.) Here, in the first two networks—with a MOVE instruction, respectively—either the floating-point number 0.0 (S1 == 0) or 700.0 (S1 == 1) is copied to the local tag `#w_fuel_tank1` (→ Instructions → Move → MOVE) (Fig. 8.31).

14. The controller block `PID_Compact` is moved to the third network. Because this block does not have multi-instance capability, it has to be assigned a data block as a single instance. It is generated automatically by Step 7 (→ Advanced instructions → PID → `PID_Compact` → OK) (Fig. 8.32).

15. Wire this block, as shown here, with the set point (local tag `#w_fuel_tank1`), actual value (global tag `X_Level_Tank1`), and manipulated variable (global tag `Y_Level_Tank1`). Then the configuration screen of the controller block can be opened (→ `#w_level_tank1` → `X_Level_Tank1` → `Y_Level_Tank1` →) (Fig. 8.33).

16. Here we have to make the Basic settings, such as controller type, and wire the internal controller structure (→ Basic settings → Controller type Volume →

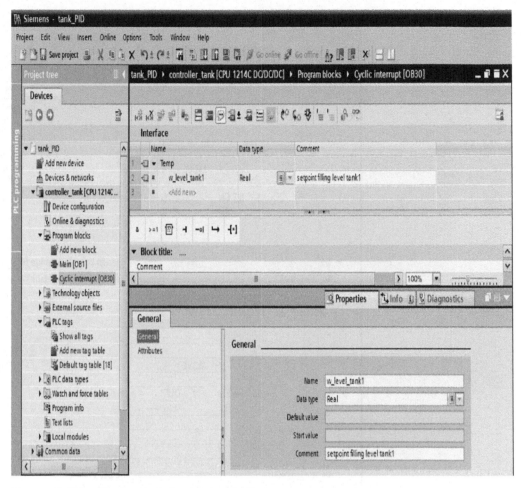

Figure 8.30 Creating/declaring local variables/tags.

l → actual value: `Input_PER` (analog) → manipulated value: `Output_PER`) (Fig. 8.34).

17. At Process value settings, we set the measuring range from 0 to 1000 L. The limits also have to be adjusted (→ Process value settings → Scaled high 1000.0 l → High limit 1000.0 l → Low limit 0.0 l → Scale low 0.0 l) (Fig. 8.35).

18. At Advanced settings, there are also Process value monitoring and a manual setting for the PID parameters. By clicking on ![Save project], the project is saved (→ Advanced settings → Process value monitoring → PID Parameters → ![Save project]) (Fig. 8.36).

19. To load the entire program to the CPU, first select the folder `controller_tank,` and then click on the symbol ![icon]. Download to device. (→ `controller_tank` → ![icon]) (Fig. 8.37).

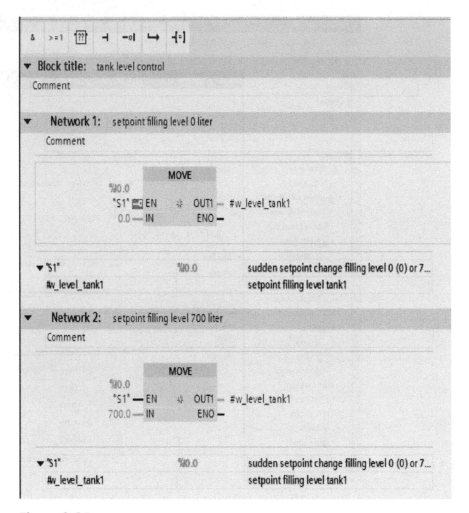

Figure 8.31 Entering the program with documentation.

20. If the PG/PC interface was not specified previously (refer to Laboratory 2.1 in Chap. 2), a window is displayed where this can still be done (→ PG/PC interface for download → Load) (Fig. 8.38).

21. Now click on Load once more. During downloading, the status is displayed in a window (→ Load) (Fig. 8.39).

22. The successful download is now displayed in a window. Click on Finish (→ Finish) (Fig. 8.40).

23. Next, start the CPU by clicking on the symbol ![symbol] (→ ![symbol]) (Fig. 8.41).

24. Confirm the question whether you are sure you want to start the CPU with OK (→ OK) (Fig. 8.42).

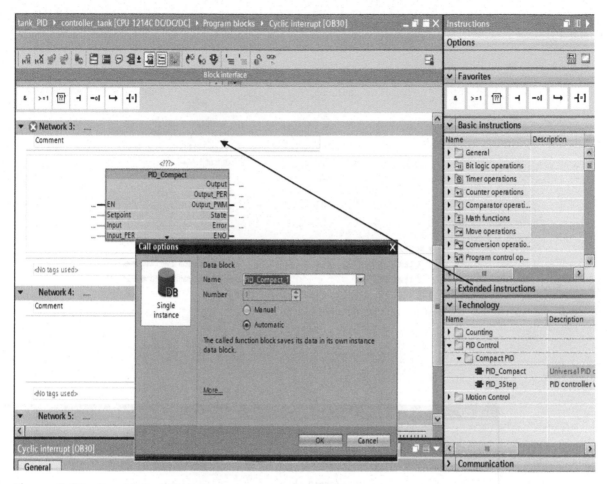

Figure 8.32 Entering the PID block.

25. By clicking on the symbol 📟 (monitoring ON/OFF), the status of the blocks and tags can be monitored while the program is tested. However, when starting the CPU, the controller **PID_Compact** is not activated. To this end, start commissioning by clicking on the symbol 📊 (→ **Cyclic interrupt[OB200]** → 📟 → **PID_Compact** → 📊 Commissioning) (Fig. 8.43).

26. On an operating screen, control variables can be displayed as shown in Fig. 8.44.

27. With Measurement Stop/Start, variables can be plotted in a diagram, including the set point, the actual value, and the manipulated variable. After the controller is loaded to the control system, it is still inactive. This means that the manipulated variable remains at 0%. Now select Tuning mode, Pretuning and then Pretuning Start (→ Measurement → Start → Tuning mode → Pretuning → Start).

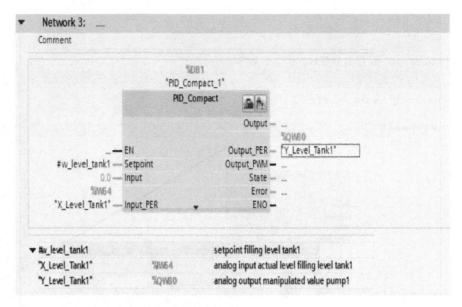

Figure 8.33 Configuring the PID block.

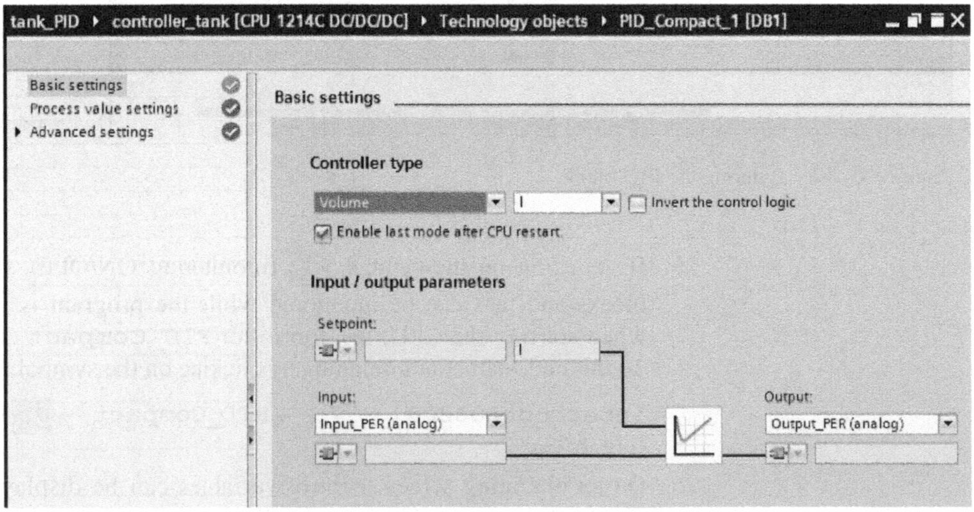

Figure 8.34 Controller structure and basic setting.

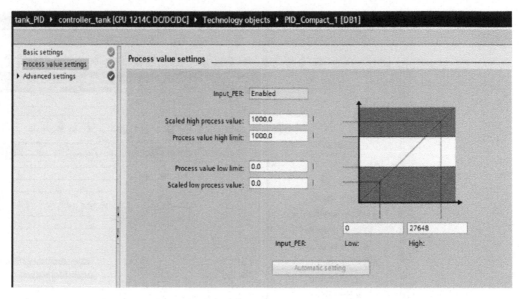

Figure 8.35 Process variable setting.

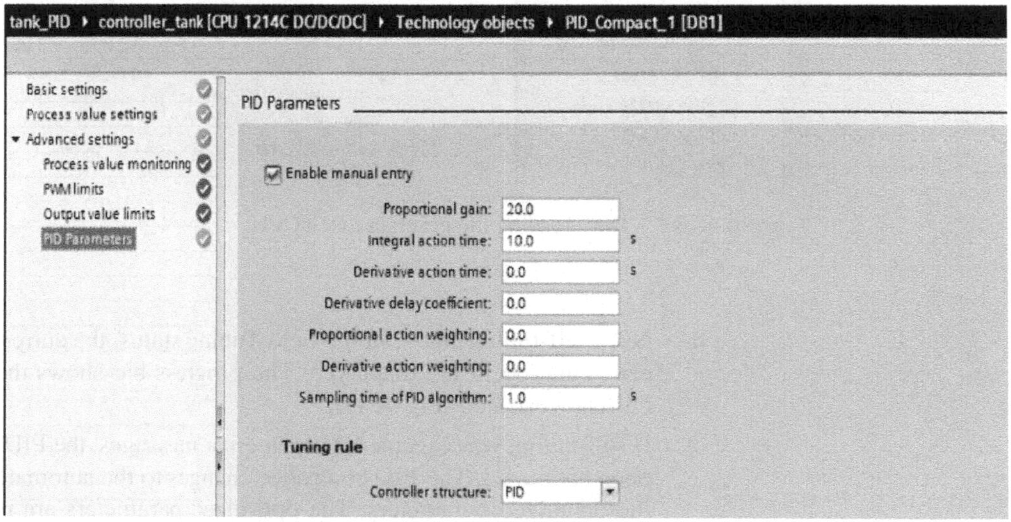

Figure 8.36 Assigning the PID closed loop control parameters.

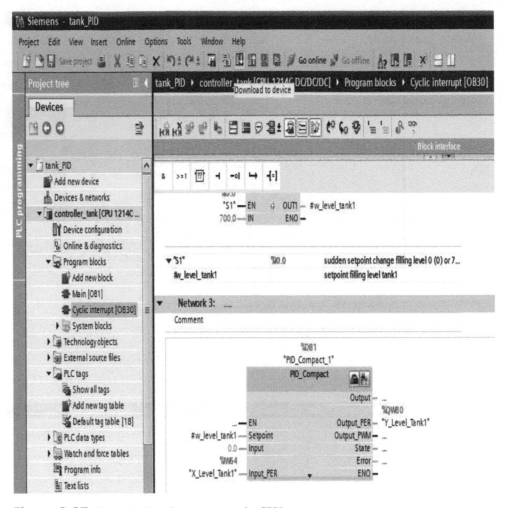

Figure 8.37 Downloading the program to the CPU.

28. Now, self-tuning starts. In the field Tuning status, the current work steps and errors that occur are displayed. The progress bar shows the progress of the current work step (Fig. 8.45).

29. If self-tuning was executed without error messages, the PID parameters have been optimized. The PID controller changes to the automatic mode and uses the optimized parameters. The optimized parameters are retained at power ON and CPU restarts. With the button ![button], the PID parameters can be loaded to the project (→ ![button]) (Fig. 8.46).

NOTE For faster processes, such as speed control, fine-tuning should be selected for optimization. A cycle is executed here that lasts several minutes, where all PID parameters are determined and set. The parameters values can be monitored in the data block after the project is loaded.

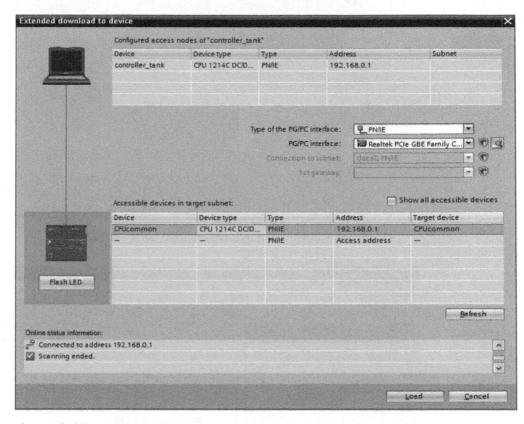

Figure 8.38 Defining the access nodes interface for download.

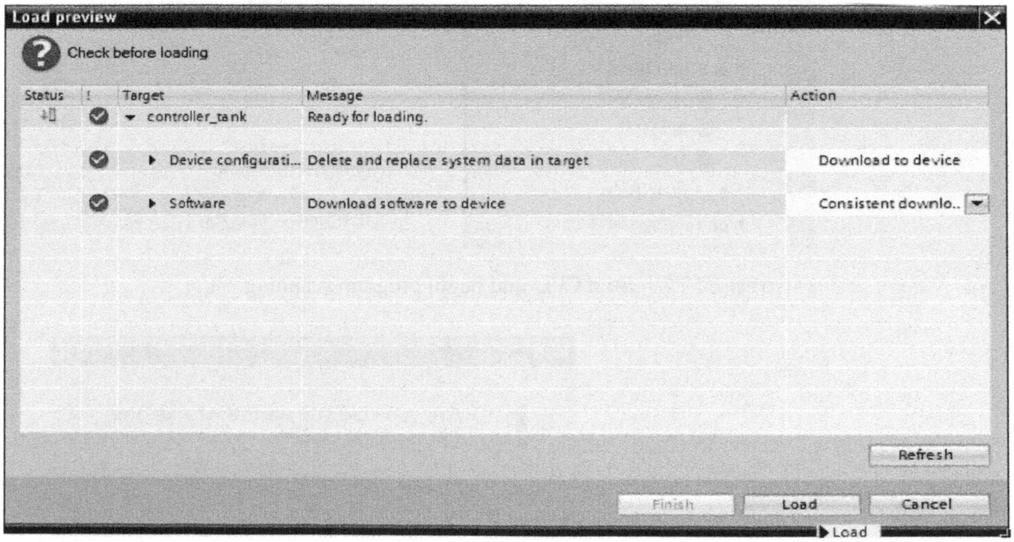

Figure 8.39 Program load and associated status.

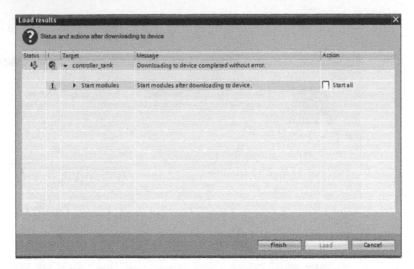

Figure 8.40 Start modules and finish the download.

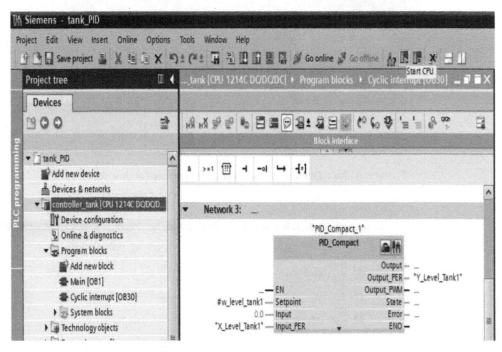

Figure 8.41 Start CPU and begin program scanning.

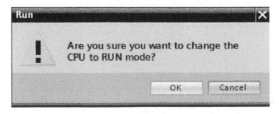

Figure 8.42 Confirm the CPU run mode.

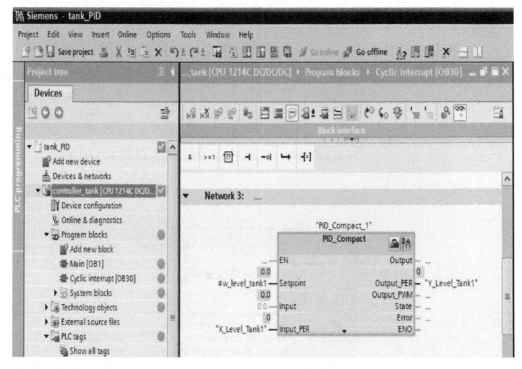

Figure 8.43 Commissioning the PID loop.

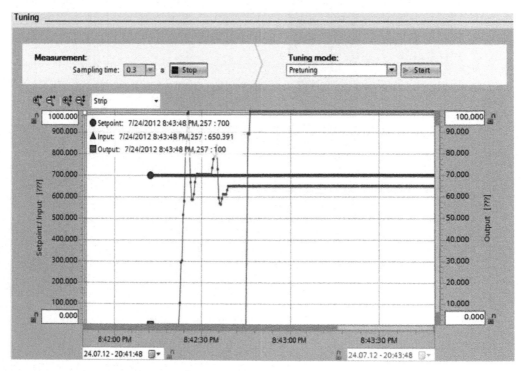

Figure 8.44 Displaying and monitoring process variables.

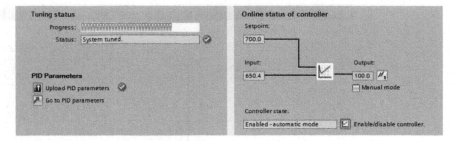

Figure 8.45   Tuning and online status of the PID controller.

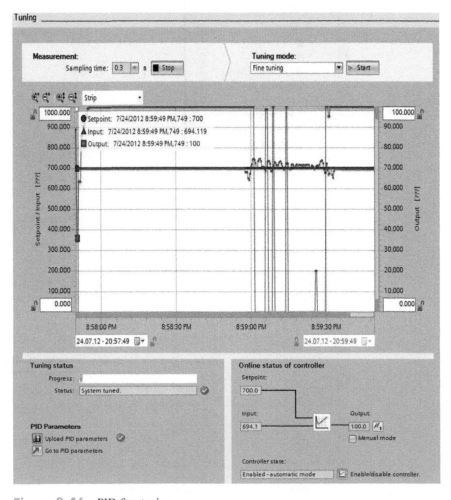

Figure 8.46   PID fine tuning.

Homework Problems and Laboratory Project

Problems

8.1 What are the actual decimal count values used by the Siemens analog I/O modules? What does it means in terms of A/D conversion resolution?

8.2 What is the function of the Wire break in scaling analog I/O signal?

8.3 Describe the function of the Normalize and Scale instructions used in analog I/O processing.

8.4 An analog input voltage signal in the range from 0 to 10 V is received in the standard form 0 to 27,648. If the signal is connected to the PLC-configured analog input module, perform the following:
   a. Show a ladder network to scale the analog input signal (0 to 10 V) to the corresponding 12-bit digital count (0 to 4095).
   b. Show another network to scale the signal to engineering units in the range from 50 to 350°F.

8.5 For Problem 8.4, show a network(s) to validate the input signal and force the equivalent standard digital count to be in the range 0 to 27,648.

8.6 Refer to Problem 8.4; if the equivalent standard digital count is from 0 to 15,300, what is the equivalent normalized count? Assuming a normalized count of 0.6257234, what is the equivalent 12-bit resolution count?

8.7 Show a network(s) to scale two analog input signals from 0 to 10 VDC to engineering units in the range of 50 to 300 lb/in². The two analog signals represent two measurements of pressure inside a boiler drum. Design ladder logic to calculate the average pressure in pounds per square foot, and then move 10 percent of the average amount to the analog output module. Use the analog addresses on the CPU module and the signal board.

8.8 Define the following terms, and give an example of each:
   a. A closed-loop control system
   b. An open-loop control system
   c. A single-variable control system
   d. A multivariable control system

8.9 Define the following terms, and quantify their influence on control-process behavior:
   a. Dead time
   b. Settling time
   c. Recovery time
   d. Delay time

8.10 Explain two methods used to determine the time response of a control system.

8.11 Draw a block diagram for a single variable *closed loop*, and describe the function of each block operation.

8.12 How is the time response of a control system determined?

8.13 Self-regulated control systems discussed are classified as three types (as shown below). Briefly explain each type.
   a. Proportional-action controlled systems
   b. First-order controlled systems
   c. Second-order delay-element controlled systems

8.14 An induction motor is a self-regulating device. The steady-state torque and speed are a linear relation in the operating range. As load increases, the motor torque increases and the speed decreases, and a new steady-state operating state is reached. Draw the relation between torque and motor speed. Show how self-regulation works under a small-motor load change.

8.15 What does non-self-regulating control system mean? Give an example.

8.16 Figure 8.47 shows the behavior of the controlled variable under transient conditions. Answer the following:
   a. If the set point is 200°F and the dead band is ±4°F, determine the maximum error in °F.
   b. What is the settling time? How much is it in this example?

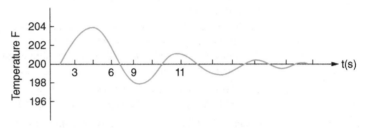

Figure 8.47 Control variable transient, homework problem 8.16.

8.17 The diagram in Fig. 8.48 shows a closed-loop system for tank outlet flow control. Redraw the diagram to include a supervisory control system, and describe the advantages of supervisory control.

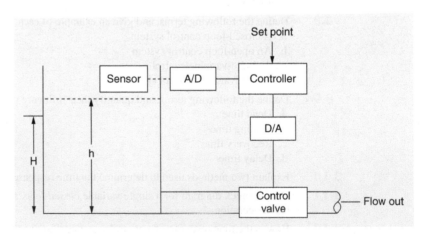

Figure 8.48 Tank level closed-loop control, homework problem 8.17.

8.18 Study the network shown in Fig. 8.49, and answer the following questions:
   a. What is the effect of the LIMIT instruction on the analog input count?
   b. Show another way of achieving the same LIMIT function.

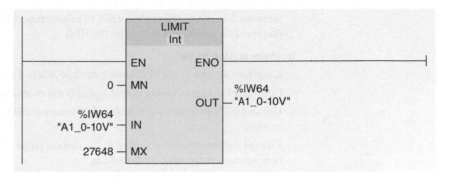

Figure 8.49 Homework problem 8.18.

8.19 What is the concept of controllability? What is the effect of dead time on controllability? Show examples of variation of controllability.

8.20 What is the effect of dead time on controller design? Which control techniques are more suitable for processes with large dead times?

Projects

Laboratory 8.1: Tank-Level Sensor Measurement Processing and Monitoring

The objective of this laboratory is to provide hands-on knowledge of the analog programming used in commercial and industrial applications.

Process Description

This lab demonstrates fundamental issues associated with analog I/O programming. Two analog signals are configured and used as tank-level measurements, and both are connected to the main CPU module. Analog input 1 (IW64) measures tank level 1, and analog input 2 (IW66) measures tank level 2. The two measurements are used to generate an average for the two tanks levels. The average is sent to an analog output module, which converts the digital count to a 0- to 10-V analog signal to be displayed on the local panel meter and in engineering units on the HMI.

Laboratory Specifications

- The first network in this lab should read an analog input voltage signal (0 to 10 V) in the standard count range from 0 to 27,648. The count is normalized in the range 0.0 to 1.0. Next, the normalized value is scaled back in the range from 0 to 4095. Use three tags in this network: **AI_0-10V**, the raw analog input (IW64); **AI_0-1**, the normalized intermediate tag (MD88); and **AI_0-4095**, the 12-bit scaled count (MD96). Only IW64 is the physical I/O address among the three used.

- Repeat the first step for the second analog input IW66 using Network 2.

- The third network calculates the average of the two tank-level scaled counts and places the result in MD 144, tag name **TANKS-AVE**.

- Network 4 normalizes the count produced by Network 3 and then scales it in the range from 0 to 27,648. The newly scaled value is stored in the **AQ_0-10V** tag, which corresponds to the physical output address QW80. This value is displayed on the local panel meter.

- Network 5 converts the average level to engineering units from 0 to 40 m. The value is displayed on the designated tag on the HMI.

Implementation Steps

- Configure the two-input PLC analog module attached to the CPU.
- Configure the single analog output signal board connected to the CPU.
- Use two 10-V potentiometers to supply the analog input signals (0 to 10 V) to the input module.
- Change the potentiometer setting in the defined range to represent a tank level in engineering units in the range of 0 to 40 m.
- Configure a Status page in the HMI to display the two tank levels and their average in meters as you change the potentiometer setting from the minimum to the maximum voltage (0 to 10 V).
- Also display the digital count in 12-bit resolution on the HMI Status page.
- Monitor the voltmeter readings, and verify the analog signal scaling and the measured values compared with the potentiometers' settings.

Laboratory Requirements

- Assign the system inputs.
- Assign the system outputs.
- Program the required networks. Use three functions: analog input validation, analog input scaling, and analog average calculation.
- Download the program, and go online.
- Record your observations for input and output values in digital count, analog signal value, and corresponding engineering units.
- Simulate the program using the training unit or the Siemens simulator. Configure the watch table, and then verify that the program is running according to the process specifications.

Ladder Program Listing (No Documentation)
See Fig. 8.50.

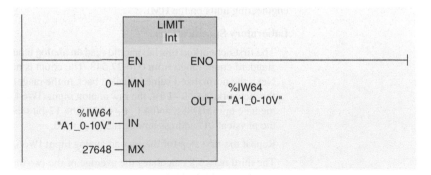

Figure 8.50 Laboratory 8.1 ladder.

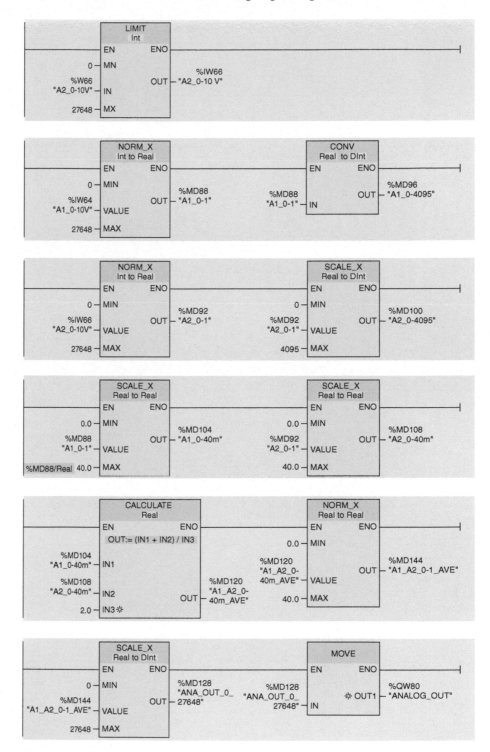

Figure 8.50 *(Continued)*

Comprehensive Case Studies

This chapter is intended to be a capstone project encompassing most of the concepts covered in this book. This project is part of a large networked multiple-sites irrigation-canal water-control process. The coverage is abbreviated to one site and is implemented using the Siemens S7-1200 PLC system.

Chapter Objectives

▶ Understand and document complete process description.

▶ Design the control system I/O and PLC memory map.

▶ Develop complete system logic diagrams from specifications.

▶ Implement ladder and HMI configuration, communication, programming, and checkout.

The two case studies selected for this chapter are intended to be capstone projects encompassing most of the concepts covered in this book. The first project is part of a large networked multiple-sites irrigation-canal water-control process that was implemented by the authors several years ago in the delta region of the River Nile in Egypt. The second project deals with a common process-control task in the wastewater treatment industry that has to do with pumping-station control. The coverage in both projects is simplified to one site and is transformed to a newer implementation using the Siemens S7-1200 PLC system.

9.1 Irrigation-Canal Water-Level Control

Irrigation water is channeled through two motorized vertical gates from an upstream to a downstream agriculture area. Motorized gates are used to regulate the water flow and thus the downstream level. Fully closed vertical gates will produce continuous reduction in the downstream water level. Low downstream water level will not allow irrigation in the downstream area. Fully raised vertical gates will result in excessive irrigation in the downstream area and thus much wasted water. Adequate regulation of the position of the vertical gates can maintain the desired downstream water levels at different times and thus support required irrigation cycles and preserve water resources at the same time.

Two identical constant-speed motors drive the two vertical gates. The motors can move the gates up or down, but only one motor can run at any time. This is done in order to reduce the overall power requirements of the regulator site. The lowest gate should be selected for a raise operation, and the highest gate must be selected for a lowering command. Both gates are equipped with position sensors, fully closed limit switches, and fully open limit switches.

Two downstream level sensors provide redundant measurements at three points downstream. These sensors must be validated because they should provide similar readings consistent with previous measurements. We will assume input values from two validated sensors for this project. An upstream flooding limit switch is used to provide the regulator with an indication of possible flooding upstream and thus command the system to raise both gates until the flooding condition is removed and the downstream level is close to the upstream level.

Each motor is equipped with an overload alarm switch that is used to trigger any unusual conditions, such as over temperature or overload. The motor provides a discrete input signal within 5 seconds from the start action indicating whether the motor is running or not. If a motor fails to start, then the other motor is selected, and an alarm is issued. Motors also can start by activating the push button located on the local panel if the AUTO/MAN switch is in manual position, and the LOCAL/REMOTE switch is on LOCAL.

A selected motor is scheduled to run for 15 seconds. This is followed by an idle period of 10 minutes. No motor is allowed to run during the idle time. This is done to prevent repetitive activation of the motors during downstream water-level transient. It is important to keep the gates close to the same height in order to minimize the loading on both gate structures. An emergency shutdown switch (ESD) is available to the operator to shut down the system in addition to the START and STOP push-button switches.

Tag Name	Address Number	Comments
VG1_ROL	I0.0	Vertical Gate1 running on line
VG2_ROL	I0.1	Vertical Gate2 running on line
VG1_RAISED	I0.2	VG1 fully raised
VG1_LOWERED	I0.3	VG1 fully lowered
VG2_RAISED	I0.4	VG2 fully raised
VG2_LOWERED	I0.5	VG2 fully lowered
ESD	I0.6	Emergency Shutdown selector switch
AUTO	I0.7	AUTO switch
UP_Flod_LS	I1.0	Upstream flooding limit switch

Figure 9.1 Irrigation-system inputs.

9.1.1 *System Input-Output (I/O) Map*

The first step in the design of a PLC control application is translation of the process specification to actual I/O resources. This is known as the *PLC I/O map*. This important step lists all I/O tags, assigned PLC addresses, and description, which are produced from the piping and instrument diagram (P&ID). P&IDs are the "schematics" used in the field of instrumentation and process control to provide the documentation needed for field implementation. The P&ID is used to better understand the process and how the instrumentation is interconnected. Figure 9.1 lists the irrigation-control-system discrete inputs, whereas Fig. 9.2 shows the corresponding PLC input tags. Figures 9.3 and 9.4 repeat the same process for the discrete outputs. Notice that none of the analog I/Os for this control process is listed. We only limited our case study to ON/OFF control based on water-level analog real-time measurements relative to a user-defined set point for the downstream water level.

VG1_ROL	Bool	%I0.0
VG2_ROL	Bool	%I0.1
VG1_RAISED	Bool	%I0.2
VG1_LOWERED	Bool	%I0.3
VG2_RAISED	Bool	%I0.4
VG2_LOWERED	Bool	%I0.5
ESD	Bool	%I0.6
AUTO	Bool	%I0.7
UP_FLOD_LS	Bool	%I1.0

Figure 9.2 Irrigation-system PLC input tags.

Tag Name	Address Number	Comments
VG1_ raise	Q0.0	Vertical Gate1 raise one step
VG1_ lower	Q0.1	Vertical Gate1 lower one step
VG2_raise	Q0.2	Vertical Gate2 raise one step
VG2_lower	Q0.3	Vertical Gate2 lower one step
VG1_FTS	Q0.4	Vertical Gate1 failed to start
VG2_FTS	Q0.5	Vertical Gate2 failed to start
DS1_FAIL	Q0.6	Down stream1 Fail
DS2_FAIL	Q0.7	Down stream2 Fail

Figure 9.3   Irrigation-system outputs.

⬛ VG1_RAISE		Bool	%Q0.0
⬛ VG1_LOWER		Bool	%Q0.1
⬛ VG2_RAISE		Bool	%Q0.2
⬛ VG2_LOWER		Bool	%Q0.3
⬛ VG1_FTS		Bool	%Q0.4
⬛ VG2_FTS		Bool	%Q0.5
⬛ DS1_FAIL		Bool	%Q0.6
⬛ DS2_FAIL		Bool	%Q0.7

Figure 9.4   Irrigation-system PLC output tags.

9.1.2   *Logic Diagrams*

Logic diagrams are recommended for sound documentation and as means of transitioning to the actual ladder programming implementation. This step, as will be demonstrated, makes ladder programming much easier. The logic diagram in Fig. 9.5 shows the implementation for the vertical gate 1 raise operation. Other operations logic diagrams can be constructed in a similar way for the vertical gate 1 lowering operation (Fig. 9.6) and both the raise and lowering for vertical gate 2 (Fig. 9.7).

The logic diagram is segmented into four parts. Part 1 (Fig. 9.5) derives VG1 selection for gate 1 control up raise operation. The control up raise command requires that the desired set point is greater than or equal to the downstream average water level, downstream average level is outside the dead band, and VG1 is the next gate to go up. VG1 is next to go up if it is available, is not fully raised, VG2 is not the next to go up, and VG1 has the lowest absolute-level position. VG1 is available to go up only if the system is on AUTO, VG1's motor is not running, VG1's motor did not fail to start, and the system is placed on REMOTE operation. The system is placed on LOCAL in order to allow motor to start only from the field using the local panel START/STOP push button.

Part 2 (Fig. 9.6) shows the logic for starting the VG1 motor for 15 seconds. This raise duration should cause an increase in the downstream water level, which

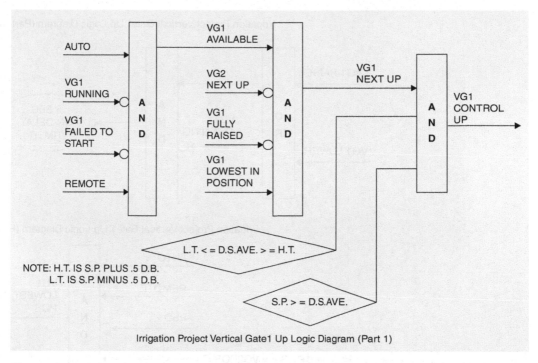

Irrigation Project Vertical Gate1 Up Logic Diagram (Part 1)

Figure 9.5 Vertical gate 1 control up raise operation.

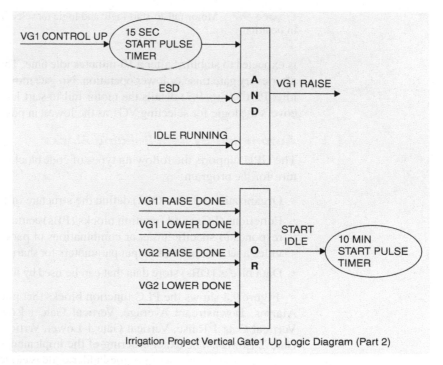

Irrigation Project Vertical Gate1 Up Logic Diagram (Part 2)

Figure 9.6 Logic for starting the vertical gate 1 motor for 15 seconds.

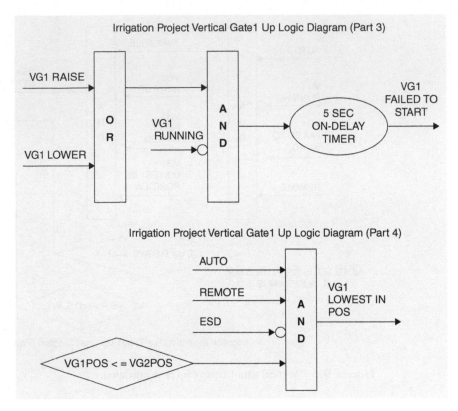

Irrigation Project Vertical Gate1 Up Logic Diagram (Part 3)

Irrigation Project Vertical Gate1 Up Logic Diagram (Part 4)

Figure 9.7 Motor fail-to-start logic and logic for selecting vertical gate 1 as the lowest in position.

is expected to stabilize after a 10-minutes idle time. The system enters the idle state after every gate raise or lower operation. No gate movement is permitted during the idle. Part 3 (Fig. 9.7) details the motor fail-to-start logic, whereas Part 4 (Fig. 9.7) covers the logic for selecting VG1 as the lowest in position among the two gates.

9.1.3 Automated-System Building Blocks

The CPU supports the following types of code blocks that allow an efficient structure for the program:

- Organization blocks (OBs) define the structure of the program.
- Functions (FCs) and function blocks (FBs) contain the program code that corresponds to specific tasks or combinations of parameters. Each FC or FB provides a set of input and output parameters for sharing data with the calling block.
- Data blocks (DBs) store data that can be used by the program blocks.

  Figure 9.8 shows the PLC function blocks (Set point Validation, Initialization, Alarms, Downstream Average, Vertical Gate 1 Lower, Vertical Gate Position, Vertical Gate 1 Raise, Vertical Gate 1 Lower, Vertical Gate 2 Raise, and Vertical Gate 2 Lower). The exact ordering of the implemented functional blocks is irrelevant because the scanning of the ladder code is continuous and repeats at a high

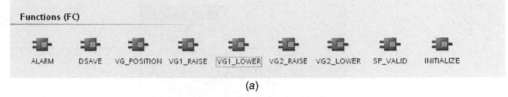

(a)

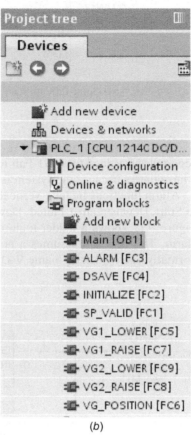

(b)

Figure 9.8 (*a*) Irrigation-system PLC function blocks (portal view). (*b*) Irrigation-system PLC function blocks (project view).

rate (at least three complete scans per second). A few time-critical tasks might require a definite ordering of some functional blocks. Most initialization tasks are executed only once on power-up or system-reset.

9.2 Irrigation-Canal Ladder Implementation

The set-point validation function (Fig. 9.9) consists of one network. It compares the set point received from the user to high and low limits. If the set point is outside the limits, a "Wrong set point, enter again" message will be displayed on the human-machine interface (HMI).

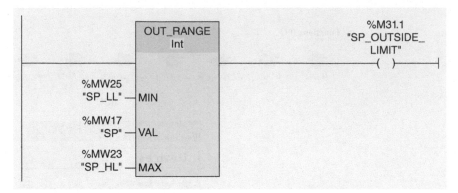

Figure 9.9 Set-point validation network.

The initialization function (Fig. 9.10) consists of one network that clears the accumulated values for VG1 Fail to Start, VG2 Fail to Start, Idle, and Common Alarm timers because the system is placed in AUTO mode.

Alarm functions include Vertical Gate 1 Failed to Start, Vertical Gate 2 Failed to Start, Downstream 1 Failed, Downstream 2 Failed, and low level float switch. Figure 9.11 shows a ladder-logic diagram for the vertical gate 1 failed-to-start alarm. This diagram assumes a normally open Q0.0 tag name VG1_RAISE, a normally open Q0.1 tag name VG1_LOWER, a normally closed contact I0.0 tag

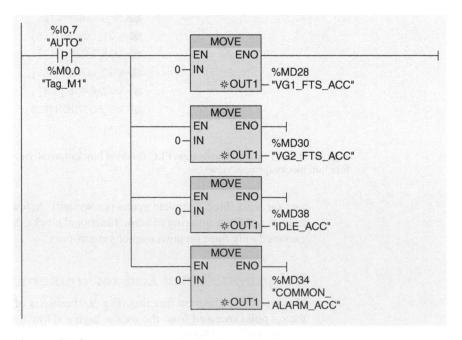

Figure 9.10 Initialization network.

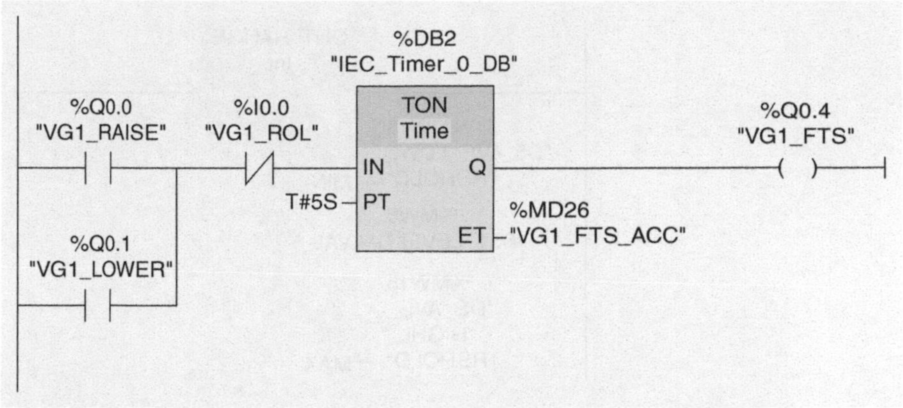

Figure 9.11 Vertical gate 1 failed-to-start network.

name `VG1_ROL`, an ON-DELAY timer (TON) with a 5-second preset time, and Q0.4 output coil tag name `VG1_FTS`. The ladder diagram works as follows:

- During the first scan, because the VG1 raise or VG1 lower output is set while I0.0 tag name `VG1_ROL` is OFF, the power flows to the timer, and the timer starts timing.

- It takes up to 5 seconds to receive the associated motor running online contact from the initiation of the motor START command.

- If motor running online signal is not received within 5 seconds, the fail-to-start coil `VG1_FTS` is energized, indicating a motor failure.

Figure 9.12 shows a ladder-logic diagram for the vertical gate 2 failed-to-start alarm. This diagram assumes a normally open Q0.2 tag name `VG2_RAISE`, a normally open contact Q0.3 tag name `VG2_LOWER`, a normally closed contact I0.1 tag

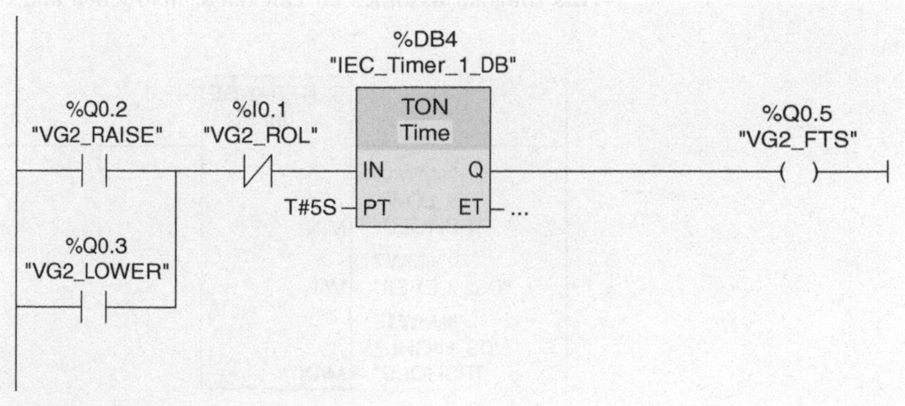

Figure 9.12 Vertical gate 2 failed-to-start network.

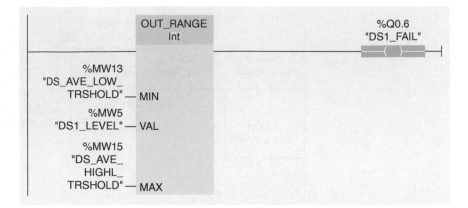

Figure 9.13 Downstream 1 transmitter fail network.

name `VG2_ROL`, an ON-DELAY timer (TON) with a 5-second preset time, and Q0.5 output coil tag name `VG2_FTS`. The ladder diagram works as follows:

* During the first scan, the VG2 raise or VG2 lower output is set while I0.1 tag name `VG2_ROL` is OFF (because it takes at least 5 seconds to receive the VG2 contact when the VG2 motor is running successfully).

* The power flows to the timer, and the timer starts timing.

* If the timer accumulated time is equal to the timer preset time (5 seconds), Q0.5 will turn ON, indicating that VG2 failed to start.

Figure 9.13 shows a ladder-logic diagram for the downstream 1 fail alarm. This diagram assumes an Out range instruction and output coil Q0.6 tag name `DS1_Fail`. The ladder diagram work as follows: If downstream 1 is greater than downstream average high threshold or less than downstream average low threshold, then power flows to the output coil, and Q0.6 is set. This is an indication of level sensor 1 failure/alarm.

Figure 9.14 shows a ladder-logic diagram for the downstream 2 fail alarm. This diagram assumes an Out range instruction and output coil Q0.7 tag name

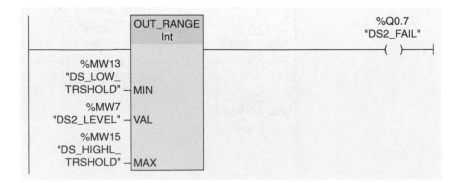

Figure 9.14 Downstream 2 transmitter fail network.

Figure 9.15 Low-low-level alarm network.

`DS2_Fail`. The ladder diagram works as follows: If downstream 2 is greater than downstream average high threshold or less than downstream average low threshold, then power flows to the output coil, and Q0.7 is set.

Figure 9.15 shows a ladder-logic diagram for the low-low-level alarm. This diagram assumes normally open contact I1.2 tag name `LL_FLOAT_SW` and output coil M29.1 tag name `LL_LVL_ALARM`. The ladder diagram works as follows: During the first scan, because the downstream low-low float switch triggers, power flows to output coil M29.1 and sets the low-low-level alarm.

Figure 9.16 shows a ladder-logic diagram for the common alarm. This diagram assumes a normally open Q0.4 tag name `VG1_FTS`, a normally open contact Q0.5 tag name `VG2_FTS`, Q0.6 tag name `DS1_FAIL`, Q0.7 tag name `DS2_FAIL`, I1.2 tag name `LL_FLOAT_SW`, and output coil Q1.0/M42.5 tag name `COMMON_ALARM`. During the first scan, because vertical gate 1 failed to start, or vertical gate 2 failed to start, or downstream1 failed, or downstream 2 failed, or low-float-level alarm is set, the power flows to the output coil common alarm, and Q1.0/M42.5 tag name `COMMON_ALARM` is set.

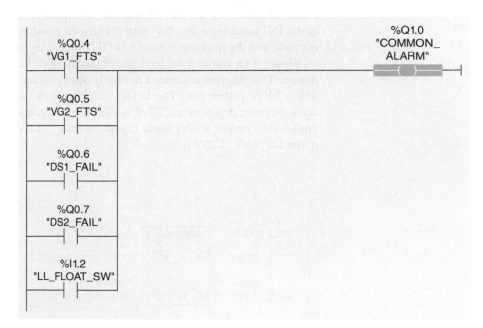

Figure 9.16 Common alarm network.

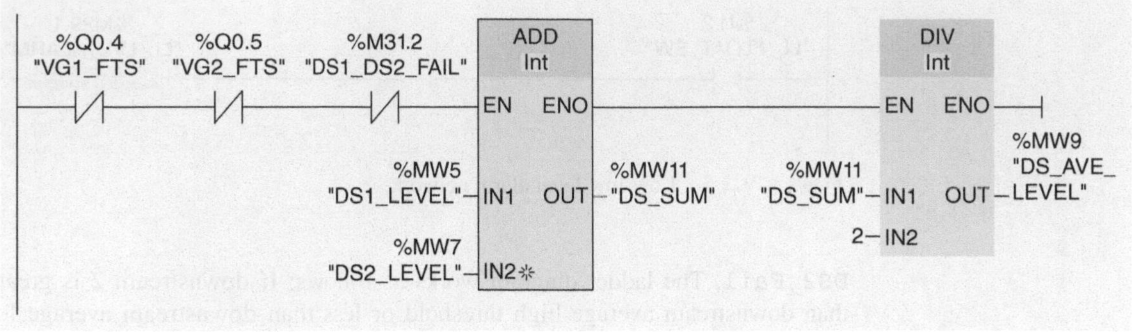

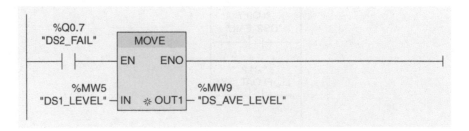

%Q0.4	%Q0.5	%M31.2	ADD		DIV	

Figure 9.17 Downstream average calculations.

Downstream average function consists of three networks, which include DS average, downstream 1 level sensor validation, and downstream 2 level sensor validation. Figure 9.17 shows a ladder-logic diagram for the downstream average. This diagram assumes a normally closed contact Q0.4 tag name VG1_FTS, a normally closed contact Q0.5 tag name VG2_FTS, a normally closed contact tag name DS1_DS2_FAIL, an ADD instruction, and a DIV instruction. The ladder diagram works as follows: If vertical gate 1 failed to start is OFF, vertical gate 2 failed to start is OFF, and downstream 1/downstream 2 fail is OFF, power flows to the ADD instruction, and the instruction is executed. The Add instruction adds the value at the IN1 input tag name DS1_LEVEL to the value of the IN2 input tag name DS2_LEVEL and outputs the sum at the OUT output tag name DS_SUM. Power also flows to the DIV instruction, and the instruction is executed. The value at the IN1 input tag name DS_SUM divides the constant value (2) at IN2 input and outputs, and the quotient is stored at OUT output tag name DS_AVE_LEVEL.

Figure 9.18 shows a ladder-logic diagram for downstream 1 level sensor validation. This diagram assumes a normally open contact Q0.7 tag name DS2_FAIL and a MOV instruction. The ladder diagram works as follows: During the first scan, because downstream 2 fail is ON, the MOV instruction is executed, and it copies the content at IN1 input tag name DS1_ LEVEL to the OUT output tag name DS_AVE_LEVEL.

Figure 9.18 Downstream average level update using sensor 1.

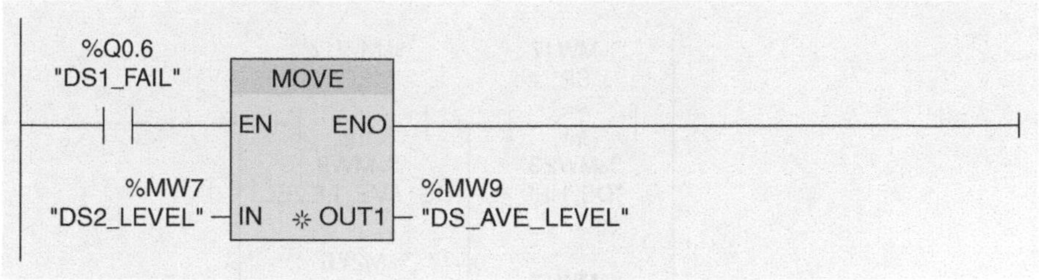

Figure 9.19 Downstream average level update using sensor 2.

Figure 9.19 shows a ladder-logic diagram for downstream 2 level sensor validation. This diagram assumes a normally open contact Q0.6 tag name `DS1_FAIL` and a MOVE instruction. The ladder diagram works as follows: During the first scan, because downstream 1 fail is ON, the MOVE instruction is executed and copies the content at IN1 input tag name `DS2_LEVEL` to the OUT output tag name `DS_AVE_LEVEL`.

Vertical gate 1 lower function consists of three networks, including Vertical Gate 1 Next Down, Control Down, and Vertical Gate 1 Lower. As shown in Fig. 9.20, if vertical gate 1 is available, highest in position, not fully lowered, and vertical gate 2 is not next to go down, then vertical gate 1 is next to go down.

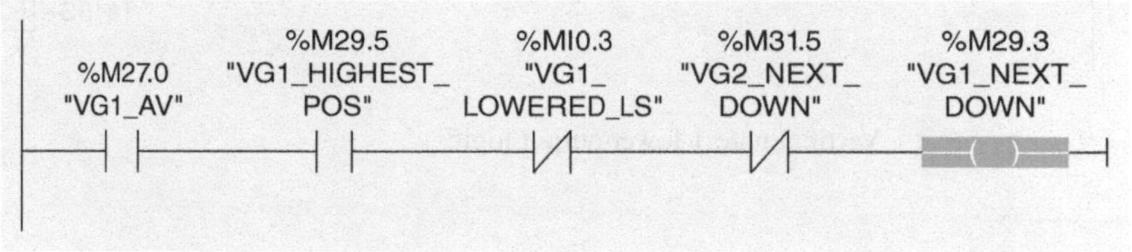

Figure 9.20 Vertical gate 1 next-down logic.

As shown in Fig. 9.21, if vertical gate 1 is next to go down, the set point is outside the dead band, and the set point is less than the downstream average level, control down is set.

As shown in Fig. 9.22, if vertical gate 1 next down is set, control down is set, no emergency shutdown, DS1/DS2 did not fail, and idle timer is not timing, then a TP instruction is executed, and vertical gate 1 lower output is ON for 15 seconds.

The vertical gate position function consists of the four networks shown in the following figures. Figure 9.23 compares VG1 position with VG2 position. When MANUAL/AUTO switch is placed in AUTO and VG1 position is greater than or equal to VG2 position, then the output coil M29.5 tag name `VG1_HIGHEST_POS` is set. Notice that if the comparison used is only greater than and the two vertical gates have equal positions, then no gate will be selected or moved.

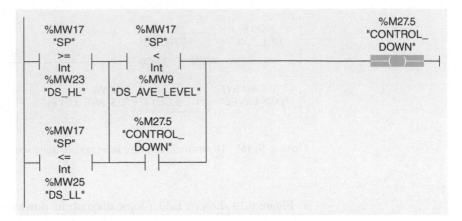

Figure 9.21   Vertical gate 1/vertical gate 2 control-down logic.

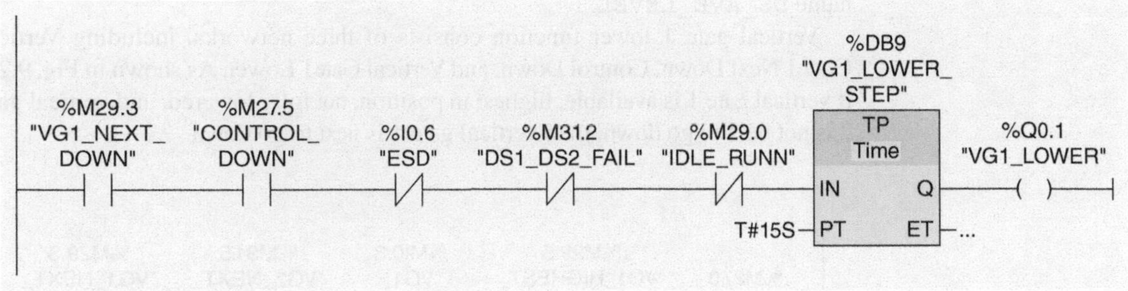

Figure 9.22   Vertical gate 1 lower-output logic.

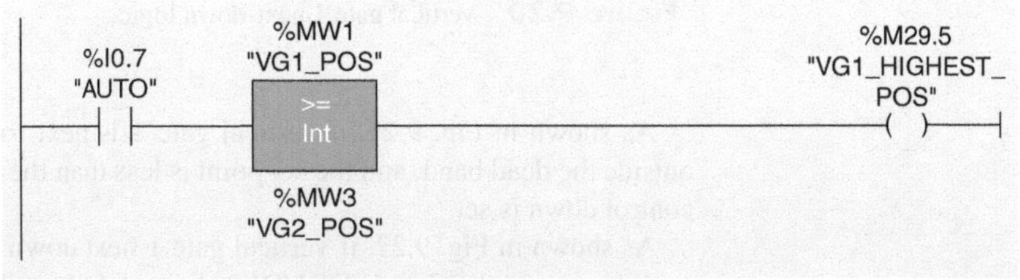

Figure 9.23   Vertical gate 1 highest-position logic.

Figure 9.24 compares VG2 position with VG1 position. When the MANUAL/ AUTO switch is placed in AUTO and the VG2 position is greater than or equal to the VG1 position, then the output coil M29.7 tag name VG2_HIGHEST_POS is set. Notice that if the comparison used is only greater than and the two vertical gates have equal positions, then no gate will be selected or moved.

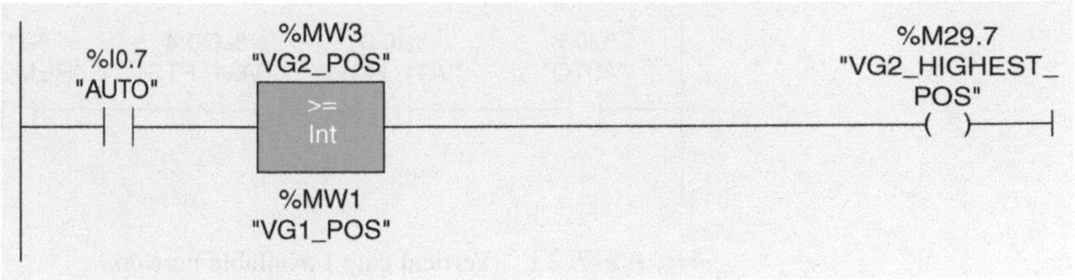

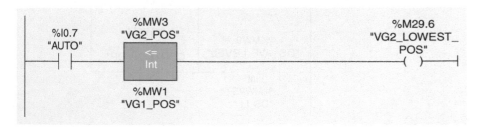

Figure 9.24 Vertical gate 2 highest-position logic.

Figure 9.25 Vertical gate 1 lowest-position logic.

Figure 9.25 compares VG1 position with VG2 position. When the MANUAL/AUTO switch is placed in AUTO and the VG1 position is less than or equal to the VG2 position, then the output coil M29.2 tag name `VG1_LOWEST_POS` is set. Notice that if the comparison used is only less than and the two vertical gates have equal positions, then no gate will be selected or moved.

Figure 9.26 compares VG2 position with VG1 position. When the MANUAL/AUTO switch is placed in AUTO, the VG2 position is less than or equal to the VG1 position, then output coil M29.6 tag name `VG2_LOWEST_POS` is set. Notice that if the comparison used is only less than and the two vertical gates have equal positions, no gate will be selected or moved.

Vertical gate 1 raise function consists of five networks: Vertical Gate 1 Available, Next Up, Control Up, Vertical Gate 1 Raise, and Idle Time. As shown in Fig. 9.27, if the AUTO/MANUAL switch is in AUTO, VG1 not running and not failed to start, and the LOCAL/REMOTE switch is in REMOTE, then VG1 is available (`VG1_AV`).

Figure 9.26 Vertical gate 2 lowest-position logic.

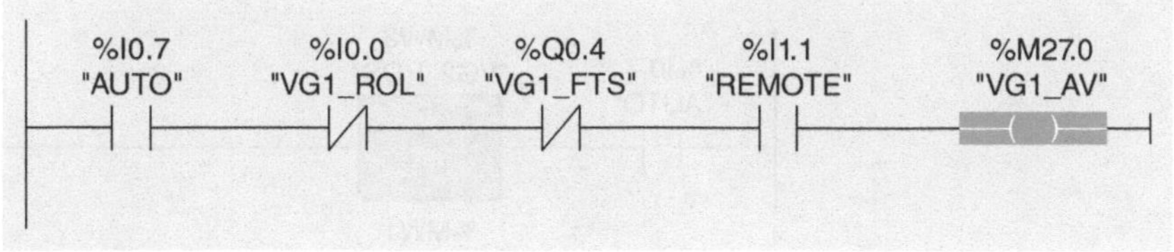

Figure 9.27 Vertical gate 1 available network.

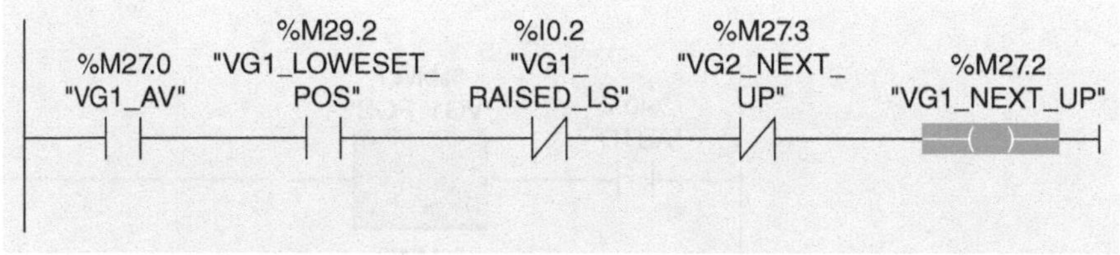

Figure 9.28 Vertical gate 2 next-up.

As shown in Fig. 9.28, if VG1 is available, lowest in position, not fully raised, and VG2 is not next to go up, then VG1 is next to go up.

As shown in Fig. 9.29, if the set point is outside the dead band and the set point is greater than the downstream average level, then CONTROL_UP is set.

As shown in Fig. 9.30, if VG1 is next to go up, Control up is set, ESD is OFF, downstream DS1_DS2_FAIL is OFF, and idle timer is not running, then the pulse timer is ON, which raises VG1 for 15 seconds.

As shown in Fig. 9.31, if VG1 raise or lower or VG2 raise or lower, then idle start pulse timer times for 10 minutes. This delay allows the control action to be realized in the downstream water level.

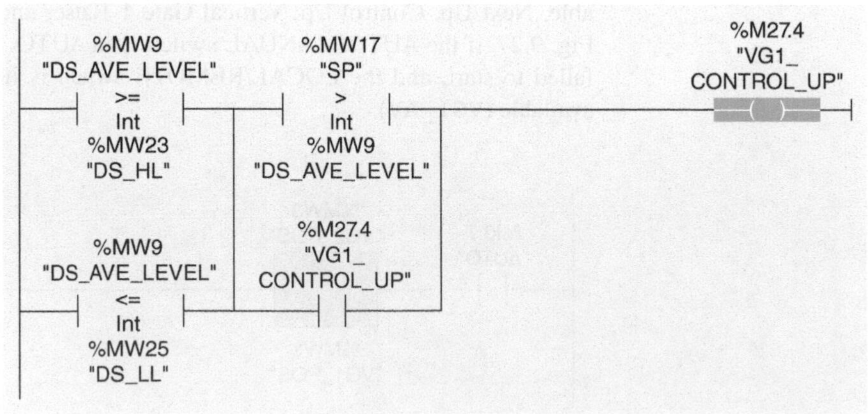

Figure 9.29 Vertical gate 1/vertical gate 2 up control logic.

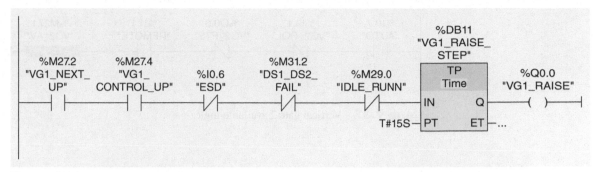

Figure 9.30 Vertical gate 1 raise network.

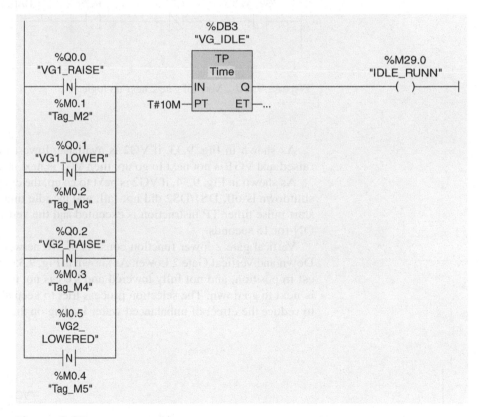

Figure 9.31 Idle timer network.

VG2 raise function consists of three networks: Vertical Gate 2 Available, Next Up, and Vertical Gate 2 Raise. As shown in Fig. 9.32, if the AUTO/MANUAL switch is in AUTO, VG2 not running and not failed to start, and the LOCAL/ REMOTE switch is in REMOTE, then VG2 is available.

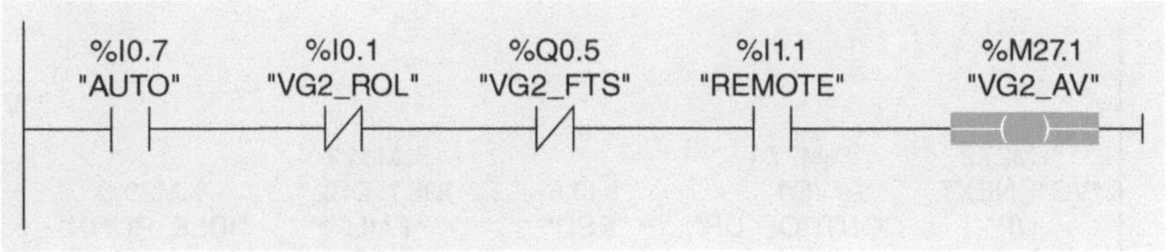

%I0.7
"AUTO"
%I0.1
"VG2_ROL"
%Q0.5
"VG2_FTS"
%I1.1
"REMOTE"
%M27.1
"VG2_AV"

Figure 9.32 Vertical gate 2 available logic.

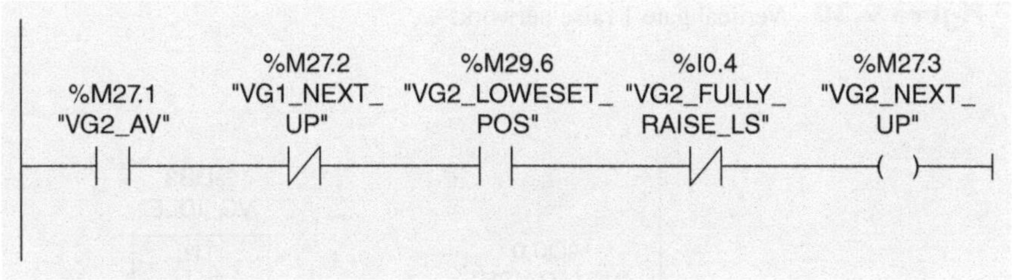

%M27.1
"VG2_AV"
%M27.2
"VG1_NEXT_
UP"
%M29.6
"VG2_LOWESET_
POS"
%I0.4
"VG2_FULLY_
RAISE_LS"
%M27.3
"VG2_NEXT_
UP"

Figure 9.33 Vertical gate 2 next-up logic.

As shown in Fig. 9.33, if VG2 is available, lowest in position, and not fully raised and VG1 is not next to go up, then VG2 is next to go up.

As shown in Fig. 9.34, if VG2 is next to go up, the control up is set, emergency shutdown is off, DS1/DS2 did not fail, and the idle timer is not timing, then the start pulse timer TP instruction is executed and the vertical gate 1 raise output is ON for 15 seconds.

Vertical gate 2 lower function consists of two networks, Vertical Gate 2 Next Down and Vertical Gate 2 Lower. As shown in Fig. 9.35, if VG2 is available, highest in position, and not fully lowered and VG1 is not next to go down, then VG2 is next to go down. The selection process tries to keep all gates at the same level to reduce the effect of unbalanced water loading on the gate structure, which can

%M27.3
"VG2_NEXT_
UP"
%M27.4
"VG1_
CONTROL_UP"
%I0.6
"ESD"
%M31.2
"DS1_DS2_
FAIL"
%M29.0
"IDLE_RUNN"
%DB12
"VG2_RAISE_
STEP"
TP
Time
IN
Q
T#15S
PT
ET
...
%Q0.2
"VG2_RAISE"

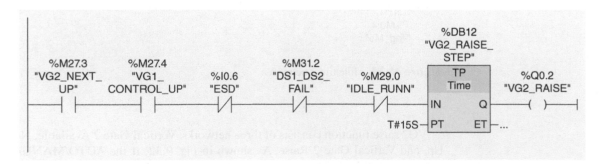

Figure 9.34 Vertical gate 2 raise network.

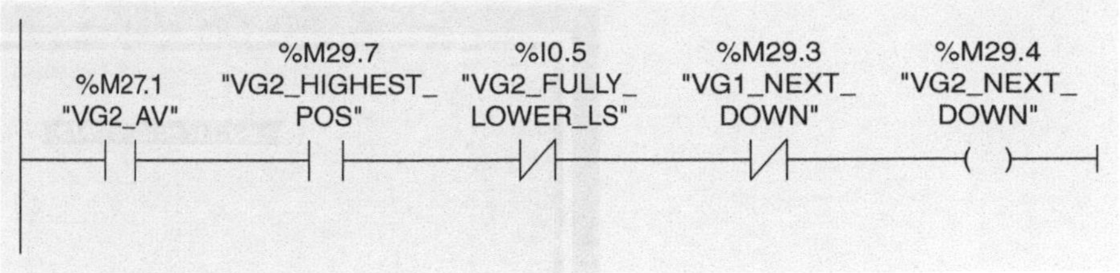

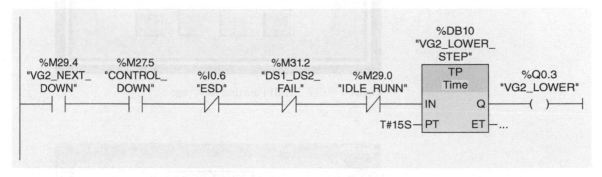

Figure 9.35 Vertical gate 2 next-down logic.

Figure 9.36 Vertical gate 2 lower network.

produce structural damage. Thus the lowest-position gate is selected when a raise operation is needed, and the highest-position gate is selected for a close operation.

If VG2 next down is set, control is down, emergency shutdown is off, DS1/DS2 did not fail, and the idle timer is not timing, then a TP instruction is executed, and VG2 lower output is ON for 15 seconds, as shown in Fig. 9.36.

9.3   Irrigation-Canal HMI Implementation

The HMI implementation of the irrigation control project is discussed in this section. A simplified user interface with only five pages is assumed in our design. The HMI Function Keys page allows the user to access STATUS, POSITION, ALARM, and CONTROL screens, as shown in Fig. 9.37. The Vertical Gate Position page in Fig. 9.38 shows the position of vertical gate 1 and vertical gate 2 in percent. Pressing fuction key F1 takes the user back to the Function Keys page.

The Alarm page in Fig. 9.39 displays Vertical Gate1 Failed to Start (VG1 FTS), Vertical Gate 2 Failed to Start (VG2 FTS), and common alarms.

The Vertical Gate Status page in Fig. 9.40 shows the status of Vertical Gate 1, Vertical Gate 2 Running/Not Running, Raise/Lower. It also displays Vertical Gate1, Vertical Gate 2 Position, Downstream 1, Downstream 2 Levels, and Set Point.

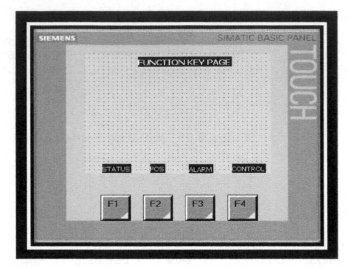

Figure 9.37 HMI Function Keys page.

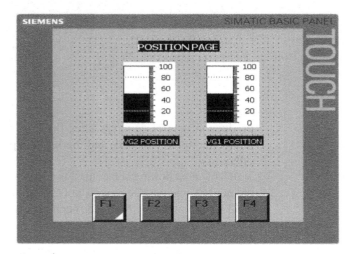

Figure 9.38 Vertical Gates Position page.

All values are displayed in engineering units. Pressing fuction key F1 takes the user back to Function Keys page.

The Control page allows the user to enter Set Point, Set Point High/Low Limits, and Downstream Dead Band. If the set point entered is outside the limits, "Wrong Set Point, Enter Again" is diplayed. Pressing Function Key F1 takes the user back to the Function Keys Page. The Control page for the irrigation downstream water level control is shown in Fig. 9.41.

The setup of HMI configuration, communication with the PLC, status/control pages, and tags is covered in Chap. 5. One or more HMI units can be used in a

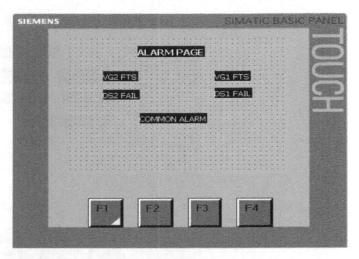

Figure 9.39 Alarm page.

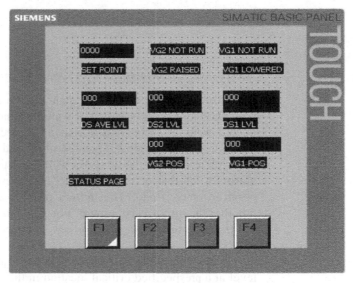

Figure 9.40 Vertical Gate Status page.

centralized control room or in remote locations. All HMIs and PLCs are integerated into one communication network where real-time information is updated and reflected on actual field devices and user interfaces. Network/communication bandwidth requirements are very minimal because a small amount of information is actually passed to connected nodes. New technology, such as with the Siemens S7-1200 system, brought huge advances in processing speeds, communication throughput, devices size, and overall cost. It allows users to implement real-time control at reasonable cost with adequate performance.

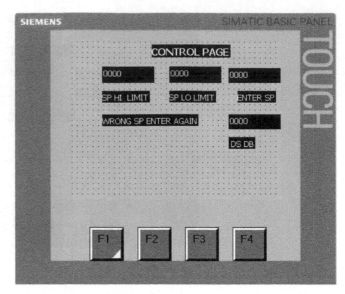

Figure 9.41   Control and Command page.

9.4 Wet-Well Pump-Station Control

High-flow-rate storm water is channeled to two large wet wells, the East wet well and the West wet well. The water is pumped to the river from the two connected wells at a constant rate using a predefined process sequence control. Two motor-driven constant-speed immersed pumps are used, one in the East wet well and one in the west wet well. Each pump is equipped with an overload alarm switch that is used to trigger any unusual conditions, such as over temperature or overload. The motors provide a discrete input signal indicating whether the motor is running or not. The motors also can be started by activating the push button located on the local panel if the AUTO/MAN switch is on MANUAL.

Three float switches are used to provide an accurate indication of the water level at a prespecified critical location in the East/West wet wells. The low-level float switch triggers the stopping of a running pump. The high-level float switch triggers the starting of a scheduled pump. If the scheduled pump fails to start within 5 seconds, the second pump is selected and started. An alarm must be issued to alert the operator of any motor failure. The very-high-level float switch triggers the starting of both pumps. If either of the two pumps fails to start, the corresponding alarm is activated by the control.

Pumps are scheduled to run according to an operator predefined calendar. This input is expected in hours of accumulated total pump run time. The two pumps must alternate while the water level is below the very high level and above the low level. The two pumps run at levels above the very high level, and cascaded timers are not altered during this condition.

Tag Name	Address Number	Comments
OFF_FLOAT	I0.0	Off Float Switch
ON_FLOAT	I0.1	On Float Switch
OVERIDE_FLOAT	I0.2	Override Float Switch
E_ROL	I0.3	East Running On Line
W_ROL	I0.4	West Running On Line
AUTO	I0.5	Auto/Manual Switch
ESD	I0.6	Emergency Shut Down selector switch
E_OVERLOAD	I0.7	East Overload Contact
W_OVERLOAD	I1.0	West Overload Contact

Figure 9.42 Pump-station-system PLC input tags.

9.4.1 *System I/O Map*

The first step in the design of a PLC control application is translation of the process specification to actual I/O resources. This is known as the *PLC I/O map*. This important step lists all I/O tags, assigned PLC addresses, and descriptions. Figure 9.42 lists the irrigation-control-system discrete inputs and the corresponding PLC input tags. Figure 9.43 repeats the same process for the discrete outputs. Notice that none of the analog I/Os for this control process are listed. We limited the case study only to ON/OFF control based on water-level analog real-time measurements relative to a user-defined set point for the downstream water level.

9.4.2 *Automated System Building Blocks*

The CPU supports the following types of code blocks that allow an efficient structure for the program:

- Organization blocks (OBs) define the structure of the program.
- Functions (FCs) and function blocks (FBs) contain the program code that corresponds to specific tasks or combinations of parameters. Each FC or FB

Tag Name	Address Number	Comments
E_PUMP	Q0.0	East Pump Output
W_PUMP	Q0.1	West Pump Output
E_FTS	Q0.2	East Pump Fail To Start
W_FTS	Q0.3	West Pump Fail To Start
COM_ALARM	Q0.4	Common Alarm

Figure 9.43 Pump-station-system PLC output tags.

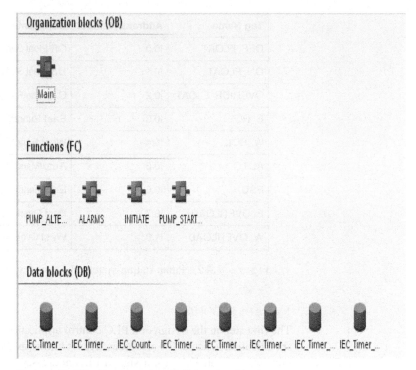

Figure 9.44 Pump-station PLC function blocks.

provides a set of input and output parameters for sharing data with the calling block.

- Data blocks (DBs) store data that can be used by the program blocks.

Figure 9.44 shows the function blocks designed and implemented for the wet-well pumping station control.

9.5 Pumping-Station Ladder Implementation

The initialization network INITIATE is shown in Fig. 9.45. A positive edge trigger instruction causes this network to execute once when the AUTO/MANUAL switch is on AUTO.

9.5.1 Pump Alarms

The Pump Alarm function includes three networks (Figs. 9.46 through 9.48). One common alarm is dedicated for the East wet well and the other for the West wet well. A common alarm is triggered from pump-motor failure to start, overload, or emergency shutdown.

- The East pump fail to start is ON if motor-running input is not received within 5 seconds from initiation of the Run command.

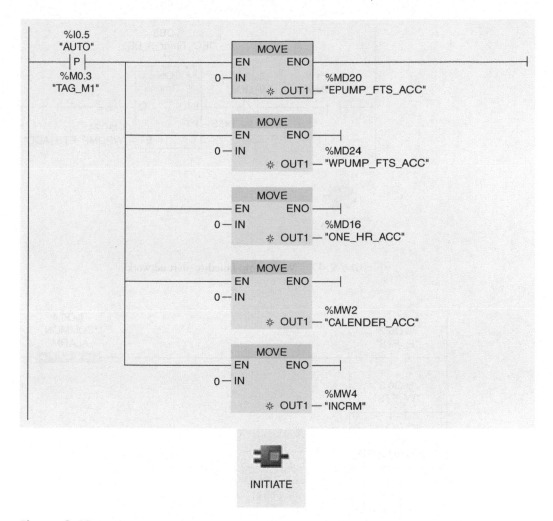

Figure 9.45 Initialization network.

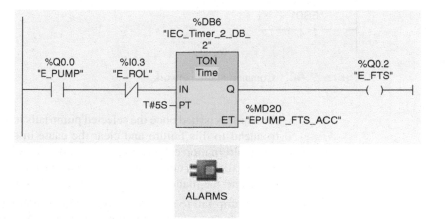

Figure 9.46 East pump failed-to-start network.

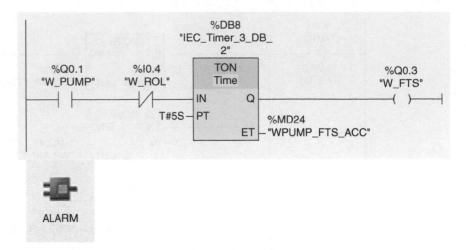

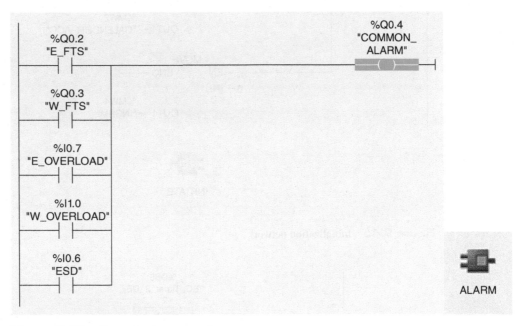

Figure 9.47 West pump failed-to-start network.

Figure 9.48 Common alarm network.

- An alarm is issued once the selected pump fails to start. The operator is expected to attend to this failure and clear the cause in order to allow and enforce the pump-alternation calendar. Having a situation where both pumps failing to start can constitute an emergency condition and must be eliminated. A third standby pump and the manual control system can eliminate this problem.

- West pump fail to start is ON if the motor-running input is not received within 5 seconds from the West pump motor-running output command.

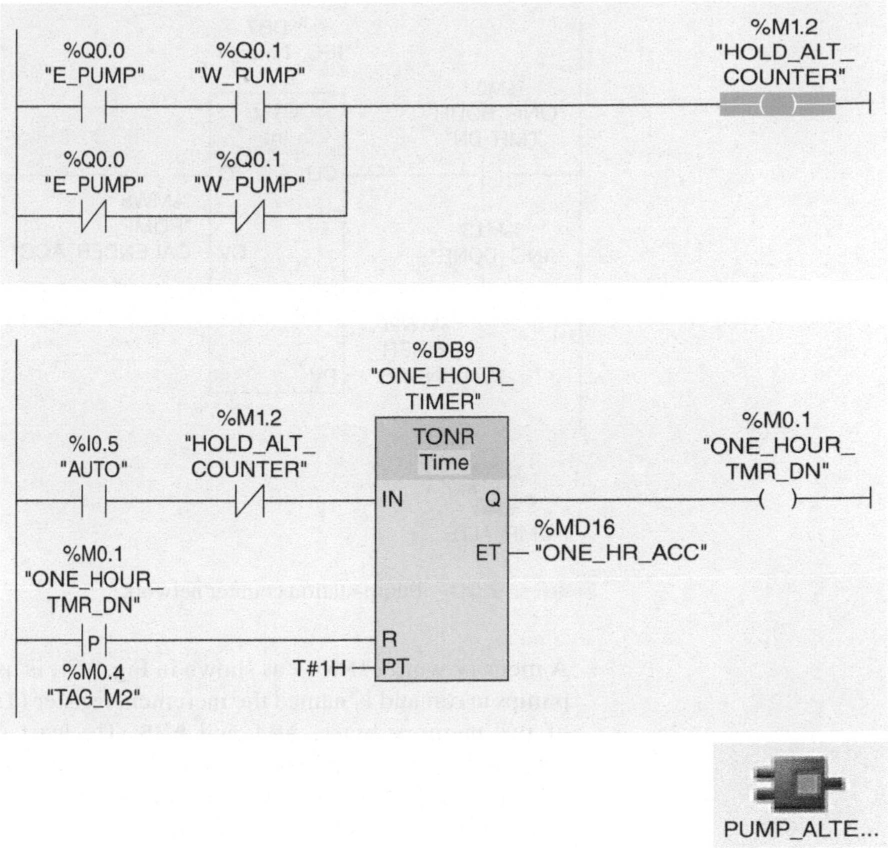

```
       %Q0.0       %Q0.1                                                    %M1.2
      "E_PUMP"    "W_PUMP"                                                "HOLD_ALT_
                                                                          COUNTER"
      ──┤ ├───────┤ ├────────┬──────────────────────────────────────────────( )──────

       %Q0.0       %Q0.1     │
      "E_PUMP"    "W_PUMP"   │
      ──┤/├───────┤/├────────┘
```

```
                                          %DB9
                                      "ONE_HOUR_
                                         TIMER"
                      %M1.2                                                %M0.1
       %I0.5       "HOLD_ALT_      │  TONR     │                       "ONE_HOUR_
       "AUTO"       COUNTER"       │  Time     │                        TMR_DN"
      ──┤ ├─────────┤/├────────────┤ IN      Q ├─────────────────────────( )──────
                                   │           │       %MD16
       %M0.1                       │        ET ├──── "ONE_HR_ACC"
      "ONE_HOUR_                   │           │
       TMR_DN"                     │           │
      ──┤P├───────────────────────┤ R         │
       %M0.4                       │           │
      "TAG_M2"              T#1H ──┤ PT        │
```

PUMP_ALTE...

Figure 9.49 Pump-station 1-hour timer networks.

- The common alarm goes ON if either the East pump fail to start or the West pump fail to start is ON. The East and the West pump overload will trigger the same common alarm. Also, an emergency shutdown (ESD) will cause the same common alarm.

- The two-pump alternation follows a defined calendar based on scheduled run time in hours. A retentive timer (RTO) second network is configured for a 1-hour preset value (PV), as shown in Fig. 9.49. This timer is halted during intervals when the two pumps are running or both are not running (XNOR logic), as shown in the first network in the figure. The calendar is updated only for the selected and running pump. The done bit **ONE_HOUR_TMR_DN** of this timer is used to trigger an up counter, which is configured to implement the desired pump-schedule calendar.

- The pump calendar is initialized by the operator in hours and indicates the time intervals for the two-pump alternations. The up counter in Fig. 9.50 is used to keep track of the accumulated pump run time in hours. The counter is incremented every hour of operation.

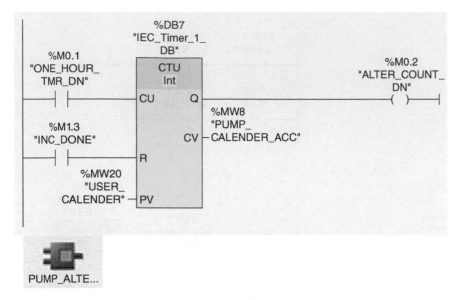

Figure 9.50 Pump-station counter network.

- A memory word (%MW4), as shown in Fig. 9.51, is used to select one of the two pumps to run and is named the increment register (INCRM). This word consists of two memory bytes, %M4 and %M5. The least significant bit of %MW4 is %M5.0, which is the least significant bit of the least significant byte of the selected word %MW4. As this word increments, the least significant bit toggles, causing the pump-selection alternation.
- The increment (INCRM) register will increment every time the user calendar expires. The even values of this register (%M5.0 is FALSE) will be used to select and start the East pump. The odd values of this register (%M5.0 is TRUE) will be used to select and start the West pump. Figures 9.52 and 9.53 implement this logic.

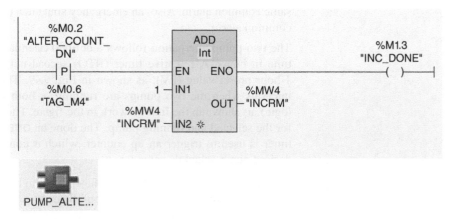

Figure 9.51 Pump-station ADD network.

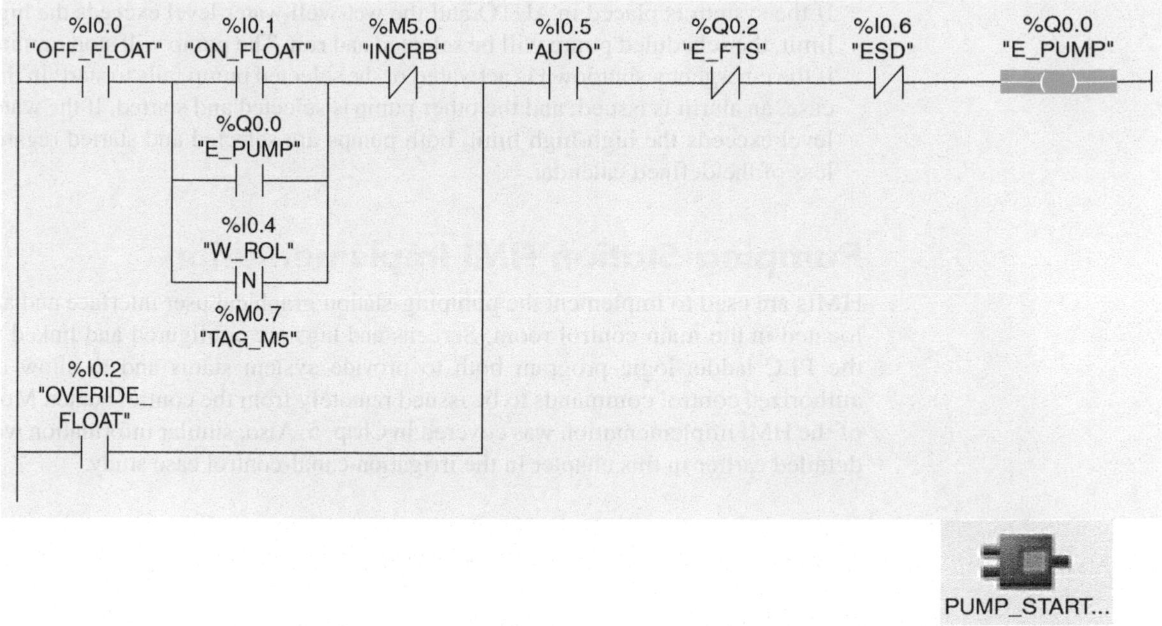

```
     %I0.0          %I0.1          %M5.0         %I0.5          %Q0.2          %I0.6                    %Q0.0
  "OFF_FLOAT"    "ON_FLOAT"       "INCRB"       "AUTO"         "E_FTS"         "ESD"                  "E_PUMP"
──────┤ ├────┬─────┤ ├──────┬──────┤/├──────────┤ ├────────────┤/├────────────┤/├──────────────────────( )──────
            │                │
            │    %Q0.0       │
            │   "E_PUMP"     │
            ├─────┤ ├────────┤
            │                │
            │    %I0.4       │
            │   "W_ROL"      │
            ├─────┤N├────────┤
            │   %M0.7        │
            │   "TAG_M5"     │
      %I0.2 │                │
   "OVERIDE_│                │
     FLOAT" │                │
      ──────┤ ├──────────────┘
```

PUMP_START...

Figure 9.52 East pump network.

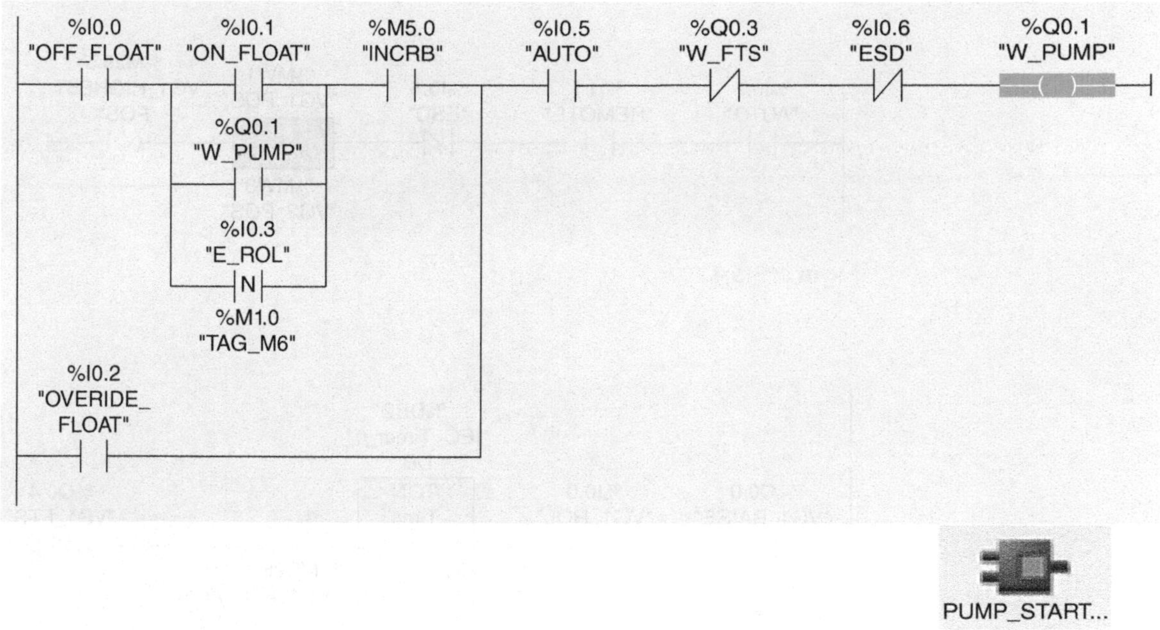

```
     %I0.0          %I0.1          %M5.0         %I0.5          %Q0.3          %I0.6                    %Q0.1
  "OFF_FLOAT"    "ON_FLOAT"       "INCRB"       "AUTO"         "W_FTS"         "ESD"                  "W_PUMP"
──────┤ ├────┬─────┤ ├──────┬──────┤ ├──────────┤ ├────────────┤/├────────────┤/├──────────────────────( )──────
            │                │
            │    %Q0.1       │
            │   "W_PUMP"     │
            ├─────┤ ├────────┤
            │                │
            │    %I0.3       │
            │   "E_ROL"      │
            ├─────┤N├────────┤
            │   %M1.0        │
            │   "TAG_M6"     │
      %I0.2 │                │
   "OVERIDE_│                │
     FLOAT" │                │
      ──────┤ ├──────────────┘
```

PUMP_START...

Figure 9.53 West pump network.

- If the system is placed in AUTO and the wet-well water level exceeds the high limit, the scheduled pump will be selected and run. The pump will stop running if the emergency shutdown is activated or the selected pump fails to start. In this case, an alarm is issued, and the other pump is selected and started. If the water level exceeds the high-high limit, both pumps are selected and started regardless of the defined calendar.

9.6 Pumping-Station HMI Implementation

HMIs are used to implement the pumping-station graphical user interface and are located in the main control room. Screens and tags are configured and linked to the PLC ladder-logic program both to provide system status and to allow all authorized control commands to be issued remotely from the control center. Most of the HMI implementation was covered in Chap. 5. Also, similar information was detailed earlier in this chapter in the irrigation-canal-control case study.

Homework Problems and Laboratory Project

Problems

9.1 What is the purpose of using piping and instrument diagram (P&ID) in industrial automation?

9.2 Document the PLC networks shown in Figs. 9.54 and 9.55 using a logic diagram:

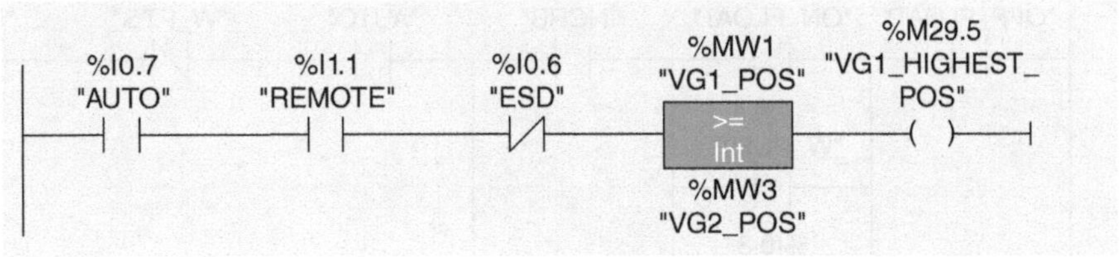

Figure 9.54

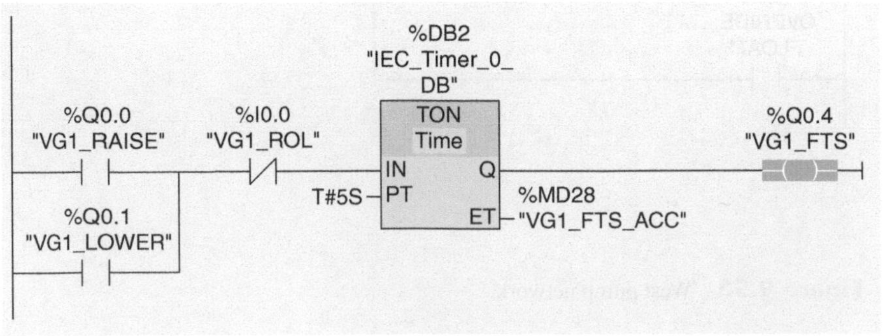

Figure 9.55

9.3   Develop the ladder network(s) for the logic diagrams shown in Figs. 9.56 to 9.58.

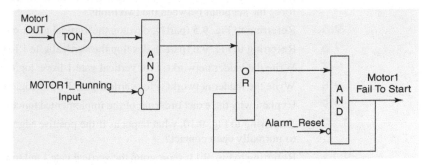

Figure 9.56

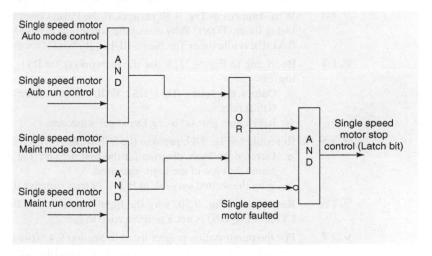

Figure 9.57

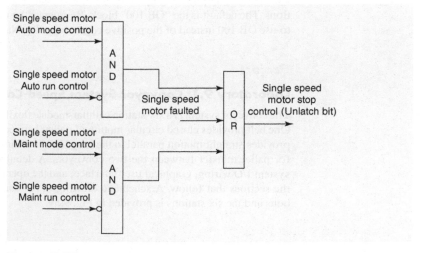

Figure 9.58

9.4   A set point is received from the operator and compared to a high limit (SP_HL) and a low limit (SP_LL). If the set point is outside the limit's range, develop a ladder network(s) to force the set point between the two limits.

9.5   Referring to Fig. 9.5 (part1), develop the vertical gate 1 control down logic diagram.

9.6   Referring to Fig. 9.6 (part2), develop the vertical gate 1 lower logic diagram.

9.7   Write the ladder network(s) for vertical gate 1 lower logic in Problem 9.6.

9.8   Write the ladder network(s) for vertical gate 1 down logic in Problem 9.5.

9.9   Explain why the exact ordering of the implemented functional blocks is irrelevant.

9.10   Referring to Fig. 9.10, what happens if the positive edge trigger AUTO switch is changed to normally open contact?

9.11   Referring to Fig. 9.11, reprogram the vertical gate 1 fail to start network using the Set Reset instructions.

9.12   Reprogram the network in Fig. 9.29, using the Out Range instruction.

9.13   What happens in Fig. 9.30 network if the Pulse Generation timer (TP) is changed to On Delay timer (TON)? Why one of the conditions to raise Vertical Gate 1, Tag Name (VG1_RAISE), is idle timer Tag Name (IDEL_RUNN) is not on?

9.14   Referring to Fig. 9.22, write the network(s) for DS1_DS2_FAIL using the following logics:
      a. Output Tag Name (DS1_DS2_FAIL) is on if downstream 1 (DS1) or downstream 2 (DS2) fails.
      b. Reprogram part (a) using Demerger's theorem.

9.15   Referring to Fig. 9.49, perform the following:
      a. Develop the logic diagram for the first network Tag Name (HLD_ALT_COUNTER). Name the type of the logic gate used.
      b. Why the second network in Fig. 9.49 uses retentive timer (TONR)?

9.16   Referring to Fig. 9.50, why the input instruction to the counter Tag Name (ALTER_COUNTER_DN) is not a positive edge trigger?

9.17   For the pump station project listed in Section 9.4, write a network(s) to sound an alarm if any of the three sensors (ON, OFF, and OVERIDE) switches fails.

9.18   Startup OBs execute one time when the operating mode of the CPU changes from STOP to RUN, including powering up in the RUN mode and in commanded STOP-to-RUN transitions. The default is the "OB 100" block. Reprogram the initialization network in Fig. 9.45 to use OB 100 instead of the positive edge trigger Tag Name (AUTO).

Project

Laboratory 9.1: Conveyor System Speed-Control Capstone Project

The conveyor system is a six-station cellular-module flexible manufacturing system (FMS). One belt provides closed circular motion of pallets between stations, whereas a second belt provides straight motion parallel to the long leg of the inner belt. Two transient belts allow for pallet transfer between the two conveyors. A detailed descriptions of the conveyor system I/O wiring, graphical user interface, and the operation of each station are given in the sections that follow. A schematic drawing of the conveyor system including the two belts and the six stations is provided in Fig. 9.59.

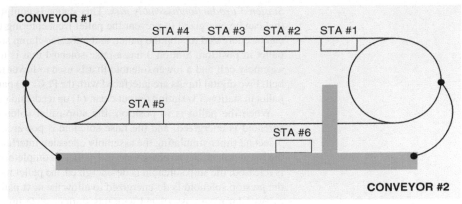

Circular Conveyor System

Figure 9.59 Circular conveyor system.

Laboratory Description

The conveyor system was configured to provide the FMS platform for this project. It also will be used to simulate and test the control algorithms and implementation of the operation, which are described next. The following is a brief description of the conveyor-system workstations and their associated control implementations.

Station 1 loading cell. This station is equipped with two solenoids, A and C. With no power applied, a pallet can enter the station and lock into position. When the two solenoids receive power, they switch position and prevent additional pallets from entering the station.

Once a pallet is detected in position at Station 1, a start pulse is output to the loading robot. The robot sends a done pulse to the PLC on completion of part loading. For this study, a 5-second timer is used to simulate the loading task. The timer timing bit is used as the Go command to the robot. The timer done bit is used as the done acknowledgment from the robot, which moves the pallet to Station 2.

Two alarms are associated with Station 1. The Station 1 loading-jam alarm is activated whenever the done pulse from robot is active and the pallet remains in position for 1 second. The Station 1 task time out alarm becomes active whenever the Go command is not acknowledged by the loading robot in 5 seconds.

Station 2 production-feed-rate regulation. This station is equipped with a pallet-advance solenoid. When this solenoid receives power and energizes, it activates and advances the pallet to Station 3. The PLC receives input signals from two limit switches indicating the advance solenoid positions (forward or backward).

When the pallet is in position, the back-limit switch is on, the Station 2 output is clear, and the Station 3 prestop is inactive. The advance solenoid will energize and move the pallet to Station 3. Otherwise, when Station 1 output is clear, the advance solenoid is de-energized and retracted.

One alarm is associated with Station 2, a production-rate-jam alarm. This alarm is activated by a malfunction in the advance solenoid's two limit switches.

Station 3 production/assembly task. This station is equipped with a prestop solenoid that can be energized to prevent the pallet from entering the station. A stop solenoid can be energized to stop the pallet, and a locator-clamp solenoid is used to clamp the pallet in position. Station 3 has a raise solenoid that is used to get the part into the assembly cell and a lower solenoid that is used to lower the pallet onto the conveyor belt. Five digital inputs are interfaced with the PLC: (1) pallet approaching station, (2) pallet in station, (3) station output clear, (4) up reed, and (5) down reed light switches.

When the pallet is in position, the stop-pallet solenoid is activated, the clamp solenoid is energized, and the raise solenoid is powered to raise the pallet. After a 5-second timer simulating the assembly operation interface, as described in Station 1, the lower solenoid energizes. Once the pallet is completely lowered, the clamp locator is released, the stop solenoid is de-energized, the pallet moves to the next station, and the prestop solenoid is de-energized to allow the next pallet to enter the station.

Two alarms are associated with this station: cycle-up failure and cycle-lower failure. At Station 3, the cycle-up-failure alarm is activated whenever the raise solenoid is on the up position and the reed switch remains off for 1 second. The cycle-down-failure alarm works the same way for the lower operation.

Station 4 testing and sorting. This station is equipped with four solenoids: raise, lower, prestop, and stop. The functions of the solenoids are similar to those explained for Station 3. The station also has six switch inputs: (1) pallet at station, (2) station up, (3) station at middle position, (4) station down, (5) pallet at output, and (6) output clear.

Station 4 can reject a pallet to a transfer-conveyor area (Continuation of Conveyor 1) or move the pallet to Conveyor 2. The reject command can be initiated from the user interface by the operator. The station has three positions: up, middle, and down. When both lower and raise solenoids are de-energized, the station is in the middle position. When the pallet is in position, the station is in the middle position. A reject signal is initiated from the user interface when the Station 6 Request-to-push command is not activated. The raise solenoid is energized to raise the pallet and move it to the transfer conveyor (pallet rejected).

Station 4 has two alarms: cycle-up failure and cycle-down failure. The operation of these alarms is exactly the same as for Station 3.

Station 5 production-output-rate regulation. This station is equipped with two solenoids: lane stop and wipe off. It also has two inputs: (1) lane clear and (2) feed-lane full.

If the operator requests a Station 4 reject or a Station 6 push, the wipe-off solenoid will energize and allow the pallet to move to Conveyor 2. Otherwise, the pallet continues on Conveyor 1. No alarms are assigned to this station.

Station 6 unloading cell. This station is equipped with four solenoids: advance pusher, prestop, pusher stop, and lane stop. The PLC receives six switch inputs from this station: (1) pusher forward, (2) pusher backward, (3) pallet present left, (4) ON/OFF enable/disable selector switch, (5) lane clear to push, and (6) output clear.

The operator can request to push from the user interface/HMI. When a pallet is present left, a request to push is made from the HMI. If a request to reject from Station 4 is inactive and the Station 6 lane is clear to push, the lane-stop solenoid will stop the pallet, and the prestop solenoid will prevent incoming pallets from entering the station. The advanced pusher solenoid will energize and unload the pallet to the transfer-conveyor area. When the pusher is back, all solenoids reverse their action, the next pallet enters the station, and the push cycle is repeated.

Two alarms are associated with this station: advance pusher forward and backward failure. The alarms operation is similar to those for Station 4.

Each of the above six stations has a separate ENABLE/DISABLE switch that can be used to bypass the station. This feature allows for FMS activities. Not only we can skip certain manufacturing operations, but the entire station also can be replaced, modified, or removed. The entire system is equipped with an emergency shutdown (ESD) master control relay (MCR). As with other systems, all hardware modules and the two conveyor motors are provided power through the ESD MCR contacts, as explained in Chap. 6.

Conveyor- System I/O Listings

The following two tables provide detailed listing of all inputs and outputs used in the conveyor-system control.

Conveyor System Inputs

Device Name	Address	Description
STA #1PIP LS	I1.0	120VAC SIGNAL LIMIT SWITCH TO INDICATE PALLET IN POSITION
STA #1 OUTCLR SEN	I1.1	120VAC SENSOR TO INDICATE STATION #1 OUTPUT IS CLEAR
STA #1 ACTIVE SEL SW	I1.2	120VAC ON/OFF SELECTOR SWITCH TO ENABLE / DISABLE STATION #1 FUNCTION
STA #2 CYL FWD LS	I1.3	120VAC LIMIT SWITCH TO INDICATE THE ADVANCE SOLENOID (SOL) FORWARD POSITION
STA #2 CYL BAC LS	I1.4	120VAC LIMIT SWITCH TO INDICATE ADVANCE SOL BACKWARD POSITION
STA #2 PIP LS	I1.5	120VAC SENSOR TO INDICATE STA #2 PALLET IN POSITION
STA #2 OUT CLR	I1.6	120VAC SENSOR TO INDICATE STA #2 OUTPUT IS CLEAR
STA #2 ACTIVE SEL SW	I1.7	120VAC ON/OFF SELECTOR SWITCH TO ENABLE / DISABLE STATION #2
STA #3 STA UP REED SW	I2.0	STA #3 120VAC SENSOR TO INDICATE RAISE SOL UPPER POSITION
STA #3 STA DOWN REED SW	I2.1	STA #3 120VAC SENSOR TO INDICATE RAISE SOL DOWN POSITION
STA #3 PAL AT IN	I2.2	STA #3 120VAC SENSOR TO INDICATE PALLET AT INPUT
STA #3 PAL AT IN STA PROX	I2.3	120VAC SENSOR TO INDICATE PALLET IN STA #3
STA #3 OUT CLR SW	I2.4	120VAC LIMIT SWITCH TO INDICATE STA #3 OUTPUT IS CLEAR
STA #3 ACIVE SEL SW	I2.5	120VAC ON/OFF SELECTOR SWICTH TO ENABLE / DISABLE STA #3
STA #4 UP LS	I2.6	120VAC LIMIT SWITCH TO INDICATE STA #4 UPPER POSITION
STA #4 MID LS	I2.7	120VAC LIMIT SWITCH TO INDICATE STA #4 MID POSITION
STA #4 DOWN LS	I3.0	120VAC LIMIT SWITCH TO INDICATE STA #4 DOWN POSITION
STA #4 PAL AT IN PROX	I3.1	120VAC SENSOR TO INDICATE STA #4 PALLET AT INPUT POSITION

Conveyor System Inputs (*Continued*)

Device Name	Address	Description
STA #4 PAL AT OUT PROX	I3.2	120VAC SENSOR TO INDICATE STA #4 PALLET AT INPUT POSITION
STA #4 OUT CLR	I3.3	120VAC SENSOR TO INDICATE STA #4 OUTPUT IS CLEAR
STA #4 TRANS AREA CLR	I3.4	120VAC SENSOR TO INDICATE STA #4 TRANSITION AREA IS CLEAR
STA #4 ACTIVE SEL SW	I3.5	120VAC ON/OFF SELECTOR SWITCH TO ENABLE/DISABLE STATION #4
STA #5 LANE CLR	I3.6	120VAC FIBER SENSOR TO INDICATE LANE IS CLEAR
STA #5 FEED LANE FULL	I3.7	120VAC SENSOR TO INDICATE STA #5 LANE IS FULL
STA #5 ACTIVE SEL SW	I4.0	120VAC ON/OFF SELECTOR SWITCH TO ENABLE/DISABLE STATION #5
STA #6 PUSHER BACK LS	I4.1	120VAC LIMIT SWITCH TO INDICATE PUSHER BACK POSITION
STA #6 PUSHER FWD LS	I4.2	120VAC LIMIT SWITCH TO INDICATE PUSHER FORWARD POSITION
STA #6 PALLET PRES PX	I4.3	120VAC PROXIMITY SWITCH TO INDICATE PALLET PRES LEFT POSTION
STA #6 LANE CLRTO PUSH	I4.4	120VAC SENSOR TO INDICATE STA #6 LANE IS CLEAR
STA #6 OUT CLR	I4.5	120VAC SENSOR TO INDICATE STA #6 OUTPUT IS CLEAR
STA #6 ACTIVE SEL SW	I4.6	120VAC ON/OFF SELECTOR SWITCH TO ENABLE / DISABLE STATION #6
STA #7 ZONE CLR FIBER	I4.7	120VAC FIBER SWITCH TO TA #7 ZONE CLEAR
CON V #1 STOP MTR	I5.0	120VAC PUSH BUTTON TO STOP CONVEYOR #1 MOTOR
CON V #1 START MTR	I5.1	120VAC PUSH BUTTON TO START CONVEYOR #1 MOTOR
CON V #2 STOP MTR	I5.2	120VAC PUSH BUTTON TO STOP CONVEYOR #2 MOTOR
CON V #2 START MTR	I5.3	120VAC PUSH BUTTON TO START CONVEYOR #2 MOTOR
STA #6 PIP	I5.4	120VAC INDICATE PALLET IN POSITION RIGHT
CONV#1 SPEED	IW64	CONVEYOR #1 0-10V DC ANALOG INPUT SIGNAL
CONV#2 SPEED	IW66	CONVEYOR #2 0-10V DC ANALOG INPUT SIGNAL

Conveyor System Outputs

Device Name	Address	Description
STA #1 CYL A	Q1.0	120VAC SIGNAL TO STA #A SOLENOID A
STA #1 CYL B	Q1.1	120VAC SIGNAL TO STA #B SOLENOID A
STA #1 CYL C	Q1.2	120VAC SIGNAL TO STA #C SOLENOID A
STA #2 ADV SOL	Q1.3	120VAC SIGNAL TO STA #2 ADVANCE SOLENOID
STA #3 RAISE STA SOL	Q1.4	120VAC SIGNAL TO STA #3 TO RAISE SOLENOID
STA #3 LOWER STA SOL	Q1.5	120VAC SIGNAL TO STA #3 TO LOWER SOLENOID
STA #3 PRE-STOP SOL	Q1.6	120VAC SIGNAL TO STA #3 PRESTOP SOLENOID
STA #3 STA STOP SOL	Q1.7	120VAC SIGNAL TO STA #3 START/STOP SOLENOID
STA #3 LOCATOR CLMP SOL	Q2.0	120VAC SIGNAL TO STA #3 TO RAISE SOLENOID
STA #4 RAISE STA SOL	Q2.1	120VAC SIGNAL TO STA #4 RAISE STATION SOLENOID
STA #4 LOWER STA SOL	Q2.2	120VAC SIGNAL TO STA #4 LOWER SOLENOID PALLET AT INPUT
STA #4 PRE-STOP SOL	Q2.3	120VAC SIGNAL TO STA #4 PRE STOP SOLENOID
STA #4 LANE STOP SOL	Q2.4	120VAC SIGNAL TO STA #4 LANE STOP SOLENOID
STA #5 LANE STOP SOL	Q2.5	120VAC SIGNAL TO STA #5 LANE STOP SOLENOID
STA #5 WIP OFF SOL	Q2.6	120VAC SIGNAL TO STA #5 LANE STOP SOLENOID
STA #6 ADANCE PUSHER SOL	Q2.7	120VAC SIGNAL TO STA #4 WIPE OFF SOLENOID
PRE-STOPS SOL	Q3.0	TO STA #6 PRE STOP SOLENOID
STA #6 PUSHER STOP SOL	Q3.1	120VAC SIGNAL TO STA #6 PUSHER STOP SOLENOID
STA #6 LANE STOP SOL	Q3.2	120VAC SIGNAL TO STA #6 LANE STOP SOLENOID
STA #7 ALLOW AIR LOGIC SOL	Q3.3	120VAC SIGNAL TO STA #7 ALLOW AIR LOGIC SOLENOID
STA #1 ALLOW	Q3.4	120VAC TO CONV #1 MOTOR
STA #2 ALLOW	Q3.5	120VAC TO CONV #2 MOTOR
CONV #1 SPEED	QW80	ANALOG OUTPUT SIGNAL TO AC DRIVE FOR CONVEYOR #1 MOTOR
CONV #2 SPEED	QW82	ANALOG OUTPUT SIGNAL TO AC DRIVE FOR CONVEYOR #2 MOTOR

Graphical User Interface (HMI)

Eight graphic screens should be implemented for control of the conveyor system. In addition to the Directory page, seven pages should be provided in the HMI. These are as follows: System Overview, System Status, Control and Trending, Alarm Graphics, Conveyor 1 Control, Conveyor 2 Control, and Alarms Summary. The following subsections will briefly describe each of the eight HMI pages. Specific details of the user-interface implementation will depend on the HMI used, for example, color or monochrome, and its capabilities.

- *Directory page.* This screen lists the seven available options and provides for the selection of individual pages. The pressing of a function key will take the user to the selected screen.

- *Overview page.* This page shows the two conveyor belts, station ENABLE/DISABLE switch status, station task busy/not busy status, Motor 1 and Motor 2 running/not running status, and color-coded status for assign station. The color code provides red for station enabled, green for station disabled, red for conveyor running, and green for conveyor not running. No user entry is assigned for this page.

- *Control and Trend page.* This page provides the status of the six stations used in the control system: ENABLE/DISABLE, alarms for stations and conveyors, acknowledge/reset of alarms, and associated color-code definition. It also provides for user input commands: Station 4 reject/not reject, Station 6 push/not push, Conveyors 1 and 2 speed, and the trending of the two conveyor speeds up or down. Three modes of control are provided on this page. In manual mode, the operator can demand changes in conveyor speed by entering either percentage speed. The automatic mode is designed to allow users to regulate the conveyor speed using PID control. The user initiates such control by entering the desired conveyor-speed set point (SP).

- *Status page.* This page displays status information for the system in numeric or symbolic form. Motor speeds are displayed in percentage value (%).

- *Alarm page.* This page provides a detailed listing and description of all system alarms. The Control and Trend page provides for an annunciation of an alarm without any details.

- *Conveyor 1 control page.* This page displays the status of the controller for the conveyor system, only for Conveyor 1.

- *Conveyor 2 control page.* This page displays the status of controller for the conveyor system, only for Conveyor 2.

- *Alarm Summary page.* This page displays all alarm transactions, including current and past alarms, and indicates whether or not they have been acknowledged.

Project Requirements

Use the preceding specifications to perform the following:

- Document and implement layer 1 of the conveyor-system speed control. Present the implementation with your project group to the course/technical training instructor. Implement recommended modifications, and secure a satisfactory layer 1 review.

- Proceed with the design of layer 2. Perform the design-review presentation, implement required changes, and secure final approval for the layer 2 document.

- Perform the final implementation and checkout, as detailed in layer 3 discussions and coverage. This includes the ladder-logic and HMI implementations, along with the communication/networks configuration and the overall system debugging/checkout. Prepare for final presentation and system operation demonstration.

- Consider valid inputs and recommendations obtained during the final presentation, and finalize your implementation. Document all tasks, and write the final project technical report and manual.

Appendix

S7-1200 Workshop Project—Simulated Bottling Project

This appendix details the steps needed to create a project and simulate it using the Siemens human-machine interface (HMI), WinCC, and the development software. It can help readers through the detailed tasks involved in the implementation, simulation, and testing of typical industrial automation tasks. The focus here is on the graphical user interface, including both command and status.

This simulated project contains notices you have to observe in order to ensure your personal safety, as well as to prevent damage to property. The notices referring to your personal safety are highlighted in the Siemens manual by safety-alert symbols; notices referring only to property damage have no safety-alert symbol.

TRADEMARKS All names identified by ® are registered trademarks of Siemens AG. The remaining trademarks in this publication may be trademarks whose use by third parties for their own purposes could violate the rights of the owner.

1. Load and Re-Store Existing Project

The project **FB_conveyor_counter** from Module 010-030 will now be opened as the model for this program (Fig. A.1).

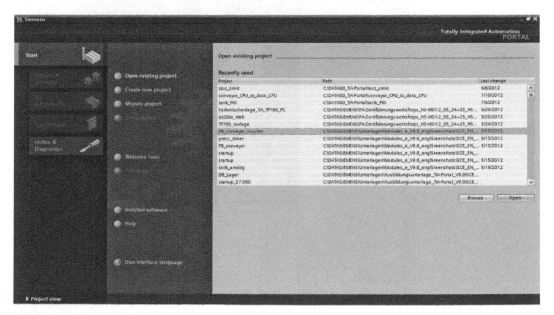

Figure A.1

Now, First steps are offered for configuring. Click on Open the project view (Fig. A.2).

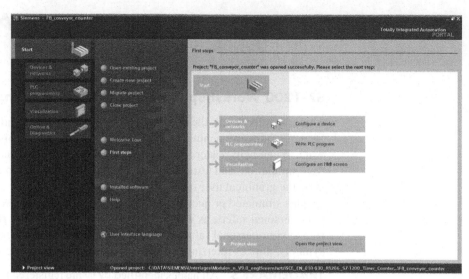

Figure A.2

First, we are going to store the project under another name. In the menu Project, click on Save As (Fig. A.3).

Figure A.3

Now, Save the project under the new name `conveyor_KTP600` (Fig. A.4).

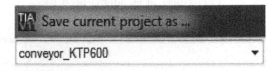
conveyor_KTP600

Figure A.4

To set up a new project, select the list box by double-clicking on Add new device. Under SIMATIC HMI, select the 6-inch display panel KTP600 Basic PN. Set the checkmark at Start device wizard. Click OK.

Under Select PLC, first select `controller_conveyor` (Fig. A.5).

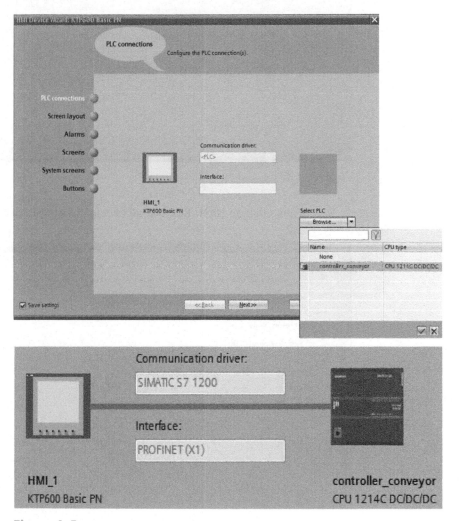

Figure A.5

Then, click on Next.

Under Screen layout, change the background color to White, and remove the checkmark at Header (Fig. A.6).

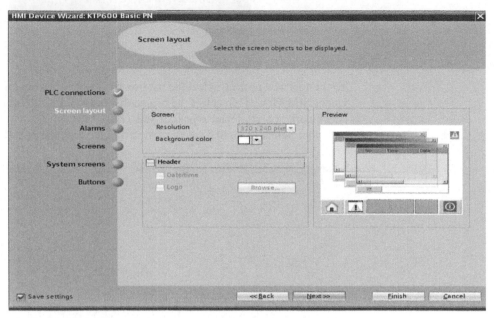

Figure A.6

Then click on Next. Remove all checkmarks at Alarms (Fig. A.7).

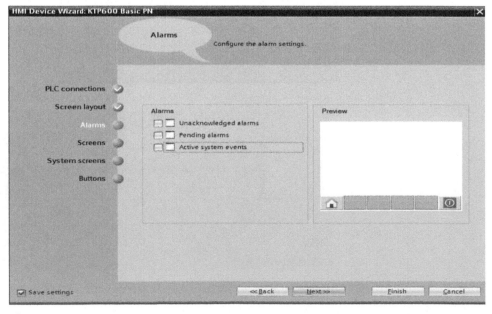

Figure A.7

Then click on Next. Under Screen navigation, a screen menu structure could be set up. For our example, the Root screen is sufficient for the time being (Fig. A.8).

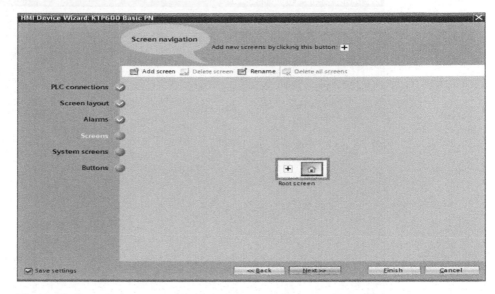

Figure A.8

Then click on Next. As System screens, select the switchover Operating modes and Stop Runtime (Fig. A.9).

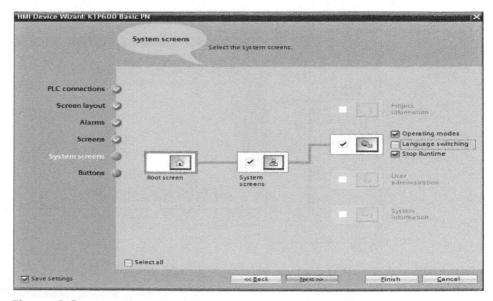

Figure A.9

Then click on Next. Finally, predefined system buttons can be added. Remove all checkmarks (Fig. A.10).

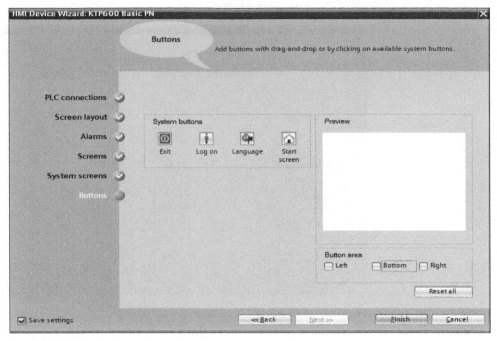

Figure A.10

Then click on Finish. The WinCC interface is now opened with the basic display (Fig. A.11).

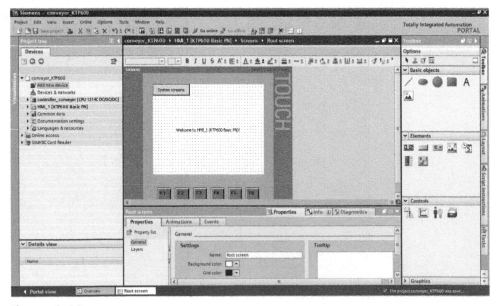

Figure A.11

2. WinCC Operator Interface (Fig. A.12)

The name SIMATIC is a registered trademark of the company Siemens, since 1958. SIMATIC connects two words "Siemens" and "Automatic" in one word. SIMATIC is a digital computer which can store and run programs. It also has some inputs and outputs. The running program on the SIMATIC controls the relation between inputs and outputs in real time. SIMATIC WinCC is a supervisory control and data acquisition (SCADA) and human-machine interface (HMI) system from Siemens. It can be used in combination with Siemens S7-1200 and SIMATIC control systems. WinCC is written for the Microsoft Windows operating system. WinCC uses Microsoft SQL Server for logging and comes with a VBScript and ANSI C application programming interface.

3. Project Navigation

The Project window is the central connection point for project editing. All components and all available editors for a project are displayed in a Project window in a tree structure and can be opened from there. Each editor is assigned a symbol with which the associated objects can be identified. Only those elements are displayed in the Project window that the selected operator panel supports.

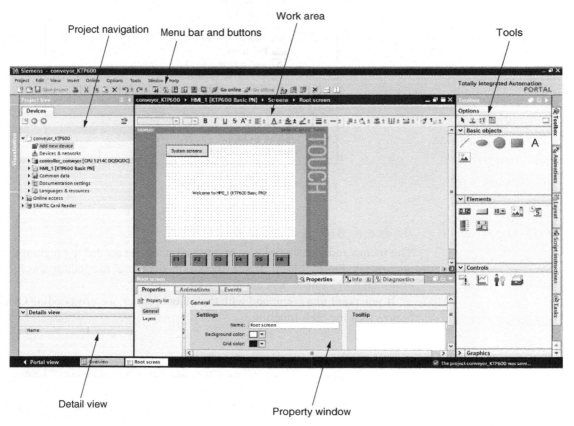

Figure A.12

In the Project window, you can access the basic settings of the operator panel (Fig. A.13).

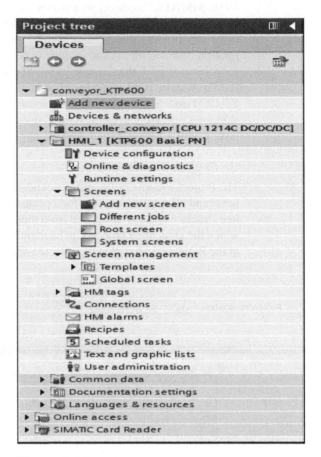

Figure A.13

4. Menu Bar and Buttons

The menus and the symbol bars provide all functions needed to program the operator panel (Fig. A.14). If a corresponding editor is active, editor-specific menu commands or symbol bars are displayed.

If you point to a command with the mouse pointer, a corresponding QuickInfo is provided for each function.

Figure A.14

Work Area

In the work area, we edit objects of the project. All WinCC elements are arranged around the work area (Fig. A.15). In the work area, we also edit the project data either in table form (e.g., tags) or graphically (e.g., a process display).

A symbol bar is located on the upper part of the work area. Here fonts, colors, or functions such as rotate, align, and so on can be selected.

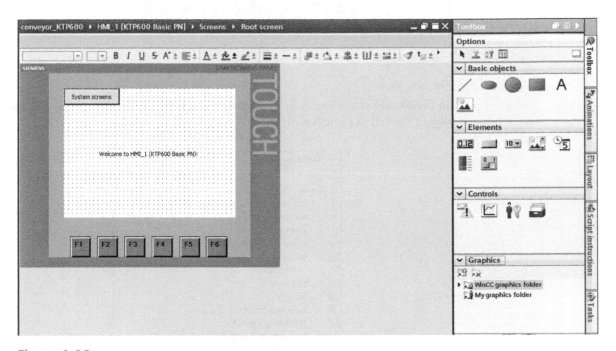

Figure A.15

Tools

In the Tools window, a selection of objects is provided that can be inserted in pictures, such as graphic objects and operating elements. In addition, the Tools window includes libraries with preassembled library objects and collections of picture blocks. Objects are moved to the work area with drag and drop.

Property Window

In the Property window, we edit the properties of objects, for example, the color of picture objects. The Property window is available only in certain editors. In the Property window, the properties of the selected object are displayed arranged according to categories. As soon as you exit an input field, the value changes become effective. If you enter an invalid value, its background is colored. Via the QuickInfo, you are then provided with information regarding the valid value range, for example. In addition, animations and events of the selected object are configured in the Property window (Fig. A.16); here, for example, a display changes when releasing the button.

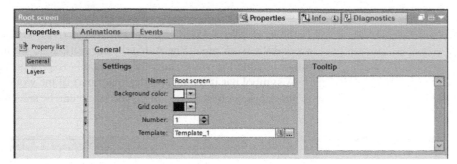

Figure A.16

Details View

In the Details view (Fig. A.17), additional information is displayed regarding the object marked in Project navigation.

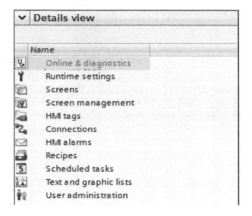

Figure A.17

5. Operating Screens and Connections

A screen can consist of static and dynamic elements. Static elements such as text and graphics are not updated by the controller. Dynamic elements are connected to the controller and visualize current values from the controller's memory. Visualization can be in the form of alphanumeric displays, curves, and bars. Dynamic elements are also inputs on the operator panel that are written to the controller's memory. *Tags* provide for the interfacing with the controller.

First, we are going to create a screen for our conveyor control.

6. Basic Screen or Start Screen

This screen is already set up automatically and defined as the Start screen (Fig. A.18). Here the entire system is represented. Buttons are provided for switching between the manual and automatic modes, for starting and stopping the conveyor motor, and for exchanging the case. The movement of the bottle on the conveyor belt and the number of bottles in the case are shown graphically. By operating the F6 key, we are jumping to the System screens.

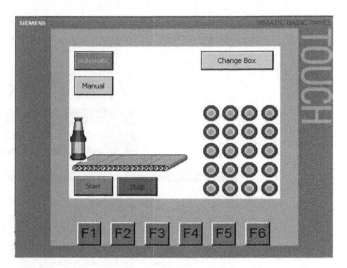

Figure A.18

7. Connections to S7 Controllers

For operating elements and display objects that access the process values of a controller, first a connection to the controller has to be configured. Here we specify how and over which interface the panel communicates with the controller (Fig. A.19). In the Project navigation, click on Connections. Through the settings in the hardware configuration, all parameters are already set.

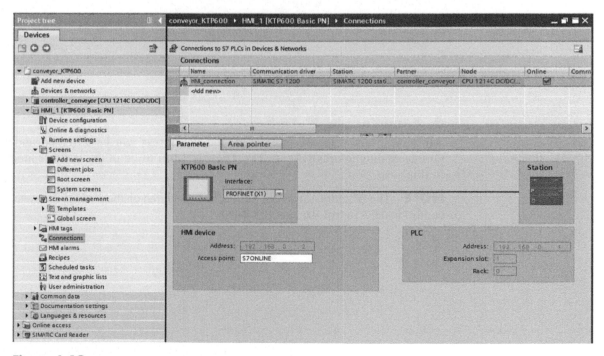

Figure A.19

The IP address still has to be assigned to the panel. With Accessible devices, read out the panel's MAC address. Then click on the button Show (Fig. A.20).

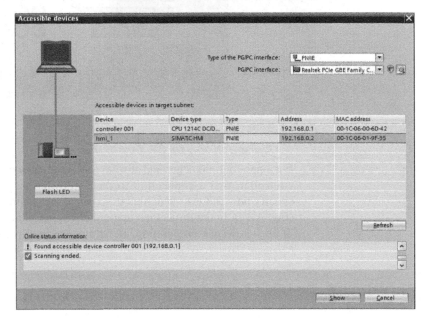

Figure A.20

8. Assigning the IP Address

After the MAC address is entered, the IP address can be assigned under Online & diagnostics (Fig. A.21). The panel has to be in the transfer mode in this case.

NOTE The IP address also can be checked or entered on the panel in the system control under Control Panel at PROFINET.

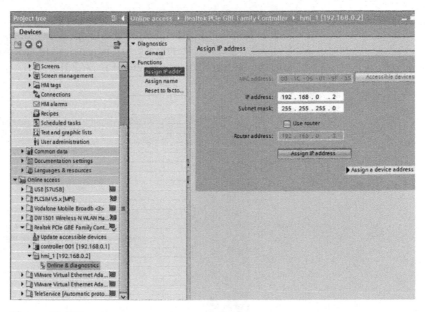

Figure A.21

9. Configuring the Basic Screen

The system screen is called using the button System screens (Fig. A.22). The function of the button System screens is to be transferred to the function key F6. Select the button System screens, and in the Property window, copy the function ActivateScreen at Events, Release key.

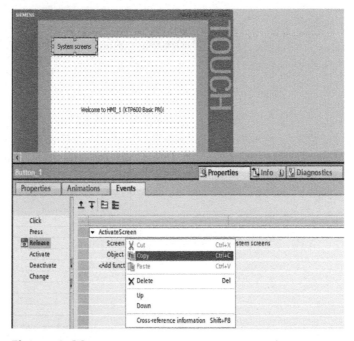

Figure A.22

10. Function Key F6

Select function key F6, and in the Property window, insert the function Activate-Screen at Events, Release key. Then delete or remove the text field in the center and the button System screens (Fig. A.23).

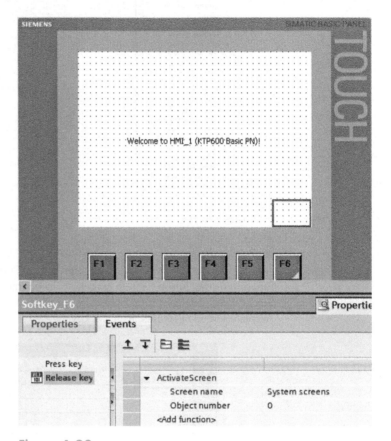

Figure A.23

The triangle on the function key F6 refers to the key having been configured.

11. Configuring the Automatic and Manual Buttons

Drag a button into the work area of the basic screen (Fig. A.24).

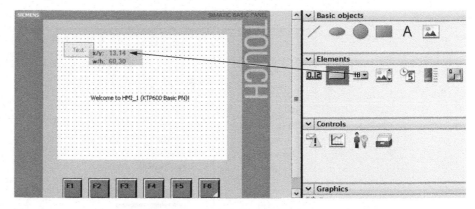

Figure A.24

As text, enter Automatic (Fig. A.25). *Caution:* Don't press the input key; otherwise, a second line is generated.

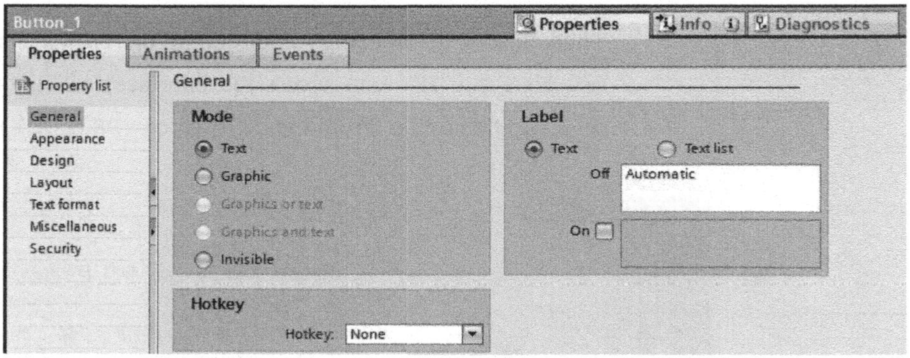

Figure A.25

Under Layout, enter position and size (Fig. A.26).

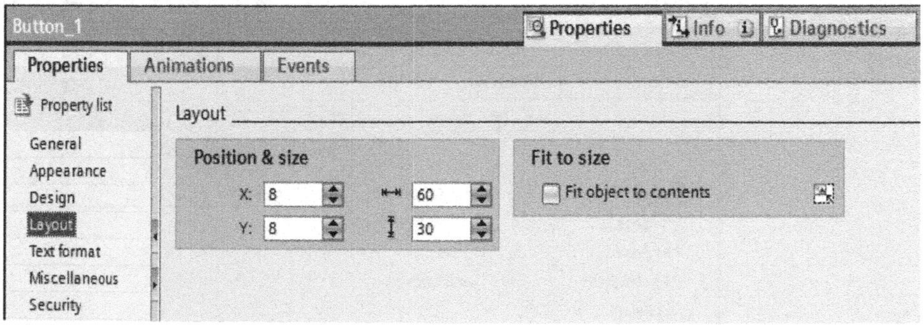

Figure A.26

Under Events, at Press, select the function under bit editing SetBitWhileKey-Pressed (Fig. A.27).

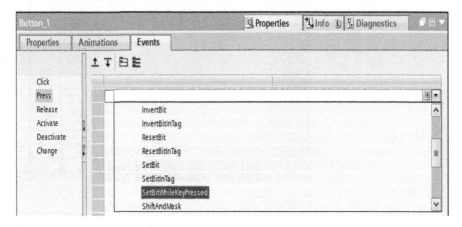

Figure A.27

Then click on the field tag (Input/output) and open the tag window with the button .... Here you also can access the interface declaration of data blocks. As tag, select `automatic` from the `conveyor_DB [DB1]` ☑ (Fig. A.28).

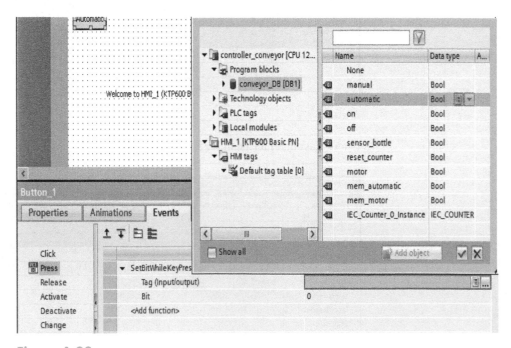

Figure A.28

In the Automatic mode, the button flashes and changes color. Under Animations, select Add new animation (Fig. A.29).

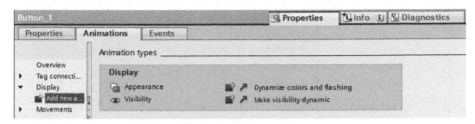

Figure A.29

Select Appearance, and confirm your selection with OK (Fig. A.30).

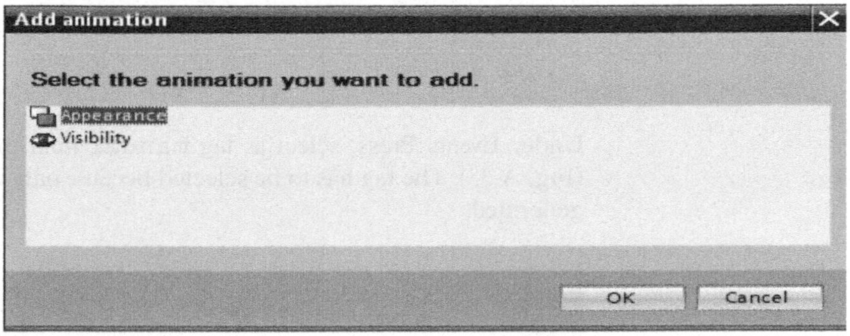

Figure A.30

As tag, select **memory_automatic** from the **conveyor_DB [DB1]**. The button changes color in the automatic mode, that is, when the tag **memory_ automatic** has the value 1. For the color change to be visible, at Appearance, change the foreground color to white and the background color to green (Fig. A.31). At Flashing, set to Yes.

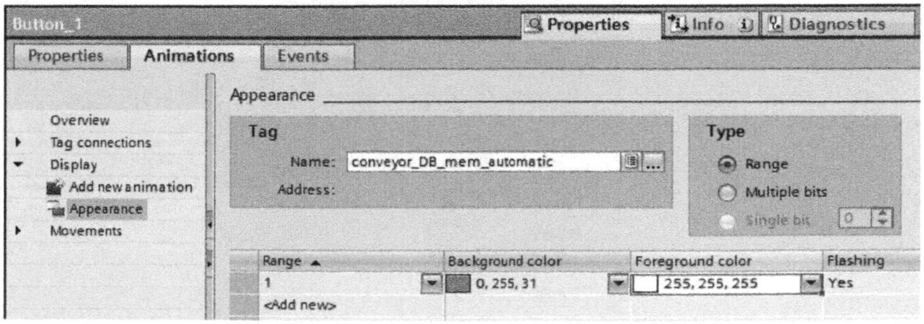

Figure A.31

Copy and add the button Automatic. Place the added button below the Automatic button. As text, enter Manual at Label (Fig. A.32). *Caution:* Don't press the input key; otherwise, a second line is generated.

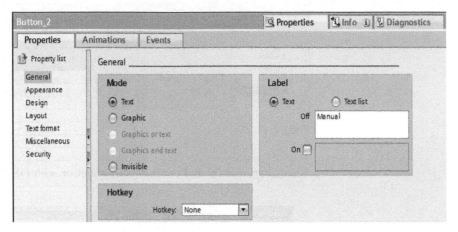

Figure A.32

Under Events Press, select as tag `manual` from the `conveyor_DB [DB1]` (Fig. A.33). The tag has to be selected because only in this way is a new HMI tag generated.

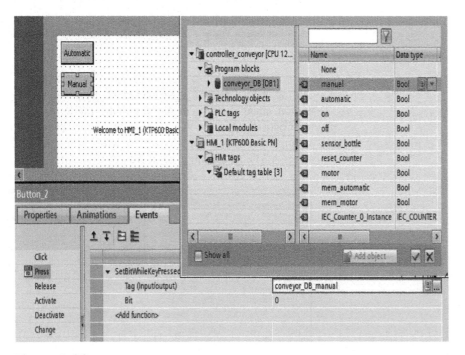

Figure A.33

The button changes color in the manual mode, that is, when the tag `memory_automatic` has the value 0. For the color change to be visible, at Appearance, change the foreground color to white and the background color to blue (Fig. A.34). At Flashing, set No.
Save the project.

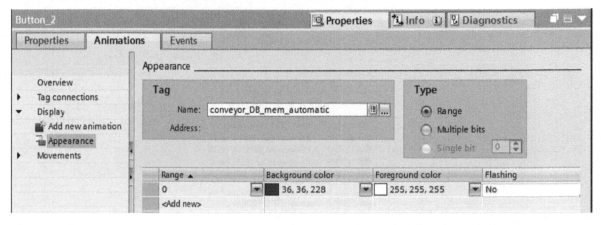

Figure A.34

12. Changes in the Step 7 Program

Before we start testing the visualization, we first have to make a change in the Step 7 program. In OB1, remove the assignments `S1_conveyor1` and `S2_conveyor1` when calling FB1 (Fig. A.35). This is necessary because otherwise the panel signals are overwritten by the process image of the inputs. Save and load the modified program.

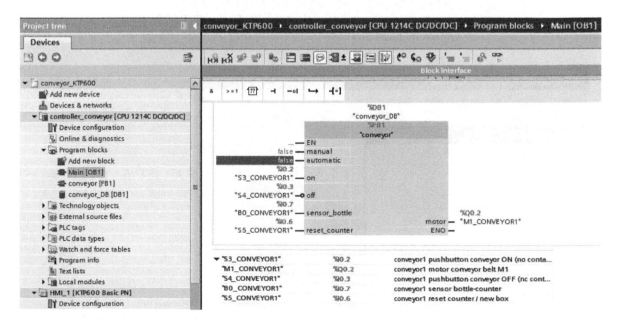

Figure A.35

13. Setting the PG/PC Interface for Run-Time Simulation

For a connection to be set up between the run-time simulation on the PG/PC and the S7-1200 CPU, first the PG/PC interface has to be set to TCP/IP (Fig. A.36).

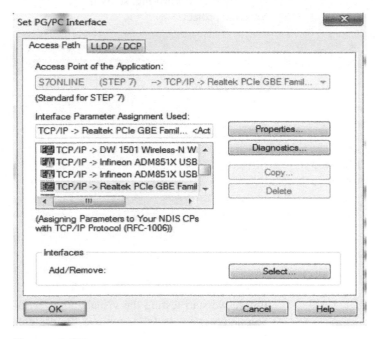

Figure A.36

Starting the Configuration in Run Time

In the Project window, select the HMI_1 [KTP600 Basic PN] panel, and click on the button Start simulation (Fig. A.37).

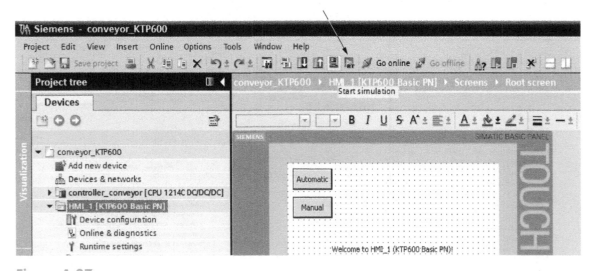

Figure A.37

Visualization is opened in the RT Simulator (Fig. A.38).

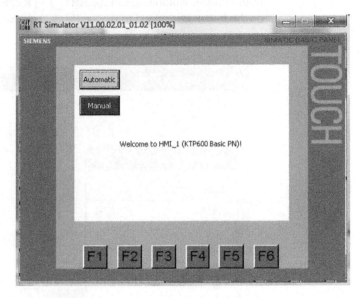

Figure A.38

Test the conveyor control project. Either the automatic or manual mode is now preselected on the panel (Fig. A.39).

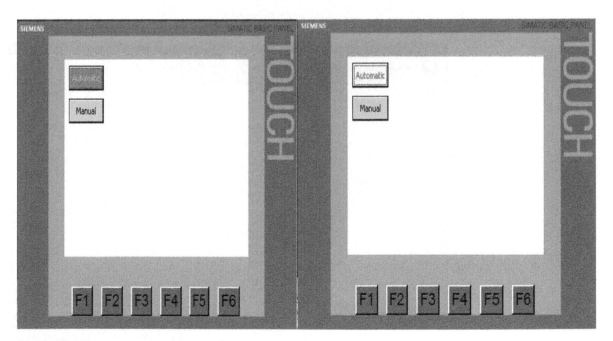

Figure A.39

14. Loading the Configuration to the Panel and Testing It

In the Project window, select the `HMI_1 [KTP600 Basic PN]` panel, and click on the button Download to device (Figs. A.40 and A.41).

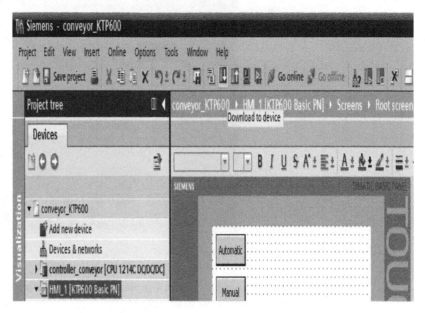

Figure A.40

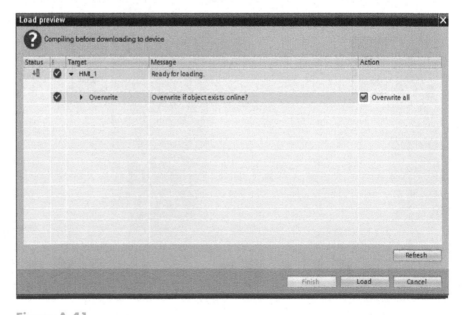

Figure A.41

Click on the button Load (Fig. A.42).

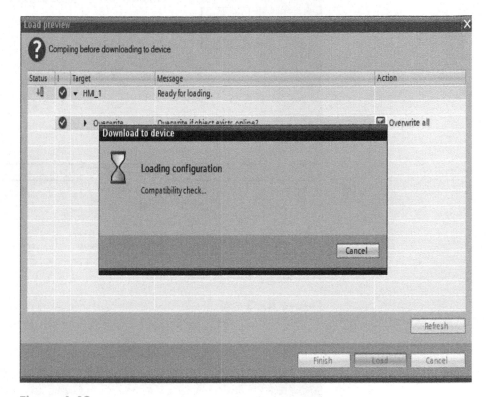

Figure A.42

If the operating system on the panel is not current, an additional window is displayed for updating the operating system. Test function key F6 as well.

15. Start and Stop Button

Now we are going to configure the Start and Stop buttons. The button Start is created in the same manner as the Automatic and Manual buttons. The button Stop has a break-contact function and has to remove the signal when operated. Generate the Start button. Set the background color to green. Under Events Press, select the function SetBitWhileKeyPressed under bit editing. Select the tag on from `conveyor_DB [DB1]` (Fig. A.43). Generate the Stop button. Set the background color to red. Under Events Press, select the function SetBitWhileKeyPressed under bit editing. Select the tag off from `conveyor_DB [DB1]` (Fig. A.44).

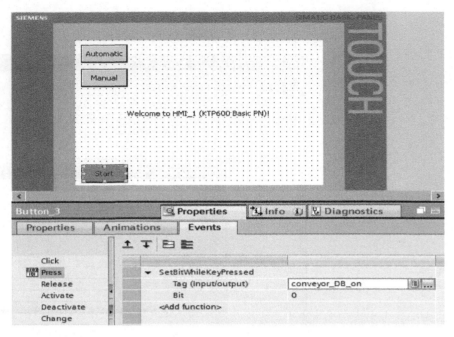

Figure A.43

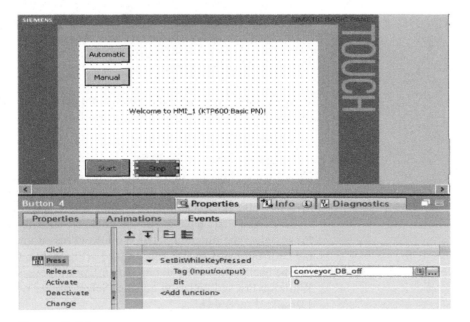

Figure A.44

Before we test visualization, another change has to be made in the Step7program. In OB1, remove the assignments `S3_Conveyor1` and `S4_Conveyor1` when calling FB1. Delete the negation at the `off` input of the block. Save and load the modified program (Fig. A.45).

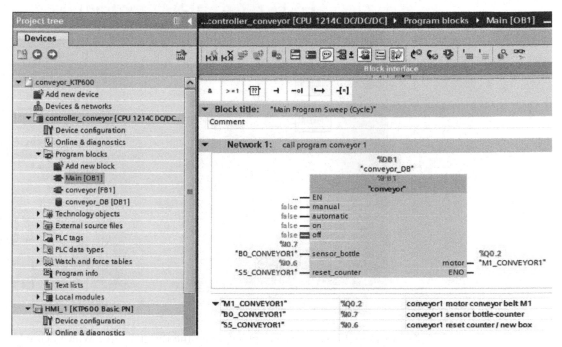

Figure A.45

Load the configuration to the panel, and test the Start and Stop buttons (Fig. A.46).

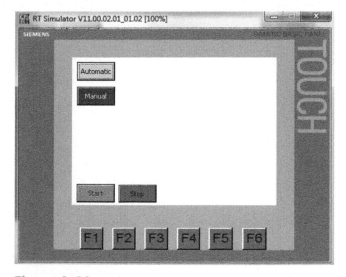

Figure A.46

16. Adding Graphics from the Graphics Folder

In the tool box under Graphics, open the directory tree WinCC graphics folder. Drag the conveyor-belt graphic to the root screen and drop it (Fig. A.47).

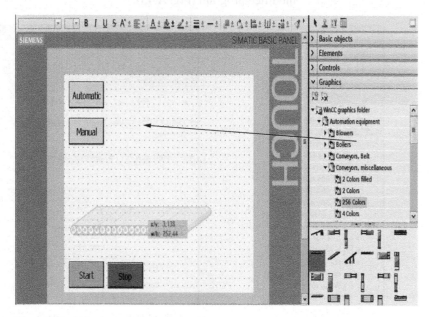

Figure A.47

In the tool box under Graphics, open the directory tree WinCC Graphics folder. Drag the graphic of the beer bottle to the basic screen and drop it. Change the size and position of the bottle (Fig. A.48).

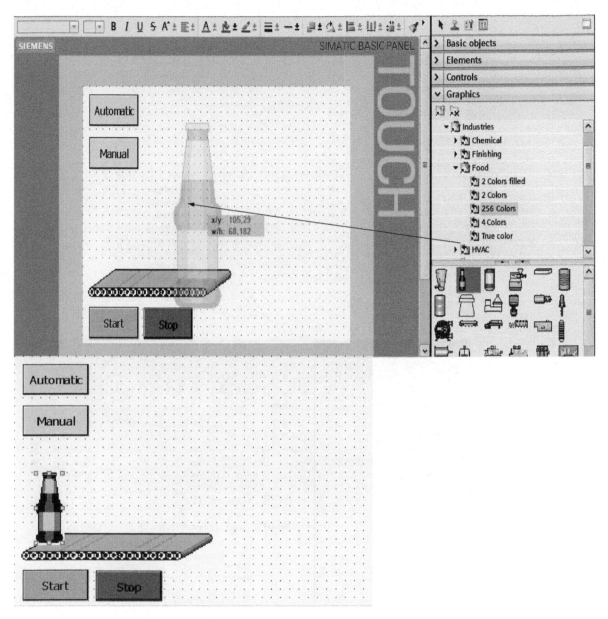

Figure A.48

NOTE All graphic objects have to be located *within* the work area (320 × 240 pixels).

17. Control Program for Simulating Bottle Movement

To simulate bottle movement and the bottle sensor, we create a new block. The FB2 (simulation), shown in Fig. A.49, with tag declaration and network consists

		Interface					
		Name	Data type	Default value	Retain	Visible in ...	Comment
1		▼ Input					
2		start	Bool	false	Non-retentive	☑	
3		pulse	Bool	false	Non-retentive	☑	
4		\<Add new\>				☐	
5		▼ Output					
6		bottle_sensor	Bool	false	Non-retentive	☑	
7		\<Add new\>				☐	
8		▼ InOut					
9		\<Add new\>				☐	
10		▼ Static					
11		▼ IEC_Counter_1	IEC_COUNTER		Non-ret... ▼	☑	
12		CU	Bool	false	Non-retentive	☑	
13		CD	Bool	false	Non-retentive	☑	
14		R	Bool	false	Non-retentive	☑	
15		LD	Bool	false	Non-retentive	☑	
16		QU	Bool	false	Non-retentive	☑	
17		QD	Bool	false	Non-retentive	☑	
18		PV	Int	0	Non-retentive	☑	
19		CV	Int	0	Non-retentive	☑	
20		\<Add new\>				☐	
21		▼ Temp					
22		count_value	Int			☐	

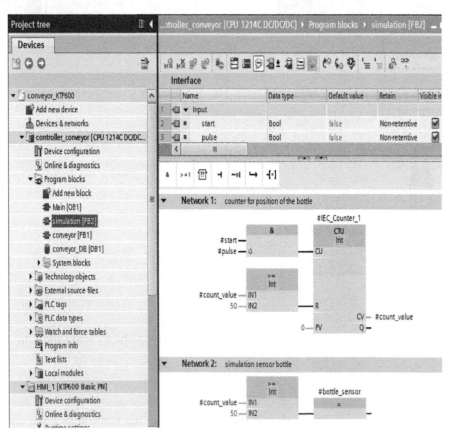

Figure A.49

of a counter that, activated by a start signal, always counts up from 1 to 51. In Network 1, the CTU (up counter) is added as a multi-instance. In Network 2, a bottle-sensor pulse signal is read out when the count reaches 50. This simulates when a bottle leaves the conveyor.

18. Activating the Clock Memory and Assigning the MB100

As clock memory, an internal CPU clock memory bit is used. Activate the clock memory bits, and assign MB100 as the address (Fig. A.50).

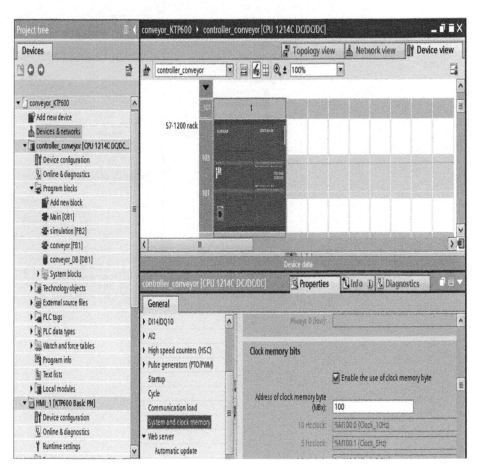

Figure A.50

19. Calling FB2 (Simulation) in OB1

Before calling FB1 (conveyor), add a new network. Call the simulation block (FB2) *before* the conveyor block (FB1). In OB1, set up the temp tag `bottle`, and wire the blocks. Then save the project and load it to the controller (Fig. A.51).

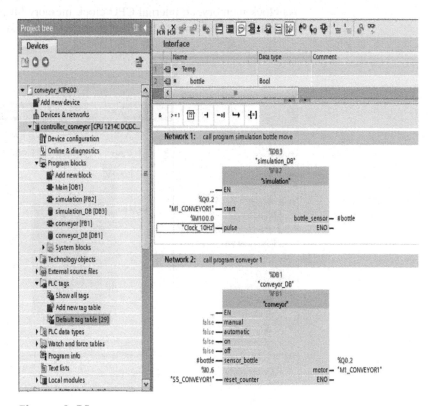

Figure A.51

20. Configuring Bottle Movement

Select the bottle, and under the tab Properties/Animations, select under Movements Add new animation (Figs. A.52 and A.53).

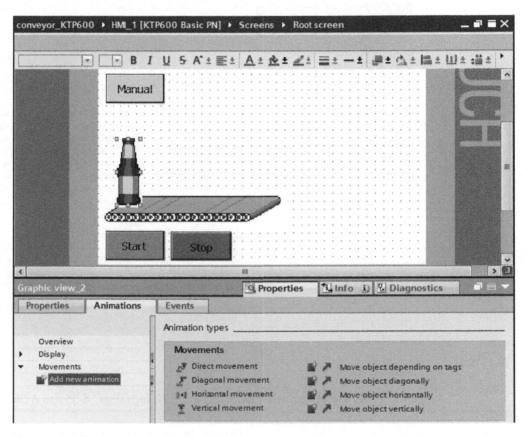

Figure A.52

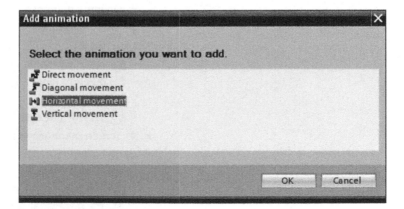

Figure A.53

As tag, select CV of IEC_Counter_1 in the simulation_DB (DB3). Under range, enter 0 to 50. Change the target position to the end of the conveyor X150 (Fig. A.54).

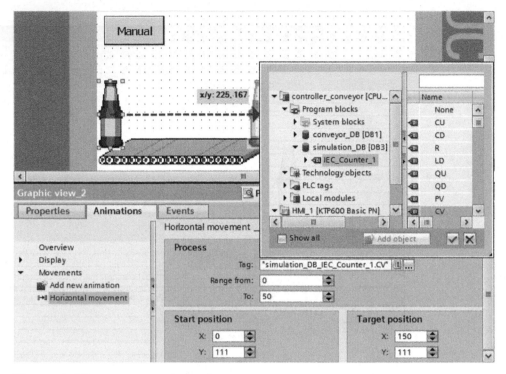

Figure A.54

In the Project window, select the HMI tags (Figs. A.55 and A.56).

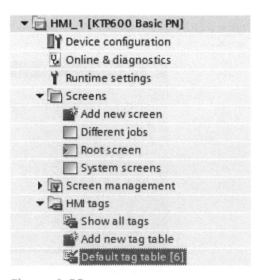

Figure A.55

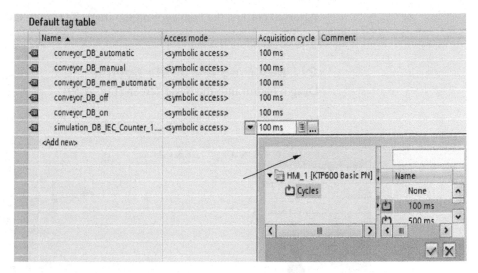

Figure A.56

Drag the slider in the window to the right to get to the column Acquisition cycle. Set the Acquisition cycle of the HMI tags to 100 ms. Then save the project, load it to the panel, and test it (Figs. A.57 and A.58).

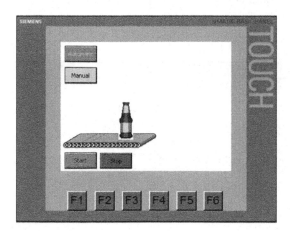

Figure A.57

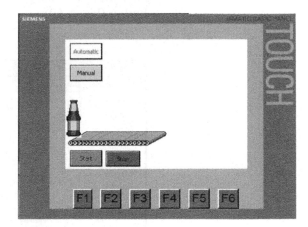

Figure A.58

After 20 bottles, the conveyor motor stops. To restart, the bottle counter has to be reset.

21. Resetting the Bottle Counter
Drag a button into the basic screen (Fig. A.59).

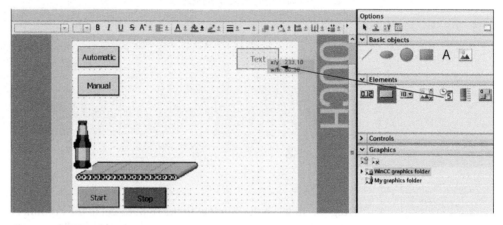

Figure A.59

As text, enter Change beer case, and adjust the color, position, and size of the button (Fig. A.60).

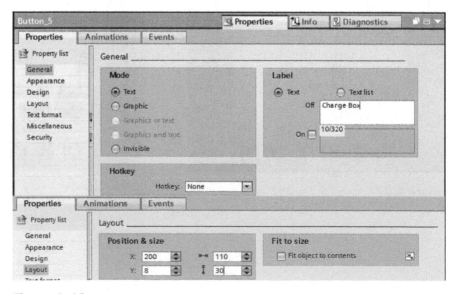

Figure A.60

Under Events Press, select under bit editing the function SetBitWhileKeyPressed. Select the tag `reset_counter` from `conveyor_DB [DB1]` (Fig. A.61).

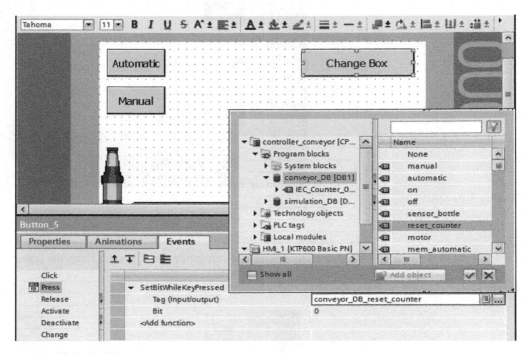

Figure A.61

Set the Acquisition cycle of the new HMI tag to 100 ms. In OB1, remove the wiring `reset_counter` when calling `conveyor_FB`. Then save the project, load it to the CPU and the panel, and test it (Figs. A.62 and A.63).

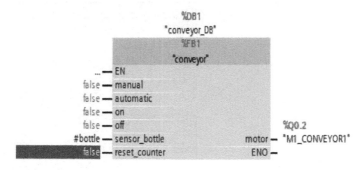

Figure A.62

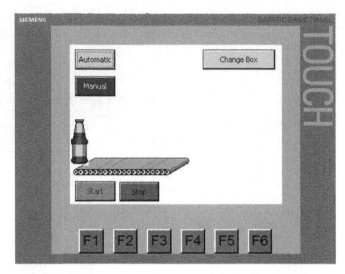

Figure A.63

22. Drawing the Beer Case

Draw a rectangle with a transparent background. Enter the frame width, the position, and the size (Fig. A.64).

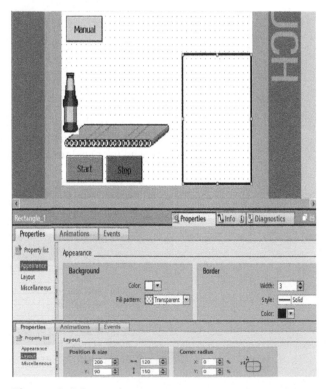

Figure A.64

Draw a vertical line with a spacing of 30 pixels (Fig. A.65).

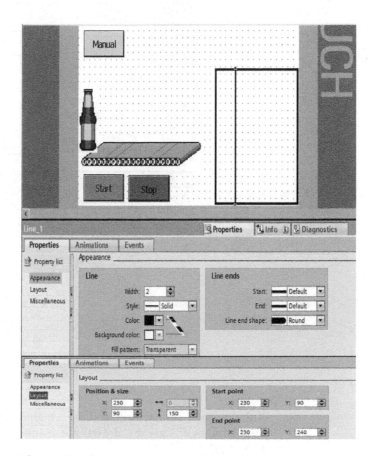

Figure A.65

Draw a horizontal line with a spacing of 30 pixels (Fig. A.66).

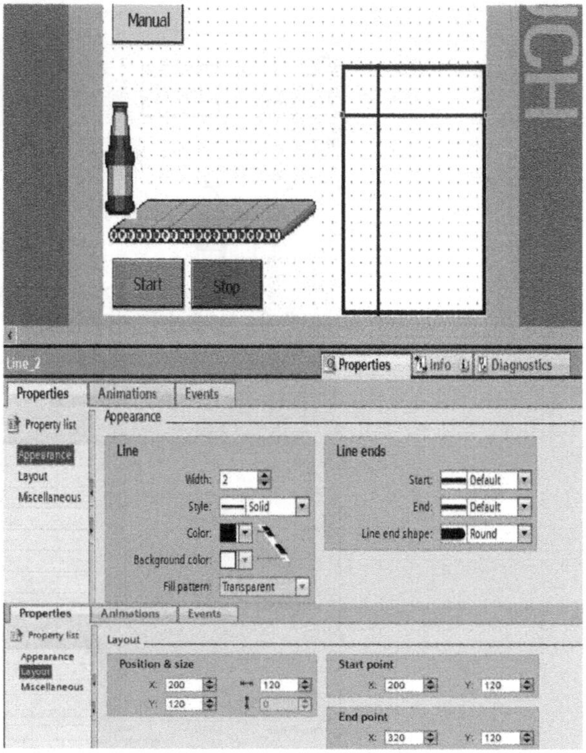

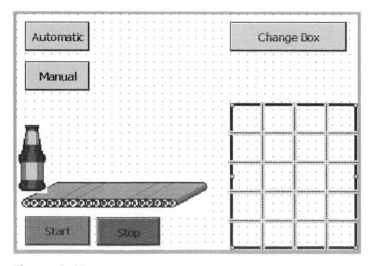

Figure A.66

With copying and inserting, add the remaining lines with a spacing of 30 pixels. Select the beer case by drawing a frame around the case with the mouse (Fig. A.67).

Figure A.67

In the Edit menu, select the function Group (Fig. A.68).

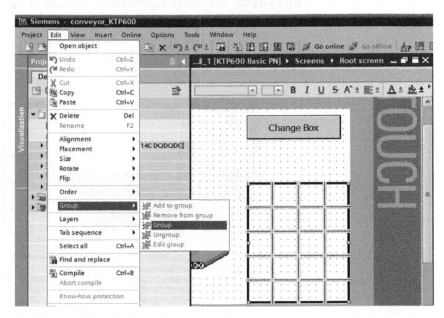

Figure A.68

The rectangle and lines are not to be displayed when the beer case is replaced. At `Rectangle_1` and at the lines, create the animation Visibility with the tag `conveyor_DB_reset_counter` at Value 1 Invisible (Fig. A.69). At the lines, the animation also can be copied and inserted.

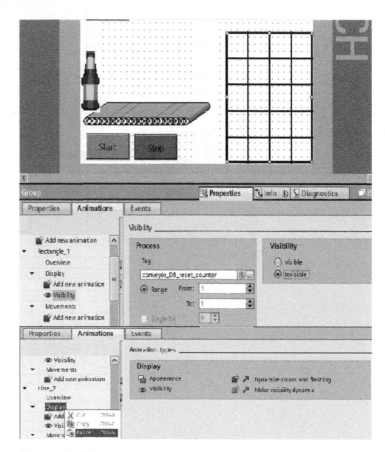

Figure A.69

Then save the project, load it to the panel, and test it.

23. Drawing the Bottles in the Case

Enlarge the view, and draw a circle in the lower-right field of the case (Fig. A.70).

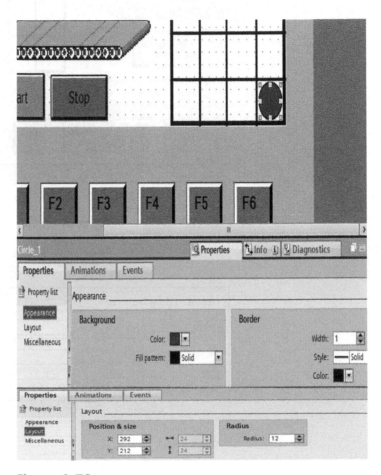

Figure A.70

Draw a second circle (Fig. A.71).

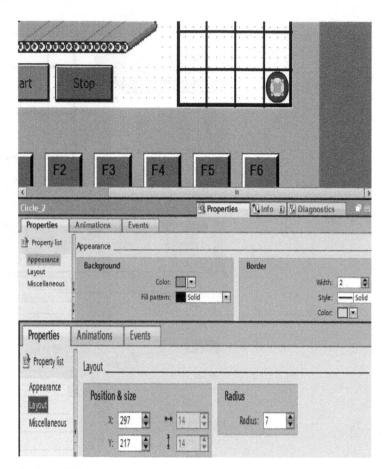

Figure A.71

Group the two inserted circles. At `Circle_1` and `Circle_2`, generate the animation Visibility with the tag `Conveyor_DB_IEC_Counter_0_Instance_CV` at Value 0 to 19 Visible (Fig. A.72).

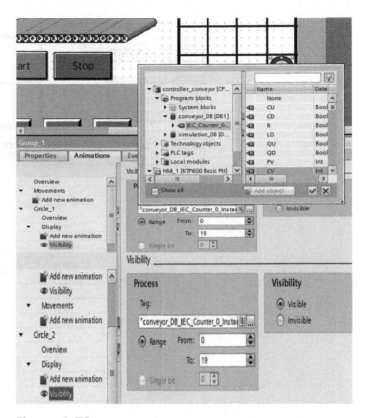

Figure A.72

Copy and insert the bottle. At the two circles under Visibility, change the value range of the tag `Conveyor_DB_IEC_Counter_0_Instance_CV` to 0 to 18 Visible (Fig. A.73).

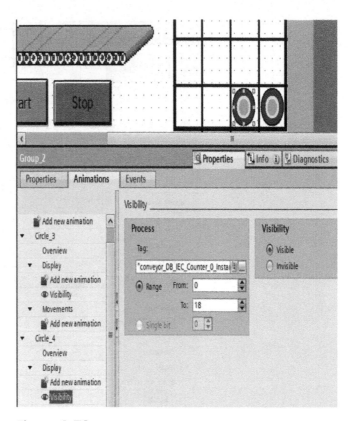

Figure A.73

Copy and insert the individual bottles. At the animation Visibility of the two circles, decrease the value at "To" by 1. The last bottle has the value range 0 to 0 (Fig. A.74).

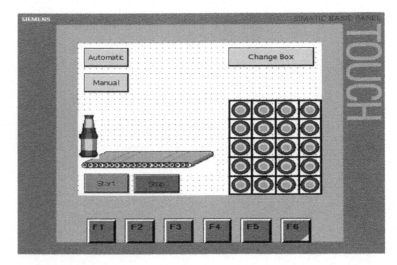

Figure A.74

Set the Acquisition cycle of the new HMI tag to 100 ms. Then save the project, load it to the panel, and test it (Fig. A.75).

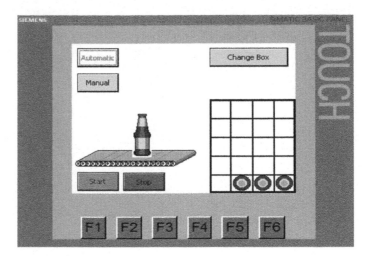

Figure A.75

Index

The Institution of Engineering and Technology

Adding value to your research

The IET is Europe's largest professional body of engineers with over 150,000 members in 127 countries and is a source of essential engineering intelligence.

We facilitate the exchange of ideas and promote the positive role of science, engineering and technology in the world. Discover the IET online to access:

- 400 eBooks
- 26 internationally renowned research journals
- 1,300 conference publications
- over 70,000 archive articles dating back to 1872
- Inspec database containing over 13 million searchable records

To find out more please visit: **www.theiet**.org/books

The Institution of Engineering and Technology is registered as a Charity in England & Wales (no 211014) and Scotland (no SC038698). The IET is located at Michael Faraday House, Six Hills Way, Stevenage, SG1 2AY, UK.

Lightning Source UK Ltd.
Milton Keynes UK
UKOW07n1123010617
302447UK00002B/2/P

9 780071 810456